Debates for the Digital Age

Debates for the Digital Age

The Good, the Bad, and the Ugly of
Our Online World

Volume I
The Good

Danielle Sarver Coombs and Simon Collister, Editors

 PRAEGER ™

An Imprint of ABC-CLIO, LLC
Santa Barbara, California • Denver, Colorado

Library of Congress Cataloging-in-Publication Data

Debates for the digital age : the good, the bad, and the ugly of our online world / Danielle Sarver Coombs and Simon Collister, editors.

volumes cm

Includes index.

Contents: Volume 1. The Good –

ISBN 978-1-4408-0123-5 (vol. 1 : alk. paper) – ISBN 978-1-4408-0124-2 (ebook)

1. Information technology–Social aspects. 2. Information technology–Moral and ethical aspects. I. Coombs, Danielle Sarver, editor. II. Collister, Simon, editor.

HM851.D4325 2016

303.48′33–dc23   2015013116

ISBN: 978-1-4408-0123-5
EISBN: 978-1-4408-0124-2

20 19 18 17 16   1 2 3 4 5

This book is also available on the World Wide Web as an eBook.
Visit www.abc-clio.com for details.

Praeger
An Imprint of ABC-CLIO, LLC

ABC-CLIO, LLC
130 Cremona Drive, P.O. Box 1911
Santa Barbara, California 93116-1911

This book is printed on acid-free paper ∞
Manufactured in the United States of America

Contents

Volume 1. The Good

Introducing the Good, the Bad, and the Ugly of Our Online World   ix
Simon Collister

PART I ACCESSIBILITY

1. New Audiences, New Markets: Accessing Music, Movies,
   Art, and Writing at Your Leisure   3
   Evan Bailey

2. The Positive Side of Social Media: Encouraging Developments
   from Sport   23
   Jimmy Sanderson and Kevin Hull

3. Reaching the World with One Song and a Few Mouse Clicks   39
   Kathryn Coduto

4. Narcissism or Self-Actualization? An Evaluation of "Selfies" as a
   Communication Tool   55
   Christina Best

5. Everyday Expertise: Instructional Videos on YouTube   77
   Jörgen Skågeby and Lina Rahm

6. Online Education, Massive Open Online Courses, and the
   Accessibility of Higher Education   91
   Kristen Chorba and R. Benjamin Hollis

PART II DEMOCRATIZATION

7. Leaks, Whistle-Blowers, and Radical Transparency:
   Government Accountability in the Internet Age   119
   Rekha Sharma

8. Rethinking Digital Democracy in a Time of Crisis:
   The Case of Spain   141
   Salomé Sola-Morales

9. Will the Revolution Be Tweeted? Activism, Politics, and the
   Internet   165
   Lázaro M. Bacallao-Pino

10. You Say You Want a Revolution? The Internet's Impact on
    Political Discussion, Activism, and Societal Transformation   183
    James D. Ponder and Rekha Sharma

11. Ground-Up Expert: Everyday People and Blogs   203
    Richard J. Batyko

12. Self-Promotion for All! Content Creation and Personal
    Branding in the Digital Age   221
    Justin Lagore

13. The Rise of Journalism Accountability   241
    Zac Gershberg

PART III COMMUNITY AND GLOBALIZATION

14. Social Media Mechanisms: A Change Agent   263
    Kiran Samuel

15. Habermas in the African E-Village: Deliberative
    Practices of Diasporan Nigerians on the Internet   287
    Farooq Kperogi

16. When Bad Timing Is Actually Good: Reconceptualizing
    Response Delays   305
    Stephanie A. Tikkanen and Andrew Frisbie

17. In Defense of "Slacktivism": How KONY 2012 Got the
Whole World to Watch   321
Christopher Boulton

18. Public Health's Courtship with the Internet: Slow but Steady   333
Samantha Lingenfelter

Index   347

About the Editors and Contributors   359

Introducing the Good, the Bad, and the Ugly of Our Online World

Simon Collister

One of the defining characteristics of contemporary society's metamorphosis from a hierarchically led, industrial, and largely national concept to a fragmented, postindustrial, and globalized space is the emergence and exponential growth of telecommunications networks and the Internet.[1]

This has radically altered the communities, politics, and media of traditionally private, localized sociocultural environments—transforming them into highly public, international networks of communication and mediation. Indeed, Livingstone goes as far as to assert that networks are "the archetypal form of contemporary social and technical organization."[2] Scholars such as Manuel Castells and Yochai Benkler conceive of this phenomenon at a macro level as the "network society"[3] or "networked information economy,"[4] respectively.

Without doubt, such developments have significantly impacted all aspects of our social and cultural domains. Yet, at the same time, the emergence of such complex factors has equally challenged the capability of scholarly research to adjust to the rapid "pace of change."[5]

Such a challenge in part accounts for creation of this text. In setting out to capture contemporary thinking on the Internet and its impact on society and culture, the authors sought to identify and address a range of issues less susceptible to change as frequently as Facebook's "Terms and Conditions" or be subject to the forces of changing consumer demand. Rather, the two volumes plot, analyze, and make sense of slowly shifting macro-themes of the sociocultural domain which—although not immediately evident—are likely to have long-term, far-reaching, and deeply profound effects on the world around us.

In making sense of such a varied and potentially complex thematic landscape, this book adapts an approach that echoes the work of other scholars
who have developed typologies for interpreting diverse literature on networked culture, media, and politics.[6] In particular, we have drawn on both
Chadwick's and Chadwick and Howard's distinction between optimistic/
positive and pessimistic/negative approaches to the Internet's impact on
society and culture. In their typology, optimistic perspectives understand the
Internet as redressing the balance of power away from dominant—often
elite—groups or spaces and returning it to networked communities of informal or amateur individuals. Conversely, pessimistic perspectives interpret the
Internet as reinforcing traditional power structures, albeit in new forms.

This typology underpins the way in which the two volumes of this text are
positioned. Optimistic perspectives are addressed in Volume 1, *The Good*, and
pessimistic perspectives are covered in Volume 2, *The Bad and the Ugly*. Each
volume is further subdivided into a series of thematic sections to enable contributors to undertake a detailed investigation into specific areas of Internet
culture. Before setting out the specific themes covered in each volume, it is
helpful to set the tone of *the Good, the Bad, and the Ugly* by providing a summary of some of the broader shifts presently occurring.

THE GOOD

Optimistic analyses of the contemporary environment created by the Internet have conventionally articulated a vision of society and culture consistent with the idea of the public sphere as identified by German philosopher
and sociologist Jürgen Habermas. For Habermas, the public sphere represents
a space within society for people to freely meet, discuss, and act on the important issues of the day.

For optimists, the Internet plays a vital role in empowering and connecting individuals in a global public discourse that facilitates "communicative
links between citizens and the power holders of society."[7] Broadly speaking,
an Internet-enabled "networked public sphere"[8] operates as the 21st century's
public space and be seen as a force for good—thanks to the expanded and
accelerated flows of information and the increased interactivity between participants it generates.[9]

Specifically, Yochai Benkler explains a networked public sphere has
"fundamentally altered the capacity of individuals . . . to be active participants . . . as opposed to passive readers, listeners or viewers" in the pursuit of
social, cultural, and political issues.[10] Benkler goes so far as to argue that such
an account of the contemporary public sphere, although aligned with Habermas's model, is *more capable* of accounting for the social and cultural complexity of modern democracies. By empowering all members of society—not

just well-educated, middle- or upper-class individuals—contemporary online communities can act collectively to perform a "watchdog" role and operate as "a source of salient observations regarding matters of public concern," and provide "a platform for discussing the alternatives open to a polity."[11]

A further reason for optimistic analyses of a networked public sphere is the Internet's distributed communications architecture combined with the low costs for producing and distributing information (all you need is a smartphone, tablet, or computer and an Internet connection). Both Benkler and Castells believe these low barriers to entry enable anyone with Internet access to shape (or reshape) the social, cultural, or political dimensions of everyday life. This "mass self-communication"[12] transforms the traditional power and influence of the mass media—typified as "more centralized, homogeneous and less pluralistic"[13]—into a decentralized, heterogeneous "social communication process"[14] characterized by a diverse and pluralistic range of participants.[15] As a result, it offers "avenues for citizen independence from mainstream news media and larger social forces."[16]

In addition to the social, political, and cultural diversity offered by the Internet, some scholars also argue that it provides a greater resilience to control by governments, states, or corporations. For example, the Internet is a communication network technically organized without any control or management by any central individual or organization. Thus, it crucially lacks any single central point of control, making it difficult to censor contentious or sensitive information.

This inevitably leads to "the emergence of multiple axes of information [that] provide new opportunities for citizens to challenge elite control of political issues"[17] and enhances "the potential for the media to exercise accountability over power."[18] Tewksbury and Rittenberg see this as a "democratization of the creation, dissemination, and consumption of news and information"[19] and Castells goes further by asserting that "mass self-communication" empowers individuals to "challenge and eventually change the power relations institutionalized in society."[20]

Such individual and collective efforts to challenge and transform social and cultural relations have been identified across a range of fields that use the Internet to share information and organize protests,[21] as well as in democratic politics for which networked communications have been used to facilitate increased engagement with and participation in democratic institutions.[22]

THE BAD AND THE UGLY

Although most scholars agree that the Internet is "bringing together individual citizens and informal networks through interconnected global webs of public communication,"[23] the idea of the networked public sphere as a force for good is not without criticism.

One of the most persistent pessimistic analyses of the Internet draws attention to the issue of access. For example, significant parts of global society are on the "wrong" side of what is called the "digital divide."[24] As a result, the universal participation in political and cultural discussion is likely to be significantly limited to those individuals who have Internet access.[25] Conversely, some scholars argue that those members of global society who do have access also are biased toward affluent "elites."[26]

Scholars also have questioned whether the society, culture, and democracy being produced by the Internet follow the same ideals imagined in Jürgen Habermas's original model. They argue that, in reality, rather than enabling civic and democratic discourse fulfilling the lofty principle of improving society, the Internet merely is facilitating broader cultural trends characterized by an increased focus on lifestyle or entertainment content.[27]

Despite its disruptive nature, for example, scholars have pointed out that the Internet merely replicates traditional media consumption habits.[28] Thus, although in theory the Internet enables people to democratically select information that matters to them, the reality is that the information consumed by the public usually excludes political information of democratic interest[29] or is limited to content that mirrors users' personal beliefs.[30]

Pessimist perspectives of the Internet's impact on society and culture also challenge the view that it can overcome censorship. Hargittai, Mansell, and Dahlberg,[31] for example, assert that the traditional dominance of commercial elites is, in fact, replicated online. This process—termed by Dahlberg as the "corporate colonization of cyberspace"[32]—weakens, rather than strengthens, the Internet's potential for free democratic, social, and cultural discussion. Despite a perceived "communicative abundance" generated by the networked public sphere's low barriers to entry,[33] corporations are outmaneuvering the public's adoption of the Internet and are "hijacking" social communication tools—such as blogs and social networks—to continue and expand economic dominance.[34]

For some scholars, the commercial adoption of the Internet represents an even more troubling aspect of the transformation of the social, political, and cultural dimensions of everyday life. Dean[35] and Terranova[36] argue that corporations and political elites co-opt public and civic discussion online and use it to create the illusion of increased individual empowerment which, in fact, instead conceals a complete removal of individual agency. For Dean, the convergence of democratic ideals—such as participation and open access—and capitalism's colonization of the networked public sphere gives rise to a "communicative capitalism"[37] which captures political power in an ever-increasing "displacement of political conflict to the terrain of networked media."[38] Moreover, every "click and interaction made in the networked media environment can be traced, capitalized and sold"[39] as "free labor" creating a "blurred territory between production and consumption, work and cultural expression."[40]

SUBTHEMES: ACCESSIBILITY, DEMOCRATIZATION, COMMUNITY, AND GLOBALIZATION

To help categorize and interpret subthemes such as accessibility, democratization, community, and globalization, the two volumes of this set are divided into three subthemes: accessibility to information; the democratization of everyday life; and community and globalization. Adopting these broad subthemes across the two volumes enables the contributing authors to isolate and perform a detailed study into—and make greater sense of—individual elements of Internet-enabled cultural phenomena.

The first subtheme, "*accessibility*," examines the opening of new markets and audiences for cultural actors, such as musicians and sports players; the freeing up of teaching and learning across informal, everyday spaces and not just in formal education settings; and the opportunities which the Internet presents for increasingly self-directed identity-formation and expression. This subtheme also challenges such constructive readings by pointing out how greater information accessibility also can lead to greater intolerance and reactionary responses by traditional elite groups, which simply undermine the Internet's potential to open up society and culture.

The next subtheme examined—"*democratization*"—encompasses a wide range of topics, including the political role of the Internet in empowering democratic engagement. The section highlights the Internet's increasingly important role in enabling and generating extra-democratic activism, particularly in the context of post-crisis Europe and the events of the Arab Spring. Crucially, it also raises important questions as to the real-world effects of such optimistic—yet largely theoretical—accounts of the Internet's democratizing power.

In the democratization section, the notion of democracy also is applied to broader, cultural topics such as the ways in which anyone with an Internet connection can create an identity (or identities) or build a commercially successful "personal brand," and what this means for self-management in an increasingly commercial space where traditional issues of privacy become challenged. Lastly, this subtheme addresses the ways in which journalism and the media increasingly are being held accountable for the ways in which they represent society and whether the Internet's democratization of news-making offers greater freedom or merely reinforces the same old problems.

The last subtheme—*"community and globalization"*—explores ways in which the Internet is used to transform communities at a local level as well as create globalized, participatory communities where specific, localized events increasingly take on a national or international significance. In doing so, the impact of these Internet-enabled transformations on community members, structures, and relations are considered from both a beneficial and a detrimental perspective. This subtheme also investigates specific features and concerns of traditional communities, such as health, education, and social mobility, and offers positive as well as problematic readings of how these phenomena are impacted by the rise of Internet-enabled individuals and groups.

VOLUME I CHAPTER SYNOPSES

Volume 1 opens with a comprehensive survey by Bailey that confidently examines the impact of the Internet on globalized audiences and cultural reception. Spanning the transformation in music consumption by a mobile-first fan base, to streaming services and the impact of wearable technology and virtual reality on art and the written word, Chapter 1 argues that despite some challenges, digitally connected audiences are capable of engaging with and producing cultural works in ways that offer a positive future for the arts.

Such positive outcomes on key forms of cultural production similarly are addressed in relation to the sports industry by Sanderson and Hull in Chapter 2. They argue that, despite the high-profile media coverage of social media sporting failures, multiple stakeholders have indeed benefitted from the development of digital technologies. They identify and discuss three key factors, including the ability of sports stars to optimize their identity and self-presentation, greater interaction with fans, and the opportunities for advocacy and activism. The chapter concludes with suggestions as to scholarly directions for further research.

In Chapter 3, Coduto offers a compelling narrative of the evolution of fan engagement and audience development in the music industry. Moving from the highly localized production of 'zines in the 1980s and 1990s, through the emergence of Napster, the rise of MySpace, and on to Radiohead's pioneering "pay what you want" approach, Coduto draws on personal experience from a transnational perspective to discuss the benefits of fan-band interaction and music industry innovation in a digital age.

In Chapter 4, Best offers a detailed analysis of a seemingly recent phenomenon, the "selfie." She provides a useful historical context to selfies, situating them within the broader theoretical framework of identity-as-performance—a notion that has become increasingly potent with the advent of social media. By plotting the origins and enduring motivations of self-presentation, Best concludes that selfies can be best understood as contemporary instantiations of a "timeless human desire."

Skågeby and Rahm argue in Chapter 5 that online video-sharing sites—in this case YouTube—empower the public to offer instructional guidance to others on a wide range of topics, from video-gaming, to cookery, to kayaking. Skågeby and Rahm offer a succinct history of this "everyday expertise" and argue that such guidance can be conceived beyond the specific applications they cover and instead be seen as an increasingly seamless integration of the material and virtual realities of everyday life.

Although Skågeby and Rahm tackle the issue of how the Internet has transformed informal learning, in Chapter 6 Chorba and Hollis explore the Internet's impact on formal education. They explain emerging virtual online learning environments, such as massive open online courses (MOOCs), and set out a comprehensive account of some of the ways people use technology to learn. The chapter also maps the benefits the new tools offer.

Sharma's contribution in Chapter 7 undertakes a robust investigation of the ways in which new communication technologies have challenged and exacerbated some of the complexities facing the media regarding the protection of sources. Sharma argues that whistle-blowing to protect civic society is a long-standing and important part of modern democracy. To help us understand what this looks like in an Internet-enabled world, Sharma identifies some of the key new actors in this digitally networked public sphere, such as bloggers and citizen journalists; discusses legal and ethical issues in relation to digital whistle-blowing; and poses important questions for the future of research in this area.

Picking up the theme of how the Internet might enrich and empower modern democratic societies, in Chapter 8 Sola-Morales focuses attention on recent developments in Spain. Drawing on examples of online activists, such as the "15 Million Movement" and more formal political parties, such as "*Podemos*," Sola-Morales offers a theoretical framework for understanding such phenomena. She explores whether the evidence suggests that the Internet is making Spain's democracy stronger or instead merely is reinforcing existing power structures.

Echoing Sola-Morales' focus but from different geographic perspectives, Ponder and Sharma explore similar topics from a U.S. perspective in Chapter 10. In Chapter 9, however, Bacallao-Pino shifts the analytical lens to a much broader perspective by examining the role of social media tools in enabling sociopolitical change at a global level.

Chapter 11 examines the issue of how the Internet, and what Manuel Castells terms "mass self-communication,"[41] has given rise to what Baytko calls "the ground-up expert." Baytko focuses his analysis on the rise of "frustrated office worker," Julie Powell, and plots how the emergence of blogging has created Hollywood stars from everyday people.

In Chapter 12, Lagore tackles a related issue: personal branding and how individuals can make use of social media tools per se to plan, create, and build a personal brand. Lagore then uses this framework for an investigation of what personal brands mean for the established, incumbent media and entertainment industries.

The impact of social media on journalism and the news industry is something explored in greater depth by Gershberg. In Chapter 13, Gershberg sets out how the growth of citizen journalists combined with the exposure of unethical practices by traditional journalists have caused a crisis of accountability in the news. But rather than seeing this as terminal shift, Gershberg argues that Internet-enabled journalism can expose poor standards and provoke a powerful "public discourse" about the state of the media.

In keeping with the theme of using the Internet to help highlight and broach social issues, in Chapter 14 Samuel explores how social media can be used as a tool enabling minority voices to enact change. By using the #CancelColbert hashtag campaign as a focal point, Samuel argues that social media was used to empower online communities to build a movement, to speak up without fear of censorship or gatekeeping by traditional media, and, moreover, to protect themselves when conventional models of social justice fail.

This notion of Internet-enabled communication prefiguring a model of "deliberative democracy" is more deeply explored by Kperogi in Chapter 15. In an analysis of a Nigerian online community, Kperogi draws on Habermas's theory of the public sphere to demonstrate how the Internet acts as a powerful tool in fostering "transnational, diasporic spheres of public discourse."

Having been introduced to the concepts of the Internet acting as a "transversal" or "transnational" platform, in Chapter 16 Tikkanen and Frisbie identify and undertake a fascinating examination of the ways in which the notion of time structures globally networked communication. Although previous chapters have looked at understanding the communicative effects of an Internet-enabled global discourse, in this chapter the authors focus on the ways the Internet—or rather the way users use the Internet to communicate—can structure meaning and influence interpretation of events. Specifically, they argue that although delays in asynchronous online discourse can be perceived as inferior to face-to-face interactions, it actually can offer opportunities for more creative and meaningful conversations.

One such criticism of online communication is that it fails to create "strong ties" between individuals and, in the particular case of social change campaigning, it is responsible for "slacktivism"—a pejorative term used to denounce ineffective, online-only activism, which often yields little or no tangible result. In Chapter 17, Boulton offers a rebuttal of such accusations.

He draws on a case study of the *KONY 2012* online campaign, which sought to highlight Joseph Kony's Lord Resistance Army as being responsible for abducting children to make them soldiers.

Finally, building on Boulton's optimistic argument for the power of the Internet to enable positive social change, Lingenfelter further explores the implications of how new technologies have—in broad terms—enabled much more effective public health communication.

CONCLUSIONS

The two volumes forming the present work gather together a range of topics, authors, and methodological approaches that the authors think will help move forward an understanding of the ways in which the Internet is changing (or not changing) the sociocultural domain. The assembled chapters have been selected to stimulate, provoke, and challenge, but also reassure scholars interested in Internet culture. Despite the rapid but arguably superficial changes in our networked society, at a macro level the transformations in the accessibility of information, the concordant shift toward the democratization of everyday life, and the effects this has on notions of local as well as globalized communities are evolving at a much more measured pace—albeit with much deeper and longer-term impacts on our sociocultural realm.

Lastly, although the authors believe that these volumes provide a good grounding for students and scholars of Internet culture, it is important to recognize that even the material contained within these two volumes eventually will be superseded by new and hitherto unthinkable changes. Returning to Chadwick and Howard's optimistic/pessimistic dichotomy, the authors also introduce a third position into their typology which they term "surprising."[42] This term is used to account for events that introduce entirely new and unrecognized ideas into the ways in which we understand society, politics, and culture in an Internet-enabled world.

This "third-way" offers a potent and constructive perspective on the ways the Internet is reshaping the everyday world and, importantly, suggests a conceptual escape route from attempts to lock scholarship into an "alternatively revolutionary or evolutionary" dichotomy[43] that "tends to treat media choice, source choice, and interactivity habits as distinct areas of inquiry."[44] Although this text focuses on arranging the contributors' analyses from a good/bad/ugly perspective, we are confident that new and "surprising" directions for future study and research into the Internet's impact on our society and culture can be glimpsed within these two volumes. Moreover, we intend for the collection to act as a springboard for the next round of enquiry into the ever-increasing and fertile domain of our online world.

NOTES

1. Castells, Manuel, *The Rise of the Network Society* (Oxford: Blackwell, 1996).

2. Livingstone, Sonia, "Critical Debates in Internet Studies: Reflections on an Emerging Field," chap. 1 in *Mass Media and Society*, edited by James Curran and Michael Gurevitch, 9–29 (London: Hodder Arnold, 2005), 12.

3. Castells, Manuel, *The Rise of the Network Society* (Oxford: Blackwell, 1996).

4. Benkler, Yochai, *The Wealth of Networks: How Social Production Transforms Markets and Freedom* (New Haven, CT: Yale University Press, 2006).

5. Castells, Manuel, *The Internet Galaxy* (Oxford: Oxford University Press, 2001), 3.

6. Wellman, Barry, "The Three Ages of Internet Studies: Ten, Five and Zero Years Ago," *New Media & Society* 6, no. 1 (2004), 123–29; Howard, Philip N., and Andrew Chadwick, "Political Omnivores and Wired States," in *Routledge Handbook of Internet Politics*, edited by Andrew Chadwick and Philip N. Howard (London: Routledge, 2009); Chadwick, Andrew, "Web 2.0: New Challenges for the Study of E-Democracy in an Era of Informational Exuberance," *I/S: A Journal of Law and Policy* 5, no. 1 (2009): 9–42. Livingstone, Sonia; "Critical Debates in Internet Studies: Reflections on an Emerging Field," chap. 1 in *Mass Media and Society*, edited by James Curran and Michael Gurevitch, 9–29 (London: Hodder Arnold, 2005), 12.

7. Dahlgren, Peter, "The Internet, Public Spheres, and Political Communication: Dispersion and Deliberation," *Political Communication* 22, no. 2 (2005): 148.

8. Benkler, Yochai, *The Wealth of Networks: How Social Production Transforms Markets and Freedom* (New Haven, CT: Yale University Press, 2006): 176–77.

9. McNair, Brian, *Cultural Chaos: Journalism, News and Power in a Globalised World* (London: Routledge, 2006): 221–23.

10. Benkler, Yochai, *The Wealth of Networks: How Social Production Transforms Markets and Freedom* (New Haven, CT: Yale University Press, 2006), 212.

11. Ibid., 271.

12. Castells, Manuel, *Communication Power* (Oxford: Oxford University Press, 2009), 65–72.

13. Beckett, Charlie, and Robin Mansell, "Crossing Boundaries: New Media and Networked Journalism," *Communication, Culture and Critique* 1, no. 1 (2008): 4.

14. Benkler, Yochai, *The Wealth of Networks: How Social Production Transforms Markets and Freedom* (New Haven, CT: Yale University Press, 2006), 181.

15. Bimber, Bruce, "The Internet and Political Transformation: Populism, Community, and Accelerated Pluralism," *Polity* 31, no. 1 (Autumn 1998), 133–60.

16. Tewksbury, David, and Jason Rittenberg, "Online News Creation and Consumption: Implications for Modern Democracies," in *Routledge Handbook of Internet Politics*, edited by Andrew Chadwick and Philip N. Howard (London: Routledge, 2009), 197.

17. Williams, Bruce A., and Michael X. Delli Carpini, "Monica and Bill All the Time and Everywhere: The Collapse of Gatekeeping and Agenda Setting in the New Media Environment," *American Behavioral Scientist* 47, no. 9 (2004): 1209.

18. McNair, Brian, *Cultural Chaos: Journalism, News and Power in a Globalised World* (London: Routledge, 2006), 229.

19. Tewksbury, David, and Jason Rittenberg, "Online News Creation and Consumption: Implications for Modern Democracies," in *Routledge Handbook of Internet Politics*, edited by Andrew Chadwick and Philip N. Howard (London: Routledge, 2009), 197.

20. Castells, Manuel, "Communication, Power and Counter-Power in the Network Society," *International Journal of Communication* 1 (2007): 248.

21. Bennett, W. Lance, "New Media Power: The Internet and Global Activism," in *Contesting Media Power: Alternative Media in a Networked World*, edited by Nick Couldry and James Curran (Oxford: Rowman & Littlefield, 2003), 17–37; Castells, Manuel, "Communication, Power and Counter-Power in the Network Society," *International Journal of Communication* 1 (2007): 238–66. Castells, Manuel, *Communication Power* (Oxford: Oxford University Press, 2009).

22. Bimber, Bruce, *Information and American Democracy: Technology in the Evolution of Political Power* (Cambridge: Cambridge University Press, 2003); Bimber, Bruce, "The Internet and Political Transformation: Populism, Community, and Accelerated Pluralism," *Polity* 31, no. 1 (Autumn 1998): 133–60; Bimber, Bruce, "The Study of Information Technology and Civic Engagement," *Political Communication* 17 (2000): 329–33; Tolbert, Caroline J., and Ramona S. McNeal, "Unraveling the Effects of the Internet on Political Participation?" *Political Research Quarterly* 56, no. 2 (2003): 175–85.

23. Curran, James, and Tamara Witschge, "Liberal Dreams and the Internet," in *New Media, Old News: Journalism & Democracy in the Digital Age*, edited by Natalie Fenton (London: Sage, 2010).

24. Van Dijk, Jan A. G. M., *The Deepening Divide: Inequality in the Information Society* (London: Sage, 2005).

25. Dahlgren, Peter, "The Public Sphere and the Net: Structure, Space and Communication," in *Mediated Politics: Communication in the Future of Democracy*, edited by W. Lance Bennett and Robert Entman (Cambridge: Cambridge University Press), 33–56; Sparks, Colin, "The Internet and the Global Public Sphere," in *Mediated Politics: Communication in the Future of Democracy*, edited by W. Lance Bennett and Robert Entman (Cambridge: Cambridge University Press, 2001).

26. Sparks, Colin, "The Internet and the Global Public Sphere," in *Mediated Politics: Communication in the Future of Democracy*, edited by W. Lance Bennett and Robert Entman (Cambridge: Cambridge University Press 2001), 83.

27. Papacharissi, Zizi, *A Private Sphere: Democracy in a Digital Age* (Cambridge: Polity, 2010), 112–25.

28. Schoenbach, Klaus, Ester de Waal, and Edmund Lauf, "Research Note: Online and Print Newspapers : Their Impact on the Extent of the Perceived Public Agenda," *European Journal of Communication* 20, no. 2 (2005): 245–58; Hargittai, Eszter, "Content Diversity Online: Myth or Reality," in *Media Diversity and Localism: Meaning and Metrics*, edited by Philip M. Napoli (London: Routledge, 2006).

29. Tewksbury, David, and Jason Rittenberg, "Online News Creation and Consumption: Implications for Modern Democracies," in *Routledge Handbook of Internet Politics*, edited by Andrew Chadwick and Philip N. Howard (London: Routledge, 2009), 194.

30. Sunstein, Cass R., *Republic.Com 2.0* (Princeton, NJ: Princeton University Press, 2007).

31. Mansell, Robin, "Political Economy, Power and New Media," *New Media & Society* 6, no. 1 (2004): 74–83. Dahlberg, Lincoln, "The Corporate Colonization of Online Attention and the Marginalization of Critical Communication." *Journal of Communication Inquiry* 29, no. 2 (2005): 160–80.

32. Dahlberg, Lincoln, "The Corporate Colonization of Online Attention and the Marginalization of Critical Communication." *Journal of Communication Inquiry* 29, no. 2 (2005): 160.

33. Karppinen, Kari, "Rethinking Media Pluralism and Communicative Abundance," *Observatorio* 11 (2009): 151–69.

34. Castells, Manuel, "Communication, Power and Counter-Power in the Network Society," *International Journal of Communication* 1 (2007): 248; Castells, Manuel, *Communication Power* (Oxford: Oxford University Press, 2009), 65–72; Stanyer, James, "Web 2.0 and the Transformation of News and Journalism," in *Routledge Handbook of Internet Politics*, edited by Andrew Chadwick and Philip N. Howard (London: Routledge, 2009).

35. Dean, Jodi, *Blog Theory: Feedback and Capture in the Circuits of Drive* (Cambridge: Polity, 2010).

36. Terranova, Tiziana, *Network Culture: Politics for the Information Age* (London: Pluto Press, 2004).

37. Dean, Jodi, *Blog Theory: Feedback and Capture in the Circuits of Drive* (Cambridge: Polity, 2010), 4–5.

38. Ibid., 124.

39. Ibid., 66.

40. Terranova, Tiziana, *Network Culture: Politics for the Information Age* (London: Pluto Press, 2004), 73–94.

41. Castells, M., *Communication Power* (Oxford: Oxford University Press, 2009), xix.

42. Chadwick, Andrew, and Philip N. Howard, eds. *Routledge Handbook of Internet Politics* (London: Routledge, 2009), 424–26.

43. Mansell, Robin, "Political Economy, Power and New Media," *New Media & Society* 6, 1 (2004): 7.

44. Howard, Philip N., and Andrew Chadwick, "Political Omnivores and Wired States," in *Routledge Handbook of Internet Politics*, edited by Andrew Chadwick and Philip N. Howard (London: Routledge, 2009), 431.

BIBLIOGRAPHY

Beckett, Charlie, and Robin Mansell. 2008. "Crossing Boundaries: New Media and Networked Journalism." *Communication, Culture and Critique* 1(1): 92–104.

Benkler, Yochai. 2006. *The Wealth of Networks: How Social Production Transforms Markets and Freedom.* New Haven, CT: Yale University Press.

Bennett, W. Lance. 2003. "New Media Power: The Internet and Global Activism." In *Contesting Media Power: Alternative Media in a Networked World*, edited by Nick Couldry and James Curran, 17–37. Oxford: Rowman & Littlefield.

Bimber, Bruce. 2003. *Information and American Democracy: Technology in the Evolution of Political Power*. Cambridge: Cambridge University Press.

Bimber, Bruce. 2000. "The Study of Information Technology and Civic Engagement." *Political Communication* 17: 329–33.

Bimber, Bruce. 1998. "The Internet and Political Transformation: Populism, Community, and Accelerated Pluralism." *Polity* 31 (1) (Autumn): 133–60.

Castells, Manuel. 2009. *Communication Power*. Oxford: Oxford University Press.

Castells, Manuel. 2007. "Communication, Power and Counter-Power in the Network Society." *International Journal of Communication* 1: 238–66.

Castells, Manuel. 2001. *The Internet Galaxy*. Oxford: Oxford University Press.

Castells, Manuel. 1996. *The Rise of the Network Society*. Oxford: Blackwell.

Chadwick, Andrew. 2013. *The Hybrid Media System: Politics and Power*. New York: Oxford University Press.

Chadwick, Andrew. 2011. "The Political Information Cycle in a Hybrid News System: The British Prime Minister and the 'Bullygate Affair'." *The International Journal of Press/Politics* 16 (3): 3–29.

Chadwick, Andrew. 2009. "Web 2.0: New Challenges for the Study of E-Democracy in an Era of Informational Exuberance." *I/S: A Journal of Law and Policy* 5 (1): 9–42.

Chadwick, Andrew. 2007. "Digital Network Repertoires and Organizational Hybridity." *Political Communication* 24: 283–301.

Chadwick, Andrew, and Philip N. Howard, eds. 2009. *Routledge Handbook of Internet Politics*. London: Routledge.

Curran, James, and Tamara Witschge. 2010. "Liberal Dreams and the Internet." In *New Media, Old News: Journalism & Democracy in the Digital Age*, edited by Natalie Fenton. London: Sage.

Dahlberg, Lincoln. 2005. "The Corporate Colonization of Online Attention and the Marginalization of Critical Communication?" *Journal of Communication Inquiry* 29 (2): 160–80.

Dahlgren, Peter. 2005. "The Internet, Public Spheres, and Political Communication: Dispersion and Deliberation." *Political Communication* 22 (2): 147–62; 176–77.

Dahlgren, Peter. 2001. "The Public Sphere and the Net: Structure, Space and Communication." In *Mediated Politics: Communication in the Future of Democracy*, edited by W. Lance Bennett and Robert Entman, 33–56. Cambridge: Cambridge University Press.

Dean, Jodi. 2010. *Blog Theory: Feedback and Capture in the Circuits of Drive*. Cambridge: Polity.

Hargittai, Eszter. 2006. "Content Diversity Online: Myth or Reality." In *Media Diversity and Localism: Meaning and Metrics*, edited by Philip M. Napoli. London: Routledge.

Howard, Philip N., and Andrew Chadwick. 2009. "Political Omnivores and Wired States." In *Routledge Handbook of Internet Politics*, edited by Andrew Chadwick and Philip N. Howard. London: Routledge.

Karppinen, Kari. 2009. "Rethinking Media Pluralism and Communicative Abundance." *Observatorio* 11: 151–69.

Livingstone, Sonia. 2005. "Critical Debates in Internet Studies: Reflections on an Emerging Field." Chap. 1 in *Mass Media and Society*, edited by James Curran and Michael Gurevitch, 9–29. London: Hodder Arnold.

Mansell, Robin. 2004. "Political Economy, Power and New Media." *New Media & Society* 6 (1): 74–83.

McNair, Brian. 2006. *Cultural Chaos: Journalism, News and Power in a Globalised World*. London: Routledge.

Papacharissi, Zizi. 2010. *A Private Sphere: Democracy in a Digital Age*. Cambridge: Polity.

Schoenbach, Klaus, Ester de Waal, and Edmund Lauf. 2005. "Research Note: Online and Print Newspapers: Their Impact on the Extent of the Perceived Public Agenda." *European Journal of Communication* 20 (2): 245–58.

Sparks, Colin. 2001. "The Internet and the Global Public Sphere." In *Mediated Politics: Communication in the Future of Democracy*, edited by W. Lance Bennett and Robert Entman. Cambridge: Cambridge University Press.

Stanyer, James. 2009. "Web 2.0 and the Transformation of News and Journalism." In *Routledge Handbook of Internet Politics*, edited by Andrew Chadwick and Philip N. Howard. London: Routledge.

Sunstein, Cass R. 2007. *Republic.Com 2.0*. Princeton, NJ: Princeton University Press.

Terranova, Tiziana. 2004. *Network Culture: Politics for the Information Age*. London: Pluto Press.

Tewksbury, David, and Jason Rittenberg. 2009. "Online News Creation and Consumption: Implications for Modern Democracies." In *Routledge Handbook of Internet Politics*, edited by Andrew Chadwick and Philip N. Howard. London: Routledge.

Tolbert, Caroline J., and Ramona S. McNeal. 2003. "Unraveling the Effects of the Internet on Political Participation?" *Political Research Quarterly* 56 (2): 175–85.

Van Dijk, Jan A. G. M. 2005. *The Deepening Divide: Inequality in the Information Society*. London: Sage.

Wellman, Barry. 2004, "The Three Ages of Internet Studies: Ten, Five and Zero Years Ago." *New Media & Society* 6 (1): 123–29.

Williams, Bruce A., and Michael X. Delli Carpini. 2004. "Monica and Bill All the Time and Everywhere: The Collapse of Gatekeeping and Agenda Setting in the New Media Environment." *American Behavioral Scientist* 47 (9): 1208–30.

Williams, Bruce A., and Michael X. Delli Carpini. 2000. "Unchained Reaction: The Collapse of Media Gatekeeping and the Clinton–Lewinsky Scandal." *Journalism* 1 (1): 61–85.

Part I

Accessibility

I

New Audiences, New Markets: Accessing Music, Movies, Art, and Writing at Your Leisure

Evan Bailey

NEW AUDIENCES AND MARKETS

Around the world, Internet users are moving toward a shared, global culture. A recent Ericsson ConsumerLab[1] survey of 23 countries found that more than three-quarters of consumers browse the Internet and half use social media daily. With the help of new technology, new opportunities are emerging for people to enjoy, engage, and share music, movies, art, and the written word.

Media usage is globalizing and audiences are evolving. Today's viewers are shifting toward inexpensive, on-demand services that allow multiplatform access to content. Regardless of where or how you receive your favorite content, the content is increasingly likely to be streamed, mobile, and wearable.

The advent of new Internet technologies and platforms also induces new debates on how artists monetize their work, copyrights are protected, and companies price content. The primary focus of this chapter, however, is the possibilities that the Internet and associated technology afford consumers, advertisers, and brand partners—and what is likely for 2015 and beyond. From nontraditional, new streaming players and content creators to the promise of new networks, the entertainment space is in full upheaval.

The global interest in streaming content in 2014 was remarkable, especially in the music sector. Although technology and Internet trends constantly evolve and are difficult to predict, streamed content looks to play a major role in how we access content for leisure in coming years.

MUSIC

According to Nielsen SoundScan,[2] streamed music grew 54 percent in 2014—from 106 billion songs streamed in 2013 to 164 billion songs streamed in 2014. The gains in streaming were in stark contrast to the diminishing number of traditional downloads of songs, which dropped off significantly from 2013 to 2014. Paid downloads for full music albums declined nearly 10 percent in 2014, and individual song downloads dropped 12 percent.

As streaming services become more important to our global culture, fans of niche music genres are sometimes overlooked. In some cases, larger streaming services can have trouble keeping up to date with the frenzied creation of releases, bootlegs, and live sets, which often leaves fans to seek music on their own.[3]

In turn, music fans have turned platforms such as SoundCloud, which has been troubled with "takedown" issues brought about by copyright crackdowns initiated by major publishers and labels. Unlike services such as Apple's iTunes or Spotify, SoundCloud doesn't own a full music catalog because what's available to consumers is limited to the content that people and music companies upload. SoundCloud, like YouTube, also has copyright infringement tool, which halts the upload of copyrighted songs, or removes them from site when copyright concerns are raised; the current process, however, isn't perfect.

Some prominent artists, such as Kaskade, have mandated that their music be removed or their profile pages be deleted from SoundCloud. This has prompted pressure from consumers and artists for the involved parties to reach a solution. According to Bloomberg, SoundCloud is negotiating with Universal Music Group, Sony Music Entertainment, and Warner Music Group by offering the labels a stake in the private company in exchange for the right to continue playing the labels' catalogues without legal disputes over copyright violations.[4]

Besides copyright issues, even "too large to fail" music-focused social networks—such as Apple's Ping—were met with challenges in the past. Apple introduced Ping in fall of 2010, but began shutting down the service in 2012, and replaced it with iTunes Facebook and Twitter integration.[5] Streaming services such as Spotify, Pandora, and iTunes Radio—which at one point represented a new hope for an ailing music business—haven't been the financial successes that the record labels expected.

NEW APPROACHES

New for 2015, some companies are creating new streaming services that attempt to address some of the music industry's current challenges, including a revenue model. A few examples of these new approaches are YouTube's Music Key, Google Chrome Cast, and Beatport's new consumer-facing strategy.

Beatport Takes Aim at Millennial Culture

Online electronic dance music retailer Beatport plans to relaunch in 2015 as a free streaming service, to create a cultural hub for the large—but niche—millennial dance music audience. Prior to the relaunch, Beatport's 50 million users mostly were professional disc jockeys that used the site as a paid download service.

In 2013, Beatport was acquired by SFX Entertainment, a company solely focused on electronic dance music culture. The SFX portfolio also includes numerous large-scale dance music festivals around the world, such as Tomorrowland, Life in Color, and Electric Zoo. The new consumer-facing Beatport aims to link the real and digital world by offering a portal of integrated live experiences, streaming, and perhaps eventually live event ticketing. SFX feels the ad-supported service will provide advertisers an efficient way of connecting with an entire audience and cultural movement.

Richard Ronstein, CFO of SFX, Beatport's parent company, in a Q&A session stated, "Historically Beatport has been a music store for professional and pro-sumer DJs, which is really a download service. However, the consumer interest in this goes beyond downloads, and in fact, many of our consumers are not interested in downloads and paying for a song, they're interested in streaming."[6]

With a new focus on streaming and on the culture itself, Beatport is vying to position itself to become the destination for the electronic music community of DJ/producers and fans alike. The Beatport model also creates a new platform for advertisers and brand partnerships, created to reach the notoriously hard-to-reach millennials.

Ronstein's point also seems to elaborate on an April 2014 release about an update to its core technology, in which Beatport announced that it had developed an API (application programming interface) to help integrate partners with Beatport and advance elements of electronic music culture.[7]

The Beatport API provides access to the millions of electronic music tracks, licensed and streaming mixes, and sounds available in the Beatport catalog, plus Beatport's database of live dance music events. The SFX database of live events also potentially holds promise to link fans in the real world to fans online with the potential for user-submitted content and live event ticketing. The API is intended to evolve to include upcoming Beatport consumer features as well as music data and analytics, which provide a framework for developers to create new web, mobile and other apps, services, and partnerships.[8]

Instead of only selling songs, the new site reportedly will feature a free, ad-supported streaming service and include the ability for users to listen to Beatport's catalog of on-demand music.[9] In 2015, Beatport is likely to seek select partners to explore and collaborate on this open API. The API will extend the next-generation Beatport platform, allowing for collaboration in line with the essence of electronic music culture.

YouTube Music Key

Google also has announced the launch of a streaming service called You-Tube Music Key, which offers "ad-free music, audio-only playback, and offline playback." Millions of users already listen to music on YouTube, but Music Key allows them to hear their favorite tunes without advertisements. The awaited service will be available by invite only. At the time of this writing, YouTube Music Key only was collecting the e-mail addresses of interested users until the service becomes more widely available. According to CNN Money,[10] those who receive an invite will be granted a six-month free trial. After six months, the beta version will cost $7.99 a month, and the future standard rate will be $9.99 a month. The service will give users access to YouTube's vast catalogue of more than 20 million tracks, albums, remixes, and live sets. The audio-only playback means that users won't use their cellular data allowance. Users also will be able to play music without ads and offline via YouTube's current iPhone and Android phone apps.

YouTube is entering the streaming market at a time when widespread disagreement exists about whether streaming positively or negatively affects the music industry as a whole. The debate was brought to the forefront when Taylor Swift pulled her entire catalog from the popular music streaming service, Spotify, a week after the release of her highly anticipated fifth album, "1989."

Other Streaming Contenders

Another potential contender in the streaming arena is Beats Music. Apple acquired Beats in August 2014, and left some parties wondering about the fate of the service. Rumors have circulated that a Beats service could be included in an iOS software update in the future. The move would pit Apple in direct competition against Spotify which, although technically profitable, has suffered setbacks—including the spat with Taylor Swift and other artists who remain skeptical about the economics of streaming. By preloading the service on its devices, however, Apple potentially would gain an edge in promoting its own service over the competitors such as Spotify and Rhapsody. Such a service also potentially would keep consumers locked into the Apple environment.

Apple's $3 billion purchase of Beats Music was the biggest acquisition in Apple's history, therefore a dedicated push to make Beats Music an integral part of Apple's hardware would come as no surprise, but such a move would mark an evident approval on the subscription-streaming-music model.[11] Apple historically has shied away from the streaming, but—given the state of slowing digital music downloads and a drop in iTunes Store sales—Apple needs to retain its positioning in the music market. To some degree, the Beats purchase would appear to be validation of the Spotify model. Apple also possesses inherent strengths to foster adoption of Beats. For instance, it could use its TouchID fingerprint reader as an easy way to subscribe to the service. The company already uses TouchID for mobile payments in stores and for in-app purchases.

Although receiving a service preloaded onto a device potentially helps stimulate consumer adoption, it doesn't guarantee success. Apple bundled iTunes Radio on its iTunes music app, for example, but the service hasn't taken notable market share away from the Internet radio giant Pandora. Integrating the service into iOS, however, could be a good entry point for first-time subscribers.

Google Chrome Cast

At the 2015 Consumer Electronics Show in Las Vegas, Google announced that it will extend its Google Cast streaming-media technology to work in audio-only devices, including speakers and home entertainment sound systems. Google Cast for Audio enables listeners to wirelessly control and send the music playing on an iOS, Android device, or Chrome browser right to new Google Cast–enabled audio gear.[12] Instead of a phone, tablet, or computer being source of the music or video, the Google Cast device pulls content from the cloud at the highest quality possible.

Spotify and Music Economics

Launched in Stockholm in 2008, Spotify has a library of more than 20 million songs and permits users to choose from millions of songs available over the Internet for free or by subscription. Spotify increasingly is seen as representing the future of music consumption. Users either pay for the premium service to stream music without interruption or they listen for free but with advertisements between songs. Per Sundin, head of Universal Music in Sweden, argues that Spotify—a service in which Universal and other big record labels have minority stakes—has saved the music industry, although not everyone agrees.

A decade after Apple revolutionized the music world with its iTunes store, the music industry is undergoing another—even more radical—digital transformation as listeners begin to move from CDs and downloads to using streaming services such as Spotify, Pandora, and YouTube. What is certain is that streaming music services are changing the economics of music.

Spotify serves its audiences legally licensed music, thus the company generally has been embraced by an industry still affected by piracy. As digital music services grow into multibillion-dollar companies, however, the proportionate trickle of money making its way to artists is concerning to some. The way streaming services pay artist royalties is fundamentally different from the way the industry has paid royalties in the past. Previously, record royalties were typically a percentage of a sale. For example, for every dollar made, a typical artist might have received $0.07 to $0.10, after paying the label, the distributor, and other fees. Under the streaming model royalties are closer to a fraction of a cent. The royalties accrue over time and leave some to wonder if the total can amount to anything substantial.[13]

TV, FILM, AND STREAMING

By 2020, the number of TV sets connected to the Internet is expected to reach 965 million, nearly nine times the number connected in 2010, and 200 million homes are expected to subscribe to a video on-demand service. The average person now spends five hours per week watching TV, video clips, and films on the Internet. The surge partially has been driven by increased use of smartphones and tablets. As screen sizes become bigger and mobile Internet speeds get faster, mobile video viewing is only going to increase. YouTube revealed in 2014 that half of its video views now originate from a mobile device.[14]

For audiences, producers, and content, greater resolutions are preferred. A display device for content using 3840 x 2160 resolution generally is referred to as "4K resolution." Several 4K resolution types exist in the fields of digital television and digital cinematography, but YouTube and the television industry have adopted Ultra HD as the 4K standard. In coming years, the 4K standard likely will have a strong influence on device construction, Internet platforms, and mobile apps. For example, although Apple didn't clearly announce the capability during its iPhone 6 release, the device actually can play 4K videos. Quality is limited by the resolution of the screen, therefore the larger size and greater resolution iPhone 6 Plus technically is a better device for viewing 4K than is the iPhone 6. Although the iPhone 6 and 6 Plus can't record 4K video, rumors already circulating about the iPhone 7 suggest that it will include a significant upgrade of the camera, which might enable it to shoot 4K video. In turn, it's possible that new versions of Apple TV might add the feature.

The emphasis on 4K in the mobile device and TV category is where new developments such as the Netflix and Amazon 4K streams, increased use of encoding, and the propagation of faster Internet connections are becoming a significant part of the Internet and home entertainment picture. Netflix also is promising to add much 4K content in 2015. Regardless of the costs, consumers are likely to be offered a much broader development of streaming entertainment in 2015 and growth in 4K-related streaming technologies and services. Consumers, broadcasters, and content creators could all potentially benefit from the developments, but some exceptions do exist. The benefits, however, are clear. With that in mind, the following are four trends in streaming 4K Ultra HD content that likely will emerge in coming years.

EXPANSION OF 4K CONTENT

To date, Netflix has offered 4K streams of various TV shows and movies since April of 2014. In December 2014, Amazon also rose to the challenge and offered 4K streams via their Amazon Prime Instant Video service. The two companies join a roster of other companies offering 4K streams, including DirecTV, Comcast, and Sony Unlimited Instant Video. These content broadcasters, and others, plan to extend their selection of 4K entertainment in 2015 and DirecTV has even launched two satellites to deliver increased 4K delivery.[15]

Financial models involving 4K are mixed. In some cases, companies are charging a premium for 4K content. New Netflix subscribers could pay $11.99 a month to access 4K TV. The "Action" service enables audiences to access four shows simultaneously, which is ideal for a family that owns multiple devices. Netflix has announced that it will use the fees to fund the production of more 4K UHD content, because filming new TV shows in 4K is a relatively expensive process. Increased costs are derived from the need for new cameras and production facilities for 4K UHD content, including for handling, storing, and streaming the huge amounts of data associated with the format.[16]

At the time of this writing, Amazon's 4K Ultra HD streaming service made much of its 4K content available to its Prime Instant subscribers in the United States at no cost. To use Amazon's 4K stream, however, viewers must have a compatible 4K TV built in 2014 (Samsung, LG, Sony, Vizio) or later and that can handle the app. Other brands almost certainly will join the list of compatible devices in 2015 and beyond.

LIVE 4K STREAMS

Live 4K streams are another technology to watch on the Internet for 2015. Live 4K sports broadcasts have already been conducted by several broadcasters in beta tests delivered to small audiences. Sony, BBC, and several other companies delivered a handful of FIFA 2014 World Cup matches to small test audiences. These companies and others are working to make live 4K feeds a reality, and in 2015 we're likely to see some of the first widespread commercial broadcasts of live sports events in 4K, especially given that DirecTV's two new satellites are in orbit—and live 4K broadcasting is one of their primary tasks.[17]

ULTRA–HIGH SPEED BROADBAND INTERNET CONNECTIVITY

Because of the massive amounts of data that 4K streams use even when compressed or encoded, truly fast high-speed Internet is crucial to wider adoption. Internet service providers and broadcasters have accepted the challenge and a race to deliver faster connectivity to consumers is under way in some larger metropolitan areas.[18] The challenge entails an expansion of the number of homes that have access to high-speed Internet, and even includes introducing connectivity of 500 Mbps or more to subscribers in some areas. Currently, Google Fiber is a leading high-speed broadband Internet provider in the United States, but the goal is to increase speeds tenfold in order and blast the Internet into living rooms across the country. Although Google Fiber offers a marked speed advantage, the number of homes it actually reaches is small. The company set up a test network in Palo Alto in 2011 before rolling out its first commercial installation in Kansas City, Missouri, after strong public demand. The cities next on the list for deployment are Austin, Texas, and Provo, Utah.[19]

By 2020, it is estimated that billions of devices will be connected to the Internet. Ultra-fast Internet could have a substantial effect on our lives in the future. New mobile networks such as 5G could be a reality by 2020, with holographic video meetings, driverless cars, and automated homes being some of many applications of the technology. The 5G mobile networks fundamentally have the potential to overhaul how we communicate using the radio spectrum. At present, the radio spectrum humans use to communicate is mostly allocated. But 5G will increase the number of available channels, thus allowing a greater number of channels to operate simultaneously at higher bandwidths, with low latency and reduced power consumption. In other words, the 5G capabilities enable significant numbers of low-power objects to speak to other Internet-ready devices with decreased delay and extremely high stability.

Although 5G still is being researched and developed, the creation of the first truly global network could play a big role in a variety of fields, including health care, transportation, and home construction.[20] Additionally, such developments also raise entirely new privacy and security concerns for consumers. What happens, for example, if hackers gain access to the "smart" products in your house?

4K STREAMING ON MOBILE DEVICES

The first wave of mobile phones featuring 4K screen displays went on sale in 2015. The capability to render UHD video will draw increased consumer interest for streaming 4K content on mobile devices.[21]

For consumers with Wi-Fi connections that enable them to download larger 4K streams, the first ultra-HD content from YouTube should be accessible for their 4K mobile phone screens at some point in 2015.[22] Additionally, 4K streams from a smartphone to an ultra–high definition TV also appear to be a possibility for 2015. In December of 2014, *PC Magazine* reported that new devices with a USB Type-C connector might support the new mobile high-definition link (MHL) 3 specification, which includes streaming 4K video. By the end of 2015, manufacturers are slated to begin replacing current USB 3.0 ports with the next-generation USB 3.1, to double the transfer rate from 4.8 Gbps to 9.6 Gbps.[23]

PERSONAL VIDEO DISTRIBUTION

With more user-submitted content being produced in 2014 than previously, 2015 will see more experimentation with how it is distributed. Applications such Snapchat and WhatsApp offer a more personal means of communication, and are ideal for reaching new and younger audiences.

In 2014, the Snapchat mobile application raised a remarkable amount of cash, putting the three-year-old company's valuation at more than $10 billion, according to Bloomberg.[24] Investors are seemingly eager to hitch a ride on Snapchat's incredible rise and continued growth. At the time of this writing, the app's monthly active user total was nearing the 200 million mark.[25] In 2014, Snapchat also announced the addition of Chat, a feature that enabled users to text and video message in real time. As personalized applications like Snapchat continue to grow and evolve, so does its user base of brands that are interested in using the platform to reach millennial audiences.

THE INTERNET, POSTMODERNISM, AND ACCESSING ART

Over the past decade, advances in technology have greatly affected the production and promotion of art. Electronic media techniques are incorporated by a majority of artists, and the use of the Internet by artists has had a substantial effect on how audiences access, interact with, and share art, as well as how artists promote and monetize their work. With advances in website technology, blogging, and social media platforms, artists have new mediums for exposing artwork. Gone are the days of traditional brick-and-mortar galleries being the primary outlet for viewing and purchasing art. Today's artists can use the Internet to directly market artworks internationally.[26] The Internet is increasingly making its way into the museum and galley space through mobile apps and interactive technologies.

Has the Internet impacted art in a positive or negative way? The answer depends partially upon your vantage point. Of course, gallery owners might not like being cut out of the deal when the public buys art online. But with these challenges come new opportunities for museums and galleries in the areas of mobile applications and wearables. For artists, the Internet has been mostly positive and has presented many opportunities, tools, and platforms that previously did not exist.

Perhaps the use of the Internet and accessing art is most usefully analyzed through a lens of ideas usually associated with postmodern theory. Postmodern theorists could argue, for example, that the goal in any conversation—including those conversations that characterize art—is not to find the "truth" but simply to further the conversation. The Internet can function as a platform for these conversations. It is a place in which there rarely is a final answer, a conclusion, a finished product, or a "truth."

Before the days of the Internet, artists often would work in relative seclusion of their studios primarily. Promotion of their work was a somewhat limited and often costly endeavor. Artists typically would converse with the public by attending art shows or joining arts organizations. With the advent of the Internet, artists easily could go online and show new projects and involve themselves in a conversation with their audiences. Today's artists can begin a blog, create a Facebook page, upload their work to an Etsy page to display, and engage in conversation. In this manner the Internet opened up a major platform for discussion—one which had not existed previously. In some ways, the notion of discussion and deeper exploration by the audience becomes central to the definition of Internet art itself.

INTERNET ART

Internet art historically has been defined as a form of digital artwork distributed via the Internet, sidestepping the traditional nation of the gallery and museum system. Internet art delivers an aesthetic experience from the use of the Internet itself. Often the audience becomes part of the work of art through interacting with the artworks, and the artists who work in this realm sometimes are referred to as "Net artists."

Internet art can be created in a variety of forms and presentations. Websites, e-mail, software projects, Internet installations, interactive pieces, streaming video, audio, networked performances, and even games, are all mediums used by Net artists. Internet art often overlaps with other computer-based art forms, such as new media art, electronic art, software art, and digital art.

POST-INTERNET ART

"Post-Internet art" is a fairly recent term circulating in the art world, and attempts to describe a growing development in the art world—one which could be one of the most noteworthy trends to emerge in some time. Contrary to how the term might seem at first glance, the key to understanding what "post-Internet" means is that the term doesn't suggest that technological advancements associated with the Internet are behind us.[27]

Instead, in the same way that postmodern artists absorbed and adapted the strategies of modernism for their aesthetics, post-Internet artists have moved beyond making work dependent on the novelty of the Web to using its tools to tackle other subjects. And although earlier Net artists often made works that existed exclusively online, the post-Internet generation—many of which have grown up with the Internet—frequently uses digital strategies to create objects that exist in the real world.

A handful of artists and galleries already are closely linked to post-Internet art, and curators are aiming to find ways to help these artists reflect our new relationship to images that are inspired by the culture of the Web.

DIGITAL MUSEUMS

For curators who wish to incorporate new technologies in museums today, the terms "online" and even "Web" have largely been replaced with "digital" as a description of such works. New technologies ranging from wearables to virtual reality are being used to enhance the museum experience—which challenges the notion that the computer monitor is central to digital experiences. Many museums have developed mobile applications, but apps can be expensive to develop and difficult to market and often "exist" primarily inside the museum itself. Some museums have developed less expensive alternatives—such as mobile tours—using WordPress and GPS technology to deliver customized experiences on a smaller budget.[28]

Along with mobile technologies, some museums are utilizing the quickly evolving field of wearable technology to interact with audiences. Although privacy can be a concern for some, the truth is that the next generation of museumgoers likely will rely upon some form of personal technology and augmented reality in their daily lives. Some museums are making use of wearables to provide personalized interpretation about art and potentially to create new types of museums. Regardless, technology and the Internet offer audiences new methods for experiencing the modern museum. These technologies also potentially offer new ways of bringing art and museums into the home.[29]

GOOGLE ART PROJECT AND DIGITAL MUSEUM INITIATIVES

The Google Art Project was released to the public in 2011 as an online platform enabling audiences to access high-resolution images of art from the comfort of their own homes. The project was especially ambitious when it launched in 2011, as high-resolution zooming on the Web was in its infancy. Seventeen museums collaborated on the project with the single goal being to come together in a central platform that audiences could use to explore art in new and dynamic ways. The project now contains tens of thousands of collections, artists, artworks, and user galleries.

Along with high-resolution imagery, Google Art Project content also featured a cadre of tools and metadata to enhance the audience's understanding and appreciation of the artwork. For example, museumgoers could view a selected work, zoom in on it, compare it with another work, and then take a virtual tour through the museum by using Google's street-view technology.[30]

Since the release of the Google Art Project, other art institutions have followed suit. In December of 2014, the Freer Gallery of Art, the Arthur M. Sackler Gallery, and the Smithsonian Institution's national museums of Asian art in the United States released their entire collections online. The artwork now is available for both viewing and downloading for noncommercial use.[31]

The Freer and Sackler galleries are the first Smithsonian entities to release their entire collections digitally, and join a small number of other U.S. museums that have undertaken similar initiatives. Although Open F | S is the galleries' largest digital initiative it is not the first; the galleries participated in the launch of the Google Art Project.

The Smithsonian created the initiative with the goal of providing the public access to the images for noncommercial educational, scholarly, artistic, and personal projects. Additionally, the Smithsonian has made available a series of works in specific sharable formats such as mobile backgrounds, desktop wallpapers, and social media headers. The general public also is invited to participate in the initiative and offer "beta tester" feedback.

VIRTUAL REALITY AND THE VISUAL ARTS

In coming years, virtual reality and the Internet potentially could offer great opportunities in the world of museums, visual arts, video games, and even marketing. One company ready for that future is Oculus VR, an American virtual-reality technology company. The company's first product was the Oculus Rift, a head-mounted display for immersive virtual reality (VR). In March 2014, Facebook agreed to acquire Oculus VR for U.S. $2 billion (in cash and Facebook stock). The Oculus Rift headset generated substantial media coverage and notable buzz in 2014, and is joined by a host of notable competitors. Google and Samsung plan to ship virtual reality sets to market in 2015.[32]

For museums, the first step is to re-create existing museums online, thus enabling audiences to visit, explore new environments, and share information and experiences. One such exhibition occurred in November 2014, when the Steven F. Udvar-Hazy Center (part of the Smithsonian National Air and Space Museum) in Virginia hosted the world-exclusive exhibition of the "Ranger" spacecraft from the Christopher Nolan film *Interstellar*. The exhibition included an immersive Oculus Rift DK2 experience, which enabled visitors to interactively explore the film's "Endurance" spacecraft via virtual-reality technology; experience weightlessness; and see views of the galaxy.

MOBILE AND STREAMING IN THE ARTS

Mobile technology has changed how the public experiences art. Anyone with a smartphone can enjoy art, either on-demand or on the go. Mobile technology can tell the artists about the participant, and vice versa.[33] Live theatre is one such area where artists are harnessing the power of Internet streaming.

In November 2014, The Globe Theater in London—the most famous Shakespeare venue in the world—launched Globe Player, a paid service that allows users to stream or download its productions onto their computers and mobile devices. Online audiences can watch more than 50 of the theater's past productions.

THE INTERNET AND THE WRITTEN WORD

Technology is constantly changing how people tell their stories. Before the invention of the printing press, people told stories orally. The printing press introduced books to the masses. Now technology has evolved into digital books. The invention of computers, tablets, and smartphones has changed the way we access the written word, read for leisure, and learn.

Today's textbooks no longer are strictly limited to flat text on a page with a few helpful images placed within the content. With the rise of technology comes a revolution in books, which are now interactive with Web-based content. While reading, students can make assessments with instant feedback, and readers can view animations that aid in understanding the content, view additional material, watch videos, and experience other content that helps with learning. eReaders enable audiences to experience content. In 2015, half of American adults own an e-reading device, but research on how e-books affect learning is scarce.

The dialog over the merits of paper books and e-readers has been heated since the first Kindle debuted in 2007. The discussion ranged from the sentimental to the practical. From how well we comprehend the digital words found in e-books, to health concerns about how safe it is to read the e-books before we go to sleep at night.

Amazon—perhaps the leader in the e-reader movement—has been particularly disruptive to these previously held notions, especially regarding the old methods of doing business. The retailer's well-documented fights with publishers—such as the Hachette Book Group—have been played out in the media. Amazon's 2014 all-you-can-read venture, Kindle Unlimited, was a catalyst in the heated debate, both for and against the service, especially among self-published writers. The subscription service gives users access to more than 600,000 e-books and audiobooks for $10 per month.[34]

Both Amazon and Hachette had been seeking a deal once Hachette's contract with Amazon expired in March 2014, only to be briefly extended by the online retailer into April. The negotiations drew protests from authors on both sides of the dispute, including some who called for Justice Department scrutiny of Amazon's business practices, which often were characterized in the media as being aggressive.

Hachette eventually won a victory against Amazon: The ability to set its own prices for e-books, which the company sees as critical to its survival. The conflict, which played out in increasingly public forums as the year progressed, caused damage on both sides. In the media Amazon was cast as a bully, and a large group of authors called for it to be investigated on antitrust grounds, although sales were hit by the dispute. Hachette showed its weakness and consumers had difficulty buying certain titles. It seemed that neither party really won.

Amazon's supporters publicly questioned the need for Hachette, a relatively large publisher, to exist at a time when authors can publish on their own, digitally. Hachette never seemed to fully respond to the accusation. Even if Amazon did receive less than it wanted, the company still controls nearly half the market for books, a previously unheard-of portion of the market captured by one retailer. Some argued that the dispute showed that Amazon was not afraid to use its power.

KINDLE UNLIMITED PROGRAM

Controversy aside, the Kindle Unlimited program offers readers access to the written word in new ways. The service grants readers the freedom to explore new genres and authors, with unlimited access to a wide and varied selection of books. Some Kindle Unlimited books also come with the free professionally narrated "Audible" audiobook. A related feature—Whispersync for Voice—enables users to switch seamlessly between reading and listening without losing their place. Using headphones, readers can continue enjoying their books when in the car, at the gym, or even in the kitchen. Readers can use the service on any device with the Kindle app installed, thus Kindle Unlimited can be used with or without a Kindle, and wherever readers might be located.

CONCLUSION

Whatever form content might take, the Internet and technology are enabling audiences to access music, art, film, and the written word for leisure in new and exciting ways. The future of this content is mobile, streamed, wearable, and increasingly connected to many facets of our lives at home, at work, in our cars, and for leisure.

Technology trends are implications of what today's consumers are looking for as they desire to work and live connected to each other via the Internet. Few aspects of our lives are unconnected. Humans constantly are connecting new objects and "smartening" every device so that it is informed of our choices and preferences. These technological advances have fostered a shared, global Internet culture with an appetite for inexpensive, on-demand services that allow multiplatform access to content.

This fundamental shift in our culture also has implications for content creators, advertisers, and brand partners. From streaming players, to the promise of new networks, to the debate over associated royalties, the entertainment space is in full upheaval. One thing is certain, however: New opportunities have emerged for people to enjoy, engage, and share music, movies, art, and the written word.

NOTES

1. ConsumerLab, "10 Hot Consumer Trends 2015," *ericsson.com*. (October 31, 2014), http://www.ericsson.com/res/docs/2014/consumerlab/ericsson-consumerlab-10 -hot-consumer-trends-2015.pdf, accessed December 10, 2014.

2. Ethan Smith, "Music Downloads Plummet in U.S., but Sales of Vinyl Records and Streaming Surge," *Wall Street Journal* (January 1, 2015), http://www.wsj.com /articles/music-downloads-plummet-in-u-s-but-sales-of-vinyl-records-and-streaming -surge-1420092579, accessed January 1, 2015.

3. Nick Jarvis, "Get Excited, Dance Fans: Beatport and YouTube Tipped to Launch Streaming Services," *In the Mix* (August 20, 2014), http://www.inthemix.com.au /industry/58769/Get_excited_dance_fans_Beatport_and_YouTube_tipped_to _launch_streaming_services, accessed December 15, 2014.

4. Adam Satariano, "SoundCloud Said to Near Deals with Record Labels," *Bloomberg* (July 10, 2014), http://www.bloomberg.com/news/2014-07-10/soundcloud -said-to-near-deals-with-record-labels.html, accessed December 2, 2014.

5. Christina Bonnington, "So Long, Ping: Apple Shuttering Failed Social Network Sept. 30," *Wired* (September 13, 2012), http://www.wired.com/2012/09/good bye-ping/, accessed December 16, 2014.

6. Seekingalpha.com, *SFX Entertainment's (SFXE) CEO Bob Sillerman on Q2 2014 Results—Earnings Call Transcript* (August 14, 2014), http://seekingalpha.com/article /2426255-sfx-entertainments-sfxe-ceo-bob-sillerman-on-q2-2014-results-earnings -call-transcript?page=7&p=qanda&l=last, accessed December 20, 2014.

7. Megan Buerger, *Beatport Aims to Be the One-Stop Streaming Shop for DJs with New Strategy* (December 19, 2014), http://www.billboard.com/biz/articles/news /digital-and-mobile/6413870/beatport-aims-to-be-the-one-stop-streaming-shop-for -djs, accessed December 19, 2014.

8. BusinessWire, "Beatport Announces Definitive API for Electronic Music Culture," *BusinessWire* (April 11, 2014), http://www.businesswire.com/news/home /20140411005642/en/Beatport-Announces-Definitive-API-Electronic-Music-Culture# .VKbAw2TF9Vo, accessed December 6, 2014.

9. Aaron Berecz, "SFX Entertainment to Re-Launch Beatport, Add Streaming Service in 2015," *Dancing Astronaut* (December 14, 2014), http://www.dancingastro naut.com/2014/12/sfx-entertainment-re-launch-beatport-add-streaming-service -2015/, accessed December 14, 2014.

10. Frank Pallotta, "YouTube to Launch Premium Music Streaming Service," *CNN Money* (November 12, 2014, http://money.cnn.com/2014/11/12/media/youtube -music-streaming-service/index.html?iid=EL, accessed December 14, 2014; Frank Pallotta, "Taylor Swift Pulls Her Music from Spotify," *CNN Money* (November 3, 2014), http://money.cnn.com/2014/11/03/media/taylor-swift-spotify/?iid=EL, accessed December 16, 2014.

11. Roger Solsman and Joan E. Cheng, "Apple Said to Be Embedding Beats Music Service into iOS," *CNET* (November 19, 2014), http://www.cnet.com/news/apple-to -embed-beats-music-service-into-ios-software-ft-says/, accessed December 17, 2014.

12. Joanna Stern, "Consumer Electronics Show Roundup," *Wall Street Journal* (January 5, 2015), http://www.wsj.com/articles/consumer-electronics-show-roundup -1420512446, accessed January 10, 2015.

13. Ben Sisario, "As Music Streaming Grows, Royalties Slow to a Trickle," *Media and Advertising: The New York Times* (January 28, 2013), http://www.nytimes.com /2013/01/29/business/media/streaming-shakes-up-music-industrys-model-for-royalties .html?pagewanted=all&_r=0, accessed December 20, 2014.

14. Stephan Jukic, "Streaming Video and 4K: Trends We Can Look Forward to in 2015," *4K* (December 29, 2014), http://4k.com/news/4k-streaming-trends-that-are -coming-in-2015-4853/, accessed December 29, 2014.

15. Ibid.

16. John Archer, "Netflix's 4K/UHD Price Hike Is Its Second Huge Mistake," *Tech: Forbes* (October 17, 2014), http://www.forbes.com/sites/johnarcher/2014/10/17 /netflix-4k-uhd-price-hike-is-second-huge-mistake/, accessed December 20, 2014.

17. Jukic, "4K"; Stephanie Mlot, "Get Ready for 4K Streams from Phone to HDTV," *PC Magazine* (November 17, 2014), http://www.pcmag.com/article2 /0,2817,2472292,00.asp, accessed December 20, 2014.

18. Jukic, "4K."

19. Iain Thomson, "Google Promises 10Gps Fiber Network to blast 4K into Living Rooms," *The Register* (February 14, 2014), http://www.theregister.co.uk/2014/02 /14/google_will_upgrade_fiber_network_to_give_10gbps_home_broadband, accessed December 21, 2014.

20. Dr. John-Paul Rooney, "5G Visionaries Make a Head Start on the Internet of Things," *The Telegraph* (December 20, 2014), http://www.telegraph.co.uk/technology /mobile-phones/11303838/5G-visionaries-make-a-head-start-on-the-internet-of -things.html, accessed January 10, 2015.

21. Jukic, "4K."

22. Ibid.

23. Mlot, "Get Ready for 4K Streams."

24. Sarah Frier, "Snapchat Raises $485.6 Million to Close Out Big Fundraising Year," *Bloomberg* (December 31, 2014), http://www.bloomberg.com/news/2015-01-01 /snapchat-raises-485-6-million-to-close-out-big-fundraising-year.html, accessed January 1, 2015.

25. Chris Welch, "Snapchat Closes Out 2014 by Raising a Ton of Cash," *The Verge* (January 1, 2015), http://www.theverge.com/2015/1/1/7477563/snapchat-closes -2014-by-raising-ton-of-cash, accessed January 2, 2015.

26. Artpromotivate, "How Is the Internet Impacting the Life of Artists?," *Artpromotivate* (December 20, 2012), http://www.artpromotivate.com/2012/12/how-is -internet-impacting-life-of.html, accessed December 18, 2014.

27. Ian Wallace, "What Is Post-Internet Art? Understanding the Revolutionary New Art Movement," *ArtSpace* (March 18, 2014), http://www.artspace.com/magazine /interviews_features/post_internet_art, accessed December 15, 2014.

28. Danny Birchall, and Mia Ridge, "Post-Web Technology: What Comes Next for Museums?," *Culture Professionals Network: The Guardian* (October 13, 2014), http://www.theguardian.com/culture-professionals-network/culture-professionals -blog/2014/oct/03/post-web-technology-museums-virtual-reality, accessed December 21, 2014.

29. Ibid.

30. Matthew Caines, "Arts Head: Amit Sood, Director, Google Cultural Institute," *Culture professionals network: The Guardian* (December 3, 2013), http://www .theguardian.com/culture-professionals-network/culture-professionals-blog/2013 /dec/03/amit-sood-google-cultural-institute-art-project, accessed December 21, 2014.

31. Jackie Dove, "Smithsonian Galleries Release Massive Asian Art Collection Online for Non-Commercial Use," *The Creativity Channel: Thenextweb.com* (January 3, 2015), http://thenextweb.com/creativity/2015/01/04/smithsonian-galleries-release -massive-asian-art-collection-online-for-non-commercial-use/, accessed January 3, 2015.

32. Christopher Heine, "How Oculus Rift Is About to Reshape Marketing Creativity," *CES 2015: AdWeek* (January 5, 2015), http://www.adweek.com/news/techn

ology/how-oculus-rift-about-reshape-marketing-creativity-162124, accessed January 6, 2015.

33. Matt Trueman, "How Mobile Tech Is Changing the Way We Make and Enjoy Art," *The Guardian* (November 11, 2014), http://www.theguardian.com/culture -professionals-network/2014/nov/11/-sp-mobile-tech-art-shakespeare-google-glass, accessed December 21, 2014.

34. David Streitfeld, "Amazon and Hachette Resolve Dispute," *Technology: The New York Times* (November 13, 2014), http://www.nytimes.com/2014/11/14/technology /amazon-hachette-ebook-dispute.html, accessed December 15, 2014.

BIBLIOGRAPHY

Archer, John. 2014 (October 17). "Netflix's 4K/UHD Price Hike Is Its Second Huge Mistake." *Tech: Forbes.* http://www.forbes.com/sites/johnarcher/2014/10/17/netflix -4k-uhd-price-hike-is-second-huge-mistake/. Accessed December 20, 2014.

Artpromotivate. 2012 (December 20). "How Is the Internet Impacting the Life of Artists?"*Artpromotivate.* http://www.artpromotivate.com/2012/12/how-is-internet -impacting-life-of.html. Accessed December 18, 2014.

Berecz, Aaron. 2014 (December 14). "SFX Entertainment to Re-Launch Beatport, Add Streaming Service in 2015." *Dancing Astronaut.* http://www.dancingastronaut .com/2014/12/sfx-entertainment-re-launch-beatport-add-streaming-service-2015/. Accessed December 14, 2014.

Birchall, Danny, and Mia Ridge. 2014 (October 13). "Post-Web Technology: What Comes Next for Museums?" *Culture Professionals Network: The Guardian.* http:// www.theguardian.com/culture-professionals-network/culture-professionals -blog/2014/oct/03/post-web-technology-museums-virtual-reality. Accessed December 21, 2014.

Bonnington, Christina. 2012 (September 13). "So Long, Ping: Apple Shuttering Failed Social Network Sept. 30." *Wired.* http://www.wired.com/2012/09/goodbye -ping/. Accessed December 16, 2014.

Buerger, Megan. 2014 (December 19). "Beatport Aims to Be the One-Stop Stream- ing Shop for DJs with New Strategy." http://www.billboard.com/biz/articles/news /digital-and-mobile/ 6413870/beatport-aims-to-be-the-one-stop-streaming-shop -for-djs. Accessed December 19, 2014.

Business Wire. 2014 (April 11). "Beatport Announces Definitive API for Electronic Music Culture." *BusinessWire.* http://www.businesswire.com/news/home/2014 0411005642/en/Beatport-Announces-Definitive-API-Electronic-Music-Culture #.VKbAw2TF9Vo. Accessed December 6, 2014.

Caines, Matthew. 2013 (December 3). "Arts Head: Amit Sood, Director, Google Cul- tural Institute." *Culture Professionals Network: The Guardian.* http://www.theguardian .com/culture-professionals-network/culture-professionals-blog/2013/dec/03/amit -sood-google-cultural-institute-art-project. Accessed December 21, 2014.

Consumerlab. 2014 (October 31). "10 Hot Consumer Trends 2015." *ericsson.com.* Ericsson. http://www.ericsson.com/res/docs/2014/consumerlab/ericsson-consumer lab-10-hot-consumer-trends-2015.pdf. Accessed December 10, 2014.

Dove, Jackie. 2015 (January 3). "Smithsonian Galleries Release Massive Asian Art Collection Online for Non-Commercial Use." *The Creativity Channel: Thenextweb .com.* http://thenextweb.com/creativity/2015/01/04/smithsonian-galleries-release -massive-asian-art-collection-online-for-non-commercial-use/. Accessed January 3, 2015.

Frier, Sarah. 2014 (December 31). "Snapchat Raises $485.6 Million to Close Out Big Fundraising Year." *Bloomberg.* http://www.bloomberg.com/news/2015-01-01/snap chat-raises-485-6-million-to-close-out-big-fundraising-year.html. Accessed January 1, 2015.

Heine, Christopher. 2015 (January 5). "How Oculus Rift Is About to Reshape Marketing Creativity." *AdWeek.* http://www.adweek.com/news/technology/how -oculus-rift-about-reshape-marketing-creativity-162124. Accessed January 6, 2015.

Jarvis, Nick. 2014 (August 20). "Get Excited, Dance Fans: Beatport and YouTube Tipped to Launch Streaming Services." *In the Mix.* http://www.inthemix.com .au/industry/58769/Get_excited_dance_fans_Beatport_and_YouTube_tipped_to _launch_streaming_services. Accessed December 15, 2014.

Jukic, Stephan. 2014 (December 29). "Streaming Video and 4K: Trends We Can Look Forward to in 2015." *4K.* http://4k.com/news/4k-streaming-trends-that-are -coming-in-2015-4853/. Accessed December 29, 2014.

Mlot, Stephanie. 2014 (November 17). "Get Ready for 4K Streams from Phone to HDTV." *PC Magazine.* http://www.pcmag.com/article2/0,2817,2472292,00.asp. Accessed December 20, 2014.

Pallotta, Frank. 2014 (November 12). "YouTube to Launch Premium Music Stream-ing Service." *CNN Money.* http://money.cnn.com/2014/11/12/media/youtube -music-streaming-service/index.html?iid=EL. Accessed December 14, 2014.

Pallotta, Frank. 2014 (November 3). "Taylor Swift Pulls Her Music from Spotify" *CNN Money.* http://money.cnn.com/2014/11/03/media/taylor-swift-spotify/?iid =EL. Accessed December 16, 2014.

Rooney, John-Paul. 2014 (December 20). "5G Visionaries Make a Head Start on the Internet of Things." *The Telegraph.* http://www.telegraph.co.uk/technology /mobile-phones/11303838/5G-visionaries-make-a-head-start-on-the-internet-of -things.html. Accessed January 10, 2015.

Satariano, Adam. 2014 (July 10). "SoundCloud Said to Near Deals with Record Labels." *Bloomberg.* http://www.bloomberg.com/news/2014-07-10/soundcloud-said -to-near-deals-with-record-labels.html. Accessed December 2, 2014.

seekingalpha.com. 2014 (August 14). *SFX Entertainment's (SFXE) CEO Bob Sillerman on Q2 2014 Results—Earnings Call Transcript.* http://seekingalpha.com/article /2426255-sfx-entertainments-sfxe-ceo-bob-sillerman-on-q2-2014-results-earnings -call-transcript?page=7&p=qanda&l=last. Accessed December 20, 2014.

Sisario, Ben. 2013 (January 28). "As Music Streaming Grows, Royalties Slow to a Trickle." *Media and Advertising: The New York Times.* http://www.nytimes.com /2013/01/29/business/media/streaming-shakes-up-music-industrys-model-for -royalties.html?pagewanted=all&_r=0. Accessed December 20, 2014.

Smith, Ethan. 2015 (January 1). "Music Downloads Plummet in U.S., but Sales of Vinyl Records and Streaming Surge." *Wall Street Journal.* http://www.wsj.com

/articles/music-downloads-plummet-in-u-s-but-sales-of-vinyl-records-and-streaming -surge-1420092579. Accessed January 1, 2015.

Solsman, Roger, and Joan E. Cheng. 2014 (November 19). "Apple Said to Be Embedding Beats Music Service into iOS." *CNET*. http://www.cnet.com/news/apple-to -embed-beats-music-service-into-ios-software-ft-says/. Accessed December 17, 2014.

Stern, Joanna. 2015 (January 5). "Consumer Electronics Show Roundup." *Wall Street Journal*. http://www.wsj.com/articles/consumer-electronics-show-roundup -1420512446. Accessed January 10, 2015.

Streitfeld, David. 2014 (November 13). "Amazon and Hachette Resolve Dispute." *Technology: The New York Times*. http://www.nytimes.com/2014/11/14/technology /amazon-hachette-ebook-dispute.html. Accessed December 15, 2014.

Thomson, Iain. 2014 (February 14). "Google Promises 10Gps Fiber Network to Blast 4K into Living Rooms." *The Register*. http://www.theregister.co.uk/2014/02/14 /google_will_upgrade_fiber_network_to_give_10gbps_home_broadband. Accessed December 21, 2014.

Trueman, Matt. 2014 (November 11). "How Mobile Tech Is Changing the Way We Make and Enjoy Art." *The Guardian*. http://www.theguardian.com/culture -professionals-network/2014/nov/11/-sp-mobile-tech-art-shakespeare-google -glass. Accessed December 21, 2014.

Wallace, Ian. 2014 (March 18). "What Is Post-Internet Art? Understanding the Revolutionary New Art Movement." *ArtSpace*. http://www.artspace.com/magazine /interviews_features/post_internet_art. Accessed December 15, 2014.

Welch, Chris. 2015 (January 1). "Snapchat Closes out 2014 by Raising a Ton of Cash." *The Verge*. http://www.theverge.com/2015/1/1/7477563/snapchat-closes -2014-by-raising-ton-of-cash. Accessed January 2, 2015.

2

The Positive Side of Social Media: Encouraging Developments from Sport

Jimmy Sanderson and Kevin Hull

The emergence of Internet and social media technologies has introduced compelling outcomes for multiple stakeholders in the sport industry.[1] Indeed, it seems that hardly a day passes during which the sport news cycle does not contain a story about an athlete or sport personality who has committed a social media miscue. In fact, some of these missteps seemingly live in perpetuity and can become standard talking points when discussing sport and social media.

In 2010, for example, University of North Carolina football player Marvin Austin began tweeting about his travels and penchant for spending money—characteristics that apparently piqued the interest of the National Collegiate Athletic Association (NCAA) given their rules regarding amateurism. The NCAA subsequently launched an investigation into the North Carolina football program and Austin ultimately lost his eligibility.[2] In another instance, former National Football League (NFL) player Chad Johnson (nicknamed "OchoCinco") gained considerable notoriety for his Twitter activity and in 2010 was fined $25,000 for carrying his phone onto the sidelines and tweeting during a preseason game—a violation of NFL rules.[3] During the 2014 NFL season, Pittsburgh Steelers player Mike Mitchell responded to criticism from fans tweeting at him about his perceived subpar performance by sending one fan the following message, "You on the other hand kill yourself."[4]

Although there are certainly issues arising from athletes' adoption of social media platforms, these incidents tend to overshadow the notion that the majority of athletes who use social media do so in a positive manner. Focusing on only the negative side of athletes' social media use also ignores the benefits that athletes can derive from social media. Indeed, the frequency with which social media issues are reported in the sport media seems likely to influence the negative view that most sport organization management personnel tend to have toward social media, particularly at the intercollegiate level.[5]

Accordingly, this chapter addresses some of the benefits that athletes obtain from using social media. These include optimizing identity expression and self-presentation; benefiting from fan advocacy; and engaging in advocacy and activism. As each of these areas is discussed, the relevant research that has explored each of these areas is reviewed. The chapter concludes by offering directions for researchers to traverse in future inquiry, along with a discussion of how sport industry practitioners can utilize this research to work with athletes to harness the power of social media and promote strategic and positive usage.

OPTIMIZING SELF-PRESENTATION

In 1959, Erving Goffman proposed that people acted differently based on the situation that they were in and the people with whom they were interacting. Goffman proposed the theory of self-presentation, in which he posited that people act in two distinct manners—one being a desirable image that they want to present to the world and the other being a relaxed image that they are more comfortable presenting only to those close to them.[6] To better explain this theory, Goffman used the idea of actors in the theater to demonstrate his propositions. Goffman posited that when people are on the front stage, their goal is to perform in such a way that the audience leaves satisfied. When those same actors are backstage with their friends and contemporaries, however, they do not have to perform and instead are more relaxed and able to reveal their true personality. Goffman suggested that people navigated between these two stages when interacting with different types of people.

Although self-presentation traditionally is linked to face-to-face communication, computer-mediated communication, including social media platforms, has enhanced people's ability to selectively self-present and to take more control over their public presentation.[7-10] For athletes, this capability is particularly important, as they are able to showcase aspects of their personality that fans do not normally see in traditional media broadcasts.[11] Previous to the emergence of social media, athletes primarily communicated with the public through traditional media channels. These interactions typically were moderated by public relations officials and media professionals who often would allow an athlete to reveal only very basic, front-stage aspects of his or her life. Social media, however, gives athletes the ability to speak directly to their fans and enables an athlete to reveal more of his or her identity. For fans this creates a scenario in which they perhaps feel as if they know the athlete better, because the fan gets a more personal view of the life of the sports star.[12] Navigating between front-stage and backstage personas, however, can be a difficult task for athletes because of the large audiences that follow the athletes' social media postings. For instance, some athletes have millions of Twitter users following their updates, and therefore tweeting messages that are of interest to all followers can be challenging.[13] Despite this struggle, as more athletes have embraced social media, many have found these channels to be an ideal place to tell millions of fans about personal events in their lives.[14, 15]

These messages can be as simple as demonstrating their support for other sport teams, their shopping habits, or wishing family members a happy birthday.[16, 17] Such disclosures can cultivate identification and parasocial interaction among fans, who can utilize these messages to find similarity with athletes that is difficult to obtain through the lens of athleticism.[18] Sanderson, for instance, examined the self-presentation on Twitter by four rookie athletes in Major League Baseball (MLB), the National Basketball Association (NBA), the National Football League (NFL), and the National Hockey League (NHL).[19] Sanderson discovered that these athletes shared topics such as their popular culture preferences, family experiences, and dedication to their sport (e.g., workout routines). One noteworthy manner in which these athletes used Twitter was to solicit information from fans, such as asking for restaurant and movie recommendations. Through sharing more of their identity with fans and going so far as to request information from them, athletes were able to appear "closer" to fans, an outcome that in addition to cultivating identification and parasocial interaction, could prompt fans to take action to defend and support athletes. In another study, Hull examined Twitter usage by professional golfers during the 2013 Masters tournament and found that golfers demonstrated more front-stage tendencies, such as promoting their endorsements, yet still revealed many behind-the-scenes stories as well.[20]

Whereas social media provide opportunities to optimize self-presentation, athletes could conform to traditional gender roles, suggesting that this capability might be underutilized. Lebel and Danylchuk investigated professional tennis players' self-presentation on Twitter and discovered that although image construction largely was similar between the two genders, male players skewed more toward a sport fan self-presentation, whereas female players tended to self-present more as brand managers.[21] Similarly, Coche explored the self-presentation of male and female professional tennis players and golfers on their Twitter accounts.[22] Coche discovered that women tended to predominantly present their femininity followed by their athletic persona, whereas males primarily self-presented as athletes.

The ability to optimize self-presentation via social media also enables athletes to counteract perceived negative media framing and to take more control over their public presentation when they perceive that it is being misrepresented in the sport media. Sanderson examined Boston Red Sox pitcher Curt Schilling's self-presentation on his blog in response to two incidents—one in which his athletic integrity was challenged by a broadcaster and another in which he publicly apologized for criticizing fellow player Barry Bonds.[23] Sanderson discovered that Schilling employed three self-presentation strategies: (1) critic; (2) committed individual; and (3) accountable person. Using the critic personality, Schilling turned the tables and lambasted sport journalists in response to baseball broadcaster Gary Throne suggesting that Schilling had staged an injury during the 2004 American League Championship Series (ALCS). Through comments such as, "Instead of using the forums they participate in to do something truly different, change lives, inspire people, you have an entire subset of people whose sole purpose in life is to actually be the news instead of report it," Schilling was able to defend himself against this allegation.[24] More specifically, he was able to assume the role of a critic—an identity aspect that often is reserved for sport media members. Schilling's blog also enabled him to express these views without any filtering, which would have been difficult to do when using traditional media channels.

Schilling utilized the committed individual self-presentation strategy to counter perceptions that he was not a team player. One poignant reminder he offered included detailing the personal sacrifices he had made to keep pitching during the Red Sox's 2004 run to a World Series title.

> Remember this, the surgery was voluntary. If you have the nuts, or the guts, grab an orthopedic surgeon, have them suture your ankle down to the tissue covering the bone in your ankle joint, then walk around for 4 hours. After that, go find a mound, throw a hundred or so pitches, run over, cover first a few times. When you're done check that ankle and see if it bleeds. It will.[25]

Schilling used the accountable person strategy to take responsibility for his terse remarks about Barry Bonds. Employing statements such as: "Regardless of my opinions, thoughts, and beliefs on anything Barry Bonds, it was absolutely irresponsible and wrong to say what I did. I don't think it's within anyone's right to say the things I said yesterday and affect other people's lives in that way."[26]

As Schilling's case demonstrates, via social media platforms athletes are endowed with the capability to optimize their self-presentation, reveal more aspects of their identity, and counteract what they perceive to be negative public portrayals. As athletes navigate between front stage and backstage they also can create potential financial benefits. In their studies of professional tennis players' self-presentation on Twitter, Lebel and Danylcuck found that female tennis players used Twitter primarily to create a positive personal brand through interactions with fans and also by promoting their sponsors.[27] Sanderson examined Florida Marlins player Logan Morrison's identity expressions on Twitter and found that Morrison's use of humor seemed to attract a large audience, which he capitalized on by promoting his corporate appearances and philanthropic efforts.[28] Thus, social media provides athletes with a platform to build a "brand" that can be strategically constructed to appeal to their organization as well as to corporate sponsors.

BENEFITING FROM FAN ADVOCACY

Whereas an athlete can directly advocate for himself or herself by countering perceived negative media framing and by promoting aspects of identity that could be silenced in traditional media, athletes indirectly receive these benefits from fans via social media. Indeed, through social media, fans can express support for athletes, combat an athlete's detractors, and advocate for the positive characteristics and attributes that an athlete possesses. Sanderson observed how Boston Red Sox pitcher Curt Schilling received social support from blog readers in response to the aforementioned incidents about which he blogged.[29] Schilling discovered that blog readers mobilized and authenticated his role as a sport media critic and proclaimed his legacy within Red Sox culture. Sanderson also noted that many fans posited that those who criticized Schilling were masquerading as Red Sox fans, and that expressing support for Schilling was an accurate indicator of whether one was a "true" Red Sox fan. Sanderson suggested that these messages from fans functioned as a form of public relations that Schilling received by merely posting his blog entries.

Athletes also could receive this benefit when they are experiencing personal adversity. For instance, Kassing and Sanderson investigated comments posted to cyclist Floyd Landis's Web site as he battled accusations that he had taken performance-enhancing drugs (PEDs) during the 2006 Tour de France, which he won. Ultimately, Landis was disqualified and stripped of the title.[30] Kassing and Sanderson found that fans conveyed support to Landis by expressing empathy and sympathy; confirming Landis's accomplishments and character; testifying about their belief in Landis; sharing that they possessed a common enemy with Landis (e.g., the United States Anti-Doping Agency, World Anti-Doping Agency); and by offering Landis support and tangible assistance such as monetary donations to his legal defense fund. Kassing and Sanderson contended that Landis's blog gave fans a mechanism to be more actively involved in his affairs, which included very elaborate and specific forms of support beyond general well-wishes.

In another study, Sanderson examined the differences in how fans framed professional golfer Tiger Woods's marital infidelity via his Facebook page as compared to how these events were framed in the mainstream media.[31] Sanderson found that mainstream media reports framed Woods's infidelity as a "tragic flaw" a characteristic that previously had been difficult to find given Woods's performance on the golf course along with his previously tightly guarded personal life. Mainstream media reports also centered on the lurid details that accompanied the reports made by many of Woods's alleged mistresses. These examples were typified by comments such as, "two more blondes and a brunette were added to Tiger Woods's sultry scorecard."[32] In contrast, although a small minority of fans conveyed that they no longer would support Woods, the overwhelming majority of participants advocated that Woods was entitled to his privacy in handling these incidents, and that the infidelity merely functioned as evidence that Woods was human—just like his fans. Sanderson observed that social media enabled fans to become more involved in Woods's media narratives and to feel as though they were assisting him in dealing with his issues. Additionally, these messages also functioned as alternative frames that ran counter to how Woods was being portrayed in the mainstream media, and Sanderson suggested that athletes benefited from fans introducing these alternative narratives.

Moreover, athletes who admit their missteps could be more likely to attain these benefits as well. Sanderson and Emmons investigated responses on the Texas Rangers' community message forum after player Josh Hamilton held a press conference to address reports that he been drinking in a Dallas-area bar.[33] Hamilton had experienced a significant drug and alcohol addiction that had resulted in major league baseball suspending him for several seasons. Hamilton made a very dramatic comeback that included a public vow that he would no longer consume alcohol. During the press conference, Hamilton acknowledged that he had indeed consumed alcohol in response to some personal problems he was experiencing and apologized for his behavior. Sanderson and Emmons found that although some fans indicated that they were unwilling to forgive Hamilton for this offense, other fans expressed support and commended Hamilton for being accountable for his actions. Still others found commonality in Hamilton's addiction problems. These fans shared their own stories of addiction and conveyed how they found strength in their own struggles by identifying with Hamilton's issues. Similar to the findings in Sanderson's study of Tiger Woods, some people suggested that Hamilton's relapse merely was evidence of his human nature and still others justified Hamilton's behavior and suggested that in a sport that had been plagued by performance-enhancing drug issues, his alcohol consumption was a minor matter.[34]

Through social media, athletes become recipients of public relations work that is willingly performed by fans. These efforts appear to stem from identification and parasocial interaction, which in some cases can be cultivated by the manner in which an athlete engages in self-presentation via social media.

As fans become advocates they defend athletes against their detractors, thereby removing athletes from what likely would be unwinnable battles if the athletes instead directly combat their detractors. Indeed, Browning and Sanderson found that one way that college athletes deal with the criticism received via Twitter is to simply retweet the offending message and let supporters attack the detractor.[35] Thus, social media becomes a key platform where support can be cultivated and where the preferred representation that athletes desire along with messages they want to promote can be widely circulated. Athletes then can extend their influence beyond the athletic arena into the social media realm. In addition to sharing their personal battles, this capability enables them to engender reinforcement for political and social causes that they support.

ENGAGING IN ADVOCACY AND ACTIVISM

In addition to allowing athletes to showcase more of their personal life, social media also provide a platform for them to engage in advocacy and activism—a characteristic that some suggest is lacking in modern athletes, particularly minority athletes.[36] Whereas athletes might be reticent to engage in advocacy via traditional media broadcasts, as illustrated by former NBA player Michael Jordan's famous quote, "Republicans buy sneakers too,"[37] it could be the case that social media presents a less threatening (although equally public) format for athletes to advocate for social and political causes.

In November 2014, for example, after the grand jury in Ferguson, Missouri, elected to not indict police officer Darren Wilson in the shooting death of Michael Brown, an 18-year-old African American male, NBA player Kobe Bryant tweeted, "The system enables young black men to be killed behind the mask of the law #Ferguson #tippingpoint #change."[38] In another noteworthy case, Schmittel and Sanderson investigated tweets from NFL players in response to the Trayvon Martin–George Zimmerman case verdict.[39] After Zimmerman was found not guilty, several NFL players received significant media attention for tweets about the verdict, including Roddy White who commented, "all them jurors should go home tonight and kill themselves for letting a grown man get away with killing a kid" and Victor Cruz who stated, "Thoroughly confused. Zimmerman doesn't last a year before the hood catches up with him."[40]

In the study's sample, which included the 15 players from each team who possessed the most Twitter followers in the 12-hour period following the verdict's announcement, the researchers found that athletes expressed their anticipation for the verdict then conveyed their disbelief when the verdict was announced. Additionally, they discovered that athletes also provided critiques of the criminal justice system and offered social commentary, while simultaneously defending themselves against fans who were displeased with their advocacy efforts. Many of these athletes elaborated that they were entitled to their views and that there were more important things than football games. Interestingly, the majority of athletes offering commentary were African American and, with one exception, the few white players who weighed in suggested that athletes needed to stay off Twitter regarding the verdict and that the American justice system, although perhaps imperfect, was the best system available. Schmittel and Sanderson observed that, via Twitter, athletes also could extend their influence into political and social arenas and the propensity for them doing so was more likely, given the convenient posting features offered by Twitter.[41] Although pushback and resistance from fans might prevent other athletes from engaging in activism via social media, they seem to be willing to get involved on these platforms. Thus, athletes' activism can become more pronounced on social media, as opposed to traditional media; constraints often exist when athletes attempt to weigh in on social and political issues via traditional media.

In addition to social justice issues, athletes also can engage in advocacy related to their sport. This is particularly pertinent in the realm of intercollegiate athletics, because monetary issues often jeopardize the sustainability of what often are considered "non-revenue" sports (at most schools this generally means every sport except football and men's basketball). Hull examined how athletes at the University of North Carolina–Wilmington used Twitter to save the men's and women's diving and swimming teams from being eliminated.[42] Hull found that these athletes used Twitter to alert as many people as possible about the planned cuts, thus becoming opinion leaders and prompting their followers to rally behind their cause. These efforts included engaging celebrities via Twitter to join them in their cause and retweeting those messages of support when they were received (Olympic swimmers Ryan Lochte and Ricky Berens were two notable celebrities who did tweet messages of support). Ultimately, these athletes were successful and social media was a key resource in their advocacy efforts. Indeed, one of the athletes noted,

> Social media was a priority. In today's world, you can spread news like wildfire online. We knew it would be the fastest and easiest way to get the word out. I've never been huge on tweeting, but this was more important to me than anything else. I just wanted to make as much noise as possible and show how much it means to me.[43]

Another athlete commented,

> I think if this had happened 15 years ago it would have taken over a month to get to the level [of support] we did in just a week or so. I am

so thankful we had social media to use as a resource. We wouldn't have been able to stir up enough trouble in the time we had without it.[44]

Via social media platforms athletes can engage in activism and advocacy and—given the public visibility of many athletes—can dramatically circulate the reach of a message as their followers and fans retransmit the message. Additionally, for cases in which advocacy causes are time sensitive, social media platforms possess significant utility to mobilize supporters and spur collective action expediently. To be sure, athletes are subject to criticism via social media for these endeavors and could find themselves in conflict with head coaches, athletic directors, general managers, and other organization personnel members. Nevertheless, this issue remains an important one for scholars to monitor and investigate, as athletes broadly might be more likely to weigh in on social and political issues via social media. Given the interactive nature that underpins social media, social and political issues might provide valuable opportunities to examine the interactions between athletes and fans—which offers unique avenues to extend the literature on this subject.

POSITIVE SOCIAL MEDIA USAGE: DIRECTIONS FOR FUTURE RESEARCH

This chapter discusses three benefits that social media provides to athletes, (1) optimizing self-presentation; (2) benefiting from fan advocacy; and (3) engaging in advocacy. There certainly are more benefits to using social media, however, and there are number of exciting directions for researchers to explore. With respect to self-presentation, much of the work to date has centered on Twitter, and although this is understandable given that Twitter appears to be the platform of choice for athletes, it is important to examine other platforms as well, particularly visual sharing platforms such as Instagram and Snapchat which seem to be growing more popular.[45] How might the self-presentation on these channels differ from the self-presentation shared on Twitter? Are there gender differences that are similar to those in the literature on self-presentation and Twitter?

Additionally, it is important to explore how amateur athletes are performing self-presentation via social media. These athletes might be subjected to more intensive monitoring than are professional athletes (at least at the intercollegiate level) and therefore they might be more scripted or "safe" in their self-presentation.[46, 47] Another area of self-presentation to consider is the manner in which acclaimed high school athletes present themselves on social media. Given the intensity that accompanies the collegiate football recruiting process, it would be fruitful to analyze how these particular athletes present themselves, in addition to determining whether certain presentations might affect player attractiveness to college programs.

Another area for researchers to pursue is determining the motivations underpinning fans engaging in supportive behaviors toward athletes via social media. Such behavior could be a function of fan identification, attachment, or a parasocial connection with the athlete, but there might be other factors that influence this behavior that also could be important to illuminate. One potential variable to consider in such an examination is whether there has been a prior interaction or acknowledgment (retweet, response) between the fan and athlete, as this could contribute to a fan's willingness to engage in advocacy on behalf of an athlete. Another area of inquiry in this topic involves the labor that is willingly performed by fans on behalf of athletes. Essentially, athletes receive this benefit at no cost. As a result, does this create implications that should be considered? In other words, is it ethical for athletes to elicit this behavior from fans, as it could be considered to be taking advantage of them? One final area to pursue in this domain is whether fans advocating on behalf of an athlete possess the capability to change dominant narratives about the athlete that are introduced by mainstream media outlets. Along those lines, it is not uncommon for an athlete to tweet at a sport journalist if they feel that they have been unfairly criticized or portrayed. How these interactions affect sport-reporting practices would be a compelling line of inquiry.

With respect to athletes engaging in advocacy, future work could investigate the response that athletes receive from fans for their advocacy and activism efforts. Are fans supportive? Do they retaliate negatively? Illuminating these reactions might shed light on barriers to athletes performing advocacy and activism via social media or could demonstrate ways that this behavior is reinforced and authenticated by fans. Social network analysis also could be utilized to examine how the messages that athletes disseminate spread across social media to better understand if these messages are more temporary or permanent in nature.

PRACTITIONER IMPLICATIONS

As noted at the beginning of this chapter, social media missteps by athletes tend to be the predominant way that athletes' social media use is framed in the mass media. Consequently, it is not surprising that many coaches and administrators possess negative attitudes regarding social media. Nevertheless, as this chapter has demonstrated, social media possesses benefits for athletes. Considering the prevalence of social media adoption and usage—particularly in the younger demographics—it is unrealistic to expect that significant numbers of athletes will abandon social media. It therefore is important that athletes receive education about the ways that social media can be utilized. Ideally, this education would begin at a young age, informing athletes that social media is a tool to optimize self-presentation and to build a personal brand that can make one attractive to relevant stakeholders. With the widespread nature of social media monitoring by employers, post-secondary admissions personnel, and internship program coordinators, helping younger individuals take control of their social media profiles likely can help to prevent negative future repercussions.

Additionally, when it comes to topics such as engaging in activism and advocacy, it could be worthwhile for sport organizations to consider the boundaries with which they are comfortable. This includes being mindful of freedom of speech elements, and potential pushback that could come from outside parties (e.g., the media) for what is perceived to be censuring an athlete's right to expression. Setting the boundaries and providing guidelines for social media usage (generally through the form of policy) might foster positive relations between athletes and administrators and transform what often is a confrontational subject into a harmonious one.

In conclusion, social media have rapidly proliferated across sport, creating a variety of positive and negative outcomes for multiple sport stakeholders, including athletes. As athletes often form the most visible stakeholder group and tend to be heavy users of social media platforms, it is important that positive behaviors are identified and cultivated. Clearly, there always will be missteps, as is the case with any new communication medium. Nevertheless, helping athletes understand "best practices" can ease the strain on administrators, coaches, parents, educators, and others. Indeed, it is important that when the social-media story is presented to athletes they are provided the proverbial "both sides."

NOTES

1. Jimmy Sanderson, *It's a Whole New Ballgame: How Social Media Is Changing Sports* (New York: Hampton Press, 2011).

2. Danah Boyd, "Taken Out of Context: American Teen Sociality in Networked Publics," PhD dissertation, University of California, Berkeley, 2008.

3. "Chad Ochocinco Fined $25K" (August 25, 2010), http://sports.espn.go.com /nfl/trainingcamp10/news/story?id=5493157, accessed December 29, 2014.

4. "Steelers Safety Mike Mitchell Tweets 'Kill Yourself' to Fan," (November 18, 2014), http://pittsburgh.cbslocal.com/2014/11/18/steelers-safety-mike-mitchell-tweets -kill-yourself-to-fan/, accessed December 29, 2014.

5. Jimmy Sanderson, "To Tweet or Not to Tweet . . . Exploring Division I Athletic Departments Social Media Policies," *International Journal of Sport Communication* (2011): 492–513.

6. Erving Goffman, *The Presentation of Self in Everyday Life* (Garden City, NY: Doubleday Anchor Books, 1959).

7. Joseph B. Walther, "Computer-Mediated Communication: Impersonal, Interpersonal, and Hyperpersonal Interaction," *Communication Research* 23 (1996): 3–43.

8. Liam Bullingham and Ana C. Vasconcelos, " 'The Presentation of Self in the Online World': Goffman and the Study of Online Identities," *Journal of Information Science* 39 (2013): 101–12.

9. Michael Hvlid Jacobsen, "Goffman through the Looking Glass: From 'Classical' to Contemporary Goffman," in *The Contemporary Goffman*, edited by Michael Hvlid Jacobsen (London: Routledge, 2010).

10. Jimmy Sanderson, "The Blog Is Serving Its Purpose: Self-Presentation Strategies on 38pitches.com." *Journal of Computer-Mediated Communication* 13 (2008): 912–36.

11. Jimmy Sanderson, "Stepping into the (Social Media) Game: Building Athlete Identity via Twitter," in *Handbook of Research on Technoself: Identity in a Technological Society*, edited by Rocci Luppicini (New York: IGI Global, 2013).

12. Ibid.

13. Boyd, "Taken Out of Context" (2008).

14. Katie Lebel and Karen E. Danylchuk, "Facing Off on Twitter: A Generation Y Interpretation of Professional Athlete Profile Pictures," *International Journal of Sport Communication* 7 (2014): 317–36.

15. Jimmy Sanderson, "Just Warming Up: Logan Morrison, Twitter, Athlete Identity, and Building the Brand," in *Sport and Identity: New Agendas in Communication*, edited by Barry Brummett and Andrew W. Ishak (New York: Routledge, 2014).

16. Kevin Hull, "A Hole in One (Hundred and Forty Characters): A Case Study Examining PGA Tour Golfers' Twitter Usage during the Masters," *International Journal of Sport Communication* 7 (2014): 245–60.

17. Lebel and Danylchuk, "Facing Off on Twitter" (2014).

18. Jimmy Sanderson, "Stepping into the (Social Media) Game: Building Athlete Identity via Twitter," in *Handbook of Research on Technoself: Identity in a Technological Society*, edited by Rocci Luppicini (New York: IGI Global, 2013).

19. Ibid.

20. Kevin Hull, "A Hole in One (Hundred and Forty Characters)" (2014).

21. Lebel and Danylchuk, "Facing Off on Twitter" (2014).

22. Roxanne Coche, "How Golfers and Tennis Players Frame Themselves: A Content Analysis of Twitter Profile Pictures," *Journal of Sports Media* 9 (2014): 95–121.

23. Sanderson, "Stepping into the (Social Media) Game" (2008).

24. Ibid., 921.

25. Ibid., 924.

26. Ibid., 926.

27. Lebel and Danylchuk, "Facing Off on Twitter" (2014).

28. Sanderson, "Stepping into the (Social Media) Game" (2014).

29. Jimmy Sanderson, " 'The Nation Stands Behind You': Mobilizing Support on 38pitches.com," *Communication Quarterly* 58 (2010a): 188–206.

30. Jeffrey W. Kassing and Jimmy Sanderson, " 'Is This a Church? Such a Big Bunch of Believers Around Here!': Fan Expressions of Social Support on Floydlandis.com," *Journal of Communication Studies* 2 (2009): 309–30.

31. Jimmy Sanderson, "Framing Tiger's Troubles: Comparing Traditional and Social Media," *International Journal of Sport Communication* 3 (2010b): 438–53.

32. Ibid., 445.

33. Jimmy Sanderson and Betsy Emmons, "Extending and Withholding Forgiveness to Josh Hamilton: Exploring Forgiveness within Parasocial Interaction," *Communication and Sport* 2 (2014): 24–47.

34. Sanderson, "Framing Tiger's Troubles" (2010b).

35. Blair Browning and Jimmy Sanderson, "The Positives and Negatives of Twitter: Exploring How Student-Athletes Use Twitter and Respond to Critical Tweets," *International Journal of Sport Communication* 5 (2012): 503–21.

36. Kwame J. A. Agyemang, "Black Male Athlete Activism and the Link to Michael Jordan: A Transformational Leadership and Social Cognitive Theory Analysis," *International Review for the Sociology of Sport* 47 (2012): 433–45.

37. L. Z. Granderson, "The Political Michael Jordan," ESPN.com (August 14, 2012), http://espn.go.com/nba/story/_/id/8264956/michael-jordan-obama-fundraiser-22-years-harvey-gantt, accessed May 20, 2015.

38. Chris Littman, "Kobe Bryant on Ferguson: System Enables Young Black Men to Be Killed," The SportingNews.Com (Charlotte, NC), (November 25, 2014), http://www.sportingnews.com/nba/story/2014-11-25/kobe-bryant-ferguson-michael-brown-darren-wilson-twitter, accessed May 20, 2015.

39. Annelie Schmittel and Jimmy Sanderson, "Talking about Trayvon in 140 Characters: Exploring NFL Players' Tweets about the George Zimmerman Verdict," *Journal of Sport & Social Issues* (Forthcoming).

40. Rich Cimini, "Victor Cruz: Tweet was 'Wrong,' " ESPNNewYork.com (New York) (July 15, 2013), http://espn.go.com/new-york/nfl/story/_/id/9497415/victor-cruz-new-york-giants-says-zimmermantweet-wrong, accessed May 20, 2015.

41. Schmittel and Sanderson, *Talking about Trayvon* (Forthcoming).

42. Hull, Kevin, "#Fight4UNCWSwimandDive: A Case Study of How College Athletes Used Social Media to Help Save Their Team," *International Journal of Sport Communication* 7 (2014): 533–52.

43. Ibid., 543.

44. Ibid., 544.

45. Browning and Sanderson, "The Positives and Negatives of Twitter."

46. Sanderson, Jimmy, "To Tweet or Not to Tweet . . . Exploring Division I Athletic Departments Social Media Policies," *International Journal of Sport Communication* 4 (2011): 492–513.

47. Sanderson, Jimmy, and Blair Browning, "Training versus Monitoring: A Qualitative Examination of Athletic Department Practices Regarding Student-Athletes and Twitter," *Qualitative Research Reports in Communication* 14 (2013): 105–11.

BIBLIOGRAPHY

Agyemang, Kwame J. A. 2012. "Black Male Athlete Activism and the Link to Michael Jordan: A Transformational Leadership and Social Cognitive Theory Analysis." *International Review for the Sociology of Sport* 47: 433–45.

"Austin's Twitter Account Sheds Light on UNC Player." http://blogs.newsobserver.com /accnow/austins-twitter-account-provides-clues-in-ncaa-probe-at-unc. Accessed on December 29, 2014.

Boyd, Danah. 2008. "Taken Out of Context: American Teen Sociality in Networked Publics." PhD dissertation, University of California, Berkeley.

Browning, Blair, and Jimmy Sanderson. 2012. "The Positives and Negatives of Twitter: Exploring How Student-Athletes Use Twitter and Respond to Critical Tweets." *International Journal of Sport Communication* 5: 503–21.

Bullingham, Liam, and Ana C. Vasconcelos. 2013. " 'The Presentation of Self in the Online World': Goffman and the Study of Online Identities." *Journal of Information Science* 39: 101–12.

"Chad Ochocinco Fined $25K." 2010 (August 25). http://sports.espn.go.com/nfl /trainingcamp10/news/story?id=5493157. Accessed December 29, 2014.

Cimini, Rich. 2013 (July 15). "Victor Cruz: Tweet Was 'Wrong.' ESPNNewYork.com (New York). http://espn.go.com/new-york/nfl/story/_/id/9497415/victor-cruz-new -york-giants-says-zimmermantweet-wrong. Accessed May 19, 2015.

Coche, Roxanne. 2014. "How Golfers and Tennis Players Frame Themselves: A Content Analysis of Twitter Profile Pictures. *Journal of Sports Media* 9: 95–121.

Goffman, Erving. 1959. *The Presentation of Self in Everyday Life*. Garden City, NY: Doubleday Anchor Books.

Granderson, L. Z. 2012 (August 14). "The Political Michael Jordan." ESPN.com (Bristol, CT). http://espn.go.com/nba/story/_/id/8264956/michael-jordan-obama -fundraiser-22-years-harvey-gantt. Accessed May 19, 2015.

Hull, Kevin. 2014a. A Hole in One (Hundred and Forty Characters): A Case Study Examining PGA Tour Golfers' Twitter Usage during the Masters. *International Journal of Sport Communication* 7: 245–60.

Hull, Kevin. 2014b. #Fight4UNCWSwimandDive: A Case Study of How College Athletes Used Social Media to Help Save Their Team. *International Journal of Sport Communication* 7: 533–52.

Jacobsen, Michael Hvlid. (2010). "Goffman through the Looking Glass: From 'Classical' to Contemporary Goffman." In *The Contemporary Goffman*, edited by Michael Hvlid Jacobsen. London: Routledge.

Kassing, Jeffrey W., and Jimmy Sanderson. 2009. "Is This a Church? Such a Big Bunch of Believers Around Here!" Fan Expressions of Social Support on Floydlandis. com. *Journal of Communication Studies* 2: 309–30.

Lebel, Katie, and Karen Danylchuk. 2012. "How Tweet It Is: A Gendered Analysis of Professional Tennis Players' Self-Presentation on Twitter." *International Journal of Sport Communication* 5: 461–80.

Lebel, Katie, and Karen E. Danylchuk. 2014. "Facing Off on Twitter: A Generation Y Interpretation of Professional Athlete Profile Pictures." *International Journal of Sport Communication* 7: 317–36.

Littman, Chris. "Kobe Bryant on Ferguson: System Enables Young Black Men to Be Killed." The SportingNews.Com (Charlotte, NC). 2014 (November 25). http:// www.sportingnews.com/nba/story/2014-11-25/kobe-bryant-ferguson-michael-brown -darren-wilson-twitter. Accessed May 19, 2015.

Sanderson, Jimmy. 2014. "Just Warming Up: Logan Morrison, Twitter, Athlete Identity, and Building the Brand." In *Sport and Identity: New Agendas in Communication*, edited by Barry Brummett and Andrew W. Ishak. New York: Routledge.

Sanderson, Jimmy. 2013. "Stepping into the (Social Media) Game: Building Athlete Identity via Twitter." In *Handbook of Research on Technoself: Identity in a Technological Society*, edited by Rocci Luppicini. New York: IGI Global.

Sanderson, Jimmy. 2011a. *It's a Whole New Ballgame: How Social Media Is Changing Sports*. New York: Hampton Press.

Sanderson, Jimmy, 2011b. "To Tweet or Not to Tweet . . . Exploring Division I Athletic Departments Social Media Policies." *International Journal of Sport Communication* 4: 492–513.

Sanderson, Jimmy. 2010a. "The Nation Stands behind You": Mobilizing Support on 38pitches.com. *Communication Quarterly* 58: 188–206.

Sanderson, Jimmy. 2010b. "Framing Tiger's Troubles: Comparing Traditional and Social Media." *International Journal of Sport Communication* 3: 438–53.

Sanderson, Jimmy. 2008. "The Blog Is Serving Its Purpose: Self-Presentation Strategies on 38pitches.com." *Journal of Computer-Mediated Communication* 13: 912–36.

Sanderson, Jimmy, and Betsy Emmons. 2014. "Extending and Withholding Forgiveness to Josh Hamilton: Exploring Forgiveness within Parasocial Interaction." *Communication and Sport* 2: 24–47.

Sanderson, Jimmy, and Blair Browning. 2013. "Training versus Monitoring: A Qualitative Examination of Athletic Department Practices Regarding Student-Athletes and Twitter." *Qualitative Research Reports in Communication* 14: 105–11.

Schmittel, Annelie, and Jimmy Sanderson. Forthcoming. "Talking about Trayvon in 140 Characters: Exploring NFL Players' Tweets about the George Zimmerman Verdict." *Journal of Sport & Social Issues*.

"Steelers Safety Mike Mitchell Tweets 'Kill Yourself' to Fan." http://pittsburgh.cbslocal.com/2014/11/18/steelers-safety-mike-mitchell-tweets-kill-yourself-to-fan/. Accessed December 29, 2014.

Walther, Joseph B. 1996. Computer-Mediated Communication: Impersonal, Interpersonal, and Hyperpersonal Interaction. *Communication Research* 23: 3–43.

3

Reaching the World with One Song and a Few Mouse Clicks

Kathryn Coduto

I found myself in the front row of Brixton Academy in London in May 2012. I was pressed against the small fence that wrapped around the front of the venue, surrounded by girls dressed in black, bouncing where they stood as they waited for the headlining act to take the stage.

We were waiting for London-based indie band the Horrors to take the stage. Previously, 4,000 miles had been between this small band I had grown to love and my hometown in Ohio. Five years after first discovering them, I was mere feet from their lead singer.

My deep love for this band came from the Internet.

THE INTERNET AND THE MUSIC INDUSTRY

You are likely familiar with the birth of the Internet; its history is now a standard part of high school and college courses. A tool that people around the world now use daily—and often constantly—began as a vague idea of electronic possibility in the early 1960s. Researchers had a notion that computers could be connected and could communicate with each other from different locations, but they had only nascent ideas as to how that interconnectedness would come to life.[1]

Fifty years later, the Internet is one of the fastest growing and widely used technological advancements in history.[2] It has evolved from a niche interest for the tech-minded to a tool necessary for a variety of jobs and endeavors. The Internet no longer is just for programmers or specialists; now it is for everyone. This means that the Internet is not used just for work; instead, the Internet now is a major source of entertainment.

In a typical day, you can log onto your computer and watch an episode of your favorite television show. If you don't want to watch the show alone, you can chat with your friends in a separate window in a social network of your choosing. Your conversation does not even have to be restricted to one person; you can broadcast your thoughts to a network that you create and cultivate. Not in the mood for television? You can check the news, read sports scores, play games, and take quizzes—whatever your interest, the Internet provide it and can deliver it to you instantly.

When talking about the Internet and the explosion of readily available entertainment, the music industry must be discussed. The music industry arguably has changed the most due to the birth of the Internet and the increasing availability of music on the Internet by artists both established and fledgling. Your typical day might involve watching a show or reading an article, but it also likely involves streaming music from one of a variety of platforms available at your fingertips, and typically at no cost to you. The Internet is divisive in the music industry, but it is revolutionary and it is here to stay.

If you are an artist, the Internet is breaking down boundaries that before would have prevented you from reaching what could be your biggest fan base. If you are a music fan, the Internet brings you music that before would have been out of your reach for months or years—if not forever. The Internet, although controversial, is important in connecting fans and musicians from around the world.

THE SLOW SPREAD OF A MUSICAL MESSAGE

In 1990 in Olympia, Washington, young kids crowded into tiny venues to see a woman scream at them. Hundreds of people jammed themselves into punk rock clubs to see Kathleen Hanna and her band, Bikini Kill, perform—knowing little about the band except that their lead singer was often ferocious and that their music was a powerful force. Hanna and her band published a small magazine, also called *Bikini Kill*, and used that magazine to spread the word about their band and the beliefs of the band. The "fanzines," as they are called, were handed out to people the band knew and to people—especially girls—who attended their shows.[3] It took considerable work for Bikini Kill to get their message out within their own city, much less their own state.

The band worked from the ground up, making connections with other bands and artists who could help spread their music and message. It took time, but the "Riot Grrrl" Movement that Bikini Kill is credited with developing eventually found its true home in Washington, DC, all the way across the country from its founders' home of Olympia. The band was not only looking to share its music but also the message behind it, specifically the idea of equality for women. Hanna said in an interview with *The AV Club* that she knew touring was important in reaching the appropriate audiences: "I felt like going out on the road and mixing it with music—which is something young people are always really interested in—would be a good way to proselytize."[4] Hanna and her group believed in what they were doing and put in the effort to reach audiences across the country. Their journey from Olympia, Washington, to Washington, DC, would take them nearly 3,000 miles to reach people with whom their music resonated.

Meanwhile, it took three more years before Hanna became fodder for the press, a woman worthy of both praise and criticism, a figure that represented hope to some and angst to others, all because of the message of her music and her performance.[5] It took nearly five years from Bikini Kill's beginning before they had both an established fan base and a position within a regular cultural discussion.

Kathleen Hanna and Bikini Kill didn't have the Internet to accelerate the spread of their message—not in their own country and definitely not internationally.

LOCAL SCENES GROW IN THE 1980s AND 1990s

Music throughout the 1980s and into the early 1990s typically was scene oriented, especially if it was music that was considered outside of the mainstream.[6] Your Madonnas and U2s and Janet Jacksons did not worry about having a local identity so much as creating something much, much larger. Their label backing and strings of hits made the world their playground after years of legwork to achieve breakthroughs. For these artists, conquering the United States through radio play was all they needed to translate their success into something greater and wider.[7]

For many independent musicians, though, the essential start came by honing a specific sound within one's hometown. Independent music often is associated with a specific region, regardless of genre. There are certain sounds ingrained within each artist or group.[8] Cities could be close physically, but one's given sound could be completely different from that of a neighboring town.

The Seattle scene is a prime example of a "local formation" in the 1990s. Grunge music is associated with Seattle, and the brand of rock music that Seattle bands including Nirvana, Soundgarden, and Stone Temple Pilots produced was unlike anything else being released at the time.[9] The bands who were playing around Seattle—including the aforementioned Bikini Kill, who had a hand in naming Nirvana's signature song, "Smells Like Teen Spirit"— were feeding off of each other and using each other as a check for what sounded good and interesting.[10] It is not to say that these bands existed completely in a bubble where they only heard each other's music. Instead, these Seattle musicians were aware of what the biggest movements in music were nationally—hair metal and boy bands seeming to be the dominant two types—and decided that that was not the sound they wanted to be associated with.[11] The goal was to create something that was distinctly their own.

Of course, radio play still mattered. It was the quickest, easiest way for a label to guarantee exposure to a new group. Writer Greil Marcus likely put it best when describing his thoughts to GQ magazine upon first hearing "Smells Like Teen Spirit," which took the Seattle scene from Seattle to the world: "I've experienced that with other records very few times, where just suddenly something new on the radio makes everything around it seem false."[12] Nirvana finally broke through, but only after connections were made through mutual friends (specifically Sonic Youth's Kim Gordon recommending Nirvana to the label) and a single made it to the radio.[13]

Would Nirvana have had the lasting impact that it does if the Internet had been available to spread their sound to more eager listeners at earlier stages of the group's career?

SOME PERSONAL EXPERIENCE

I joined MySpace during my freshman year of high school in August 2006. At the time, MySpace (which made its online debut in 2004) was an exciting proposition—it was one of the first ways I could stay in touch with my friends outside of school that didn't involve sitting on the phone with them and re-hashing the day's events. It was a good way to combine blogging, messaging, and quick comments. MySpace helped with the earliest versions of personal brands online.[14]

One of the biggest benefits of MySpace came from its integrated music service, which even in its fledgling days was a powerful tool for new bands to reach audiences previously unheard of. It wasn't just that you could discover bands on MySpace—which was clearly huge—but MySpace users could endorse bands by including one of their songs as a "profile" song. For me, at age 14, including a band's song on the page was one of the most appealing aspects. I loved associating myself with different musicians and what they represented.

When I discovered the Horrors—a post-punk band based in London and originally from Essex, England—on MySpace later in 2007, it was important not just because I could listen to them but because I could show my small world who this band was. A casual MySpace search through "punk" music brought me to the Horrors, but my unyielding interest exposed them to even more people in a short amount of time. MySpace created a bridge that spanned nearly 4,000 miles to London, and then I expanded that bridge by incorporating songs from the group's debut album into my MySpace page in the coming months. If I had not joined MySpace, then I never would have heard of the Horrors in 2007 while I was a freshman in a U.S. high school.

The Internet evolved from MySpace into other social platforms, however, especially for music, so there is a chance that the Horrors and I still would've been brought together.

THE IMPORTANCE OF MUSIC BEFORE THE ADVENT OF THE INTERNET

So why does this history lesson matter? What difference does it make that Bikini Kill had to travel thousands of miles just to spread a single message within their own country? Why does the formation of a local music scene matter at all—especially if the birth of one scene ultimately led to success for both the band behind the hit and for those musicians around them?

These events happened without the Internet. Music fans had to work just as hard to hear their choice of music as the musicians worked to expose their music to audiences. Additionally, if a band had a message it wanted to share—as in the case of Bikini Kill—then it needed an audience to receive that message. For music fans at that time, it also was much harder to find a band whose message mattered. The value in knowing how things were in the past is that it enables today's listeners to fully appreciate what they have before them—a world full of music that can be played almost completely uninterrupted if a listener so desires, and the ability to access music that the listener specifically wants to hear. Before music was a regular commodity online there existed a strong physical limitation to finding it. Whether physically attending a show to hear a group, or going to a record store and searching through piles of records to find one that *looked* like it might sound good, music fans had to be much more involved—far beyond keystrokes and mouse clicks.

The Internet has been good to both musicians and fans, especially in connecting fans around the world with musicians that they can truly invest in. In doing so, the Internet has opened fans to new cultures, new ideas, and new messages—much like Kathleen Hanna envisioned in her travels in the early 1990s.

DOWNLOADS READILY AVAILABLE—FOR BETTER OR FOR WORSE

A large part of the online music revolution came with the invention of Napster, a product that frightened labels and enticed music fans equally. Napster came online in 1999[15]—just a few years after the dissolution of both Bikini Kill and Nirvana. Napster was the pinnacle of peer-to-peer software; as such, it also became the target of the wrath of label heads and the Recording Industry Association of America (RIAA).[16] Napster and its ilk were different from Apple's iTunes Store, which offered the first legally available downloads in 2003.[17] Peer-to-peer downloading not only provided listeners with a free option for music to come to them, but it put the power in the hands of those doing the sharing. Music came from users and was delivered to users—all free of charge. Peer-to-peer sharing had even stronger selling point: Users could download entire discographies, not just albums or songs. If a user liked one song, it was easy to safely acquire everything the artist had done without having to purchase any of it.[18]

As the RIAA fretted over what peer-to-peer downloading would do to the industry and brought numerous lawsuits against specific users who were caught engaged with the technology, consumers had been given a taste of what the future could be for avid music listeners. Suddenly, listeners did not have to rifle through import albums in a store or wonder what their counterparts in Australia were listening to. Napster made everything readily available—right at the user's fingertips. There was no turning back; consumers now could connect with new artists and with each other.

THE BENEFITS OF DIGITAL FOR LISTENERS

The availability of digital files eradicated two roadblocks to buying and experiencing music: Sampling and importing.

Sampling

For centuries people have loved music. But it takes trial and error to figure out which music makes a listener the happiest, to decipher which rhythms and words will have the longest lasting effect. Listeners grow up hearing the music that their parents, siblings, older friends, and relatives hand down through time; they are exposed to musicians and albums by people who hope that the sharing will bring something exciting to the life of the new listener. The listener's—and consumer's—earliest experiences with music come from other people and their musical tastes.

There comes a time, however, when a listener discovers music on his or her own, and begins defining a new taste outside of what he or she grew up hearing. Instead of rifling through racks of records or bins of compact discs, though, such discovery now takes place on the Internet. When a listener is ready for a new sound, all he or she must do is go online and search for whatever it is he or she is trying to experience.

The idea of sampling a song or album before buying it is enticing to listeners. It is a safety net; it makes consumers feel that they are making an educated and, ultimately, safe choice. Instead of buying an album at random because of the artwork or name recognition from a chance listen to the radio, a consumer can buy music that already feels familiar.[19] Peitz and Waelbroeck conducted research on the idea of consumer sampling in 2006 and found that sampling did not just benefit consumers; it could, in fact, lead to greater profits for the music industry. This is because fans are operating in a "multiproduct" setting that provides greater product diversity and therefore more options that are likely to fit with what a fan wants to hear.[20]

In 2006, research also was conducted on digital rights management (DRM), a technology that could be embedded into songs that were legally purchased online to prevent them from being reproduced. Digital rights management is defined as "the technical systems and technologies that digital publishers and copyright holders use to exert control over how consumers may use digital works."[21] A song that was embedded with DRM could be played using specified products and operating systems—for example, early iTunes store purchases were embedded with DRM and could be played on an iPod. Purchases from iTunes would not work with other hardware or operating systems, however, and it was illegal to try to save the purchases if a listener wanted them to work with a new operating system. From 2006 through 2009, DRM was a regular discussion in the music industry, as it hardly battled piracy and only made it more difficult for those who legally wanted to share music with others.[22] If a listener's friend used Zune and the listener had purchased an iTunes song, the legally purchased track still only could be loaded on the purchaser's iPod. The research in 2006 showed that files that were easier to download could still help legally sell music—in part because listeners might illegally download some tracks but, once familiar with the music, would purchase an entire album because it was familiar and usage was unlimited.

However, the discussion around DRM-free music went beyond the ease of download for listeners who were looking to sample music before adding it to their collection. DRM-free music also brought awareness to the social-networking effects music could have on people, just as the earliest versions of social networks were being unveiled and adopted by the general public.[23] With DRM-free music widely available and songs easier than ever to download and share, conversations about that music were happening across platforms. The discussion was not limited to a record store or coffee shop with the usual suspects; instead, a user on MySpace based in Connecticut could connect with a user in Brazil over shared love of a given song that both users could access easily and equally. Digital rights management enabled easier interactions and discussions about music that inevitably bridged cultural gaps.

Imports

Beyond sampling, digital music encourages adoption of imported music. With the distribution of vinyl records and then compact discs, imported music had to be pressed in a given country of origin and then shipped world-wide—meaning that, in many cases, demand determined what product was sent where. For a store to carry specific albums, people had to be asking for those albums. This meant that lesser-known bands from around the world often were relegated to their city to sell their records and spread the word.[24]

Research conducted at the University of New Orleans in 2006 showed that digital file sharing did, in fact, impact sales of compact discs, determining that digital downloads could decrease physical sales over time.[25] As digital recordings became more prevalent in the early to mid-2000s, physical sales of compact discs dropped. The increase in digital downloads, however, means that consumers were finding the songs and albums they wanted with greater ease—it also means that more artists were gaining more exposure to new fans. Risk-taking music consumers were gone, with more assured consumers taking their place.

What this meant for imports, though, is that albums that originated overseas no longer were risky purchases. Instead of having to spend more money for an album that was unfamiliar to the listener, imports became equal with music that was made in the consumer's home country. An artist in India could release an album to a peer-to-peer network, the iTunes Store, or another online entity and garner the same exposure as artists in Japan and the United Kingdom. All the artist need do is upload a song and let fans find the music.

Two of the most impactful benefits of digitalization in the music industry are sampling and the normalization of imported music. These two benefits go hand in hand; sampling enables listeners to know what they are buying before spending any money, and the normalization of imports means that international artists have a fair chance at exposure in countries that they previously never would have been able to traverse. Digitalization is a win for fans and musicians.

OVERCOMING GEOGRAPHY

Knowing the history of bands and the music industry before the Internet became a factor, and understanding two key benefits of digital music and some of the story of music going online, we can draw this conclusion: The greatest achievement of the Internet's marriage with music is how it overcomes geographical boundaries that previously never would have been broken down. This is how cultures come together and people can find similarities instead of differences; as borders melt away, greater connectivity occurs. Fans have more options of what music to listen to than ever before—and they can exercise their options on the same technology that they use to call their friends and take pictures. At the same time, musicians can gain exposure to new audiences, determine early on where they should tour, and know where their best chances are for the greatest exposure.

Before the Internet provided such ease of accessibility, bands such as Sonic Youth—the same band that referred Nirvana to their record label—had to travel the world to raise awareness of their music. When the band formed in the 1980s, it began traveling the United States with like-minded bands to show people who they were.[26] The group of bands struggled as it traveled throughout the Midwest, however, rarely finding receptive crowds and often met with confusion. Performance attendees knew only what little they might have read in (small) magazines, and the musicians were rarely paid for their performances.[27]

After touring the United States, the Sonic Youth finally went to Europe. The band found that it had to do much of the same work it had done in the United States: Jump in a van, play tiny venues to unsuspecting audiences, and hope to be paid for the performance.[28] As the band toured Europe, what Sonic Youth eventually found was that audiences were growing to enjoy them. It took rounds of touring, but it the music started to click with "tastemakers" and college kids alike.

In their time touring, Sonic Youth easily covered more than 3,000 miles across the United States; 3,500 miles across the Atlantic Ocean; and more than 1,000 miles around Europe. In just a few years, Sonic Youth traveled nearly 8,000 miles to make people aware of their music. In time, Sonic Youth became one of the most successful bands of their ilk, honing their sound and releasing the genre-defining *Daydream Nation*. But it took years of legwork and travel to reach that goal.

Not every band has the support and the ability that Sonic Youth had. Sonic Youth was backed by small records labels and savvy business friends that believed in what the band was doing, plus a network of bands with similar interests and goals.[29] Even today, touring musicians are not guaranteed easy access to certain locations. Artists have their visas revoked, and flight cancellations frequently leave musicians stranded and unable to make it to their shows. The difference today is that—because of the growth of the Internet and digitalization—artists do not have to rely solely on travels or tours to raise awareness. Instead, a show is just one component strongly complemented by what is happening in the digital space; thus fans still have something to interact with, and the promise of the show does not die if an artist cannot make it on schedule or waits to put on a performance.

The digitalization of music allows musicians to flourish in ways they simply could not before. Applications such as Pandora and Spotify do the legwork that bands previously had to do on tour. Music videos on YouTube can deliver an experience that formerly could be had only by being in the same country as a particular artist. Times have changed, and it is good to be a musician and a fan now. A fan can travel around the world with the right search terms and just a few clicks of a mouse button.

GENRES AND SOUNDS EXTEND BEYOND THEIR HOME

With geographic barriers broken down, listeners able to sample whatever music they choose, and imports now being an affordable and viable option, new sounds can be heard anywhere a listener desires. The possibilities are endless as to what can be experienced and what will resonate with listeners around the world.

Remember the Madonnas and U2s from earlier times? Artists who found success with some airplay in the United States and then traversed the world on massive tours and were guaranteed sell-out crowds of thousands no matter where they went? Well, they are not the only ones who can experience that now. Granted, many performers will not sell out arenas on their first trek around the world. But it is becoming more and more likely—and much, much easier—for artists with massive success on the Internet to translate that success into bodies in stands, stalls, and seats. What is beautiful with the growth of the Internet and music's ability to reach more people than ever before is that genres and sounds that previously would have had difficulty selling even a few tickets now can sell to massive crowds.

A perfect example of an artist who first found success on the Internet and translated it into international acclaim and fandom is Harlem rapper A$AP Rocky (born Rakim Mayers).[30] A$AP Rocky first earned attention by posting a video of his song "Peso" on YouTube. In just a few months the video had more than three million views and Rocky was destined for international recognition—aided by considerable hype from those in rap's innermost circles.[31] Rocky followed "Peso" with a highly acclaimed mixtape that was distributed online and made easily accessible to a public growing more and more interested in the young rapper. Over the course of two years Rocky built a following through videos and free releases that guaranteed him an excited and engaged fan base when his official debut album, *Long.Live.A$AP*, finally was released in 2013.[32]

Rocky embracing the Internet did not just help his debut album sales or the skyrocketing success of his early singles. He, along with his A$AP Mob cohorts, were able to go on tour and play sold-out shows around the world. Rocky and his crew brought the sounds of their Harlem to kids in places as distant as London and Sydney. His first tour in support of *Long.Live.A$AP* went to 47 cities, with an average of 3,000 tickets sold per venue.[33]

As of 2013, Peter Schwartz is Rocky's tour manager as well as the vice president and head of Urban Music at The Agency Group in New York. In an interview about Rocky's rapid and Internet-based success, Schwartz spoke about a number of factors that make Rocky so appealing as a performer and why he resonates with his audiences. In his discussion, Schwartz spoke candidly about Rocky's international success. "We are definitely impressed to see how well he is selling tickets internationally. 4,000 capacity venues blowing out in a matter of days in Australia and Rocky has never been there before. [That's] definitely impressive."[34] The excitement around Rocky that started on the Internet translated into actual ticket sales for him in places he previously never had visited.

Schwartz also gives credit to social media for how awareness of Rocky's music was raised by his fans around the world. As a musician, it was easy for Rocky to put his songs online and make them available for anyone who was interested. But for fans, it is easy for them to discover and share Rocky just as quickly. Social media did a lot of the early talking for Rocky, and as Schwartz said, "Word of web is far stronger than word of mouth."[35]

The rap music that A$AP Rocky makes evokes Harlem specifically. His sound is one of many within the realm of rap and hip-hop, and with just one song he brought that specific sound to millions of people. Without the Internet, people in London likely would not know or care about a Harlem rapper's thoughts, much less why they matter. Instead, listeners with thousands of miles between them—with oceans and mountains and other physical barriers blocking them from each other—can hear A$AP Rocky and his distinct sound, and then find a meaning within the music that matters to them. These listeners certainly would hear rap music but it would not evoke Harlem as a place and an experience, as life in Harlem is likely unknown to listeners in the United Kingdom.

Bikini Kill had to travel throughout the United States to spread the message of female equality. It took radio play for Nirvana's biggest single to take off first in the United States and then around the world. Sonic Youth drove a van through the entirety of the United States, and then they drove another van through Europe in the hopes of having even 20 people come to a performance. A$AP Rocky had three million listeners, a record deal, and almost 50 sold-out shows just by building his image on the Internet. A$AP Rocky brought the sound of his neighborhood to listeners internationally with one video. A$AP Rocky covered the world with the release of one video.

Music on the Internet goes everywhere, and in doing so it makes the world a little smaller and brings people a little closer. And it is only going to continue growing, evolving, changing, exploding.

RADIOHEAD AND THE RECORD LABELS

Thom Yorke is a crazy man.

As the front man of Radiohead for nearly 30 years, Tom Yorke has had a consistent platform to offer his take on the world. Radiohead formed in 1986 in Oxfordshire, England[36]; the group was able to witness both the success of major pop stars and the explosion of various independent music scenes in the United States.

Radiohead first found massive success with "Creep," the band's breakthrough single from the 1993 album *Pablo Honey*. Following the release of that album, Yorke and crew continued churning out albums—many critically acclaimed—through a six-album record deal with Columbia.[37]

Radiohead's Capitol/EMI deal ended with the release of 2003's *Hail to the Thief*. When the time came to renew with Capitol or find another label, Radiohead chose neither option.[38] Instead, the group went to work on a seventh album for the next four years, never committing to a label and never promising a release date for whatever the next album would be.

Then, in October 2007, Radiohead announced via blog post that the band's next album, *In Rainbows*, would be released later that month. The plot twist, though, was that Radiohead had not committed to a new label. Instead, the group recorded and produced the album on its own, and then planned to release it online—and let consumers determine the price. As *Time* magazine explained, "It's the first major album whose price is determined by what individual consumers want to pay for it. And it's perfectly acceptable to pay nothing at all."[39] When *In Rainbows* was officially released, interested listeners could log on to the designated Web site and fill in their chosen payment amount. For many, this meant putting "0.00" in the box. Radiohead did not care, and Thom Yorke made his point: He did not need a label to share the music he was creating for his fans.

Of course, the story is even better because *In Rainbows* ended up being massively successful both online and on CD. Industry reports a year later, in October 2008, showed that the album sold 3 million copies online, as well as 1.75 million physical copies worldwide.[40] Special edition "discboxes" were available as well, and more than 100,000 of those special editions also were sold. For Radiohead, the online-first, pay-what-you-want release translated into some of the group's biggest sales. Previously, Radiohead usually sold hundreds of thousands of albums. With *In Rainbows*, millions became the standard.[41]

Radiohead's experiment is just one way that the digital space is going to continue growing and improving for musicians and their fans worldwide. All a listener needs is an Internet connection to hear new music; and now, backed by innovative artists such as Radiohead, music is available in exciting and affordable forms. A user does not just have to log in to a designated store or account to buy music; instead, artists understand just how important it is to make their music accessible to anyone who might be interested. The implications of the Radiohead experiment in 2007 are crucial, as it paves the way for further experimentation from artists who, prior to the *In Rainbows* album release, might have been uncertain just what the Internet—and an affordable album—could do for them.

MY STORY, CONTINUED

When the Horrors released *Strange House* in 2007,[42] I did almost all of my listening via MySpace and the tracks that were available on the social network. Four years later, when the Horrors released their third album, *Skying*, I found out about it not through MySpace, but through Pitchfork.com. By 2011, I was a sophomore in college and I shed my MySpace habits and instead immersed myself in two worlds, new social networks (Facebook, Twitter) and the music blogosphere, which for me hinged on Pitchfork's coverage of new music and releases.

When the Horrors released *Skying*, I followed the coverage on Pitchfork. Again, the Internet brought me news of a band that—even 20 years ago—it would've taken me years to hear about. Instead, though, I already knew the band's whole backstory and learned about their previous releases from other Web sources. I never had to buy import copies of their albums at high costs; I didn't have to rely on domestic radio to play a few of their songs for me to be interested. Instead, when *Skying* was released, I found out about the album through Pitchfork, and then listened to it on Spotify. I never even had to leave my dorm room to listen to a band producing music across the ocean. Equally important for me in my college years, I didn't have to spend a dime to experience any of it.

My journey with the Horrors is just one of many experiences with musicians that began online. The Internet's power to connect people around the world with new music is unbelievable. The connections happen, and happen *rapidly*. As those connections are made around the world, a physical component comes into play. When a listener knows he or she is invested in an artist, buying physical copies of the artist's music becomes an exercise in more than consumerism. It becomes a way to show how much the artist is a part of the listener's life; the artist earns a physical spot.

That same trip to London found me conducting research in record shops throughout the city. While doing some observational research at one of the Rough Trade record shops, I had the opportunity to pick up *Skying* on vinyl. I already had played the album compulsively on Spotify, so I knew exactly what I was getting—and I was even willing to spend a little more to buy a special version of it. Without the Internet, I never would have been able to explore London's indie rock scene so easily, and I definitely would not have known when I arrived in London the exact bands that I would support financially.

This experience is common today. With access to albums, mixtapes, songs, remixes, and EPs, listeners can experience any artist they want at any time they choose. Music brings us together. Throughout time people have fought each other; strife is common. As people, we find reasons to be divided and to be argumentative—especially in the preservation and exultation of our own cultures and heritages. But music can break those boundaries. For as long as we have fought, we have also sung. Music can break barriers that have been built over centuries—one note and one brick at a time. If music can break boundaries established for centuries between people and their cultures, then why would music be stopped by physical geographical boundaries? We become stronger as people when we learn about other cultures and embrace them; we better understand our own histories when we have the context of others. By letting new music into our daily lives with the help of the Internet and the ease with which we can access the new and unusual, we can become that much closer, our world can become that much smaller—and our artists, the ones who bring us the relief from the everyday, can thrive in previously unheard of ways.

NOTES

1. "Internet Society," Brief History of the Internet, http://www.internetsociety.org /internet/what-internet/history-internet/brief-history-internet (January 1, 2014), accessed January 2, 2015.

2. Ibid.

3. "Kathleen Hanna Biography," Bio.com (January 1, 2014), accessed January 3, 2015. http://www.biography.com/people/kathleen-hanna-17178854.

4. Marah Eakin (November 20, 2012), "Kathleen Hanna on Bikini Kill, Growing Up, and Being a Feminist Icon," *The A.V. Club.*

5. Ibid.

6. Holly Kruse, "Local Identity and Independent Music Scenes, Online and Off," *Popular Music and Society* 33 (5) (2010): 625–39.

7. Larry King, "Interview with Bono," CNN.com (January 1, 2002), accessed January 2, 2015.

8. Holly Kruse, "Local Identity and Independent Music Scenes, Online and Off," *Popular Music and Society* 33, no. 5 (2010): 625–39.

9. Nathaniel Penn, "Nirvana and Kurt Cobain: An Oral History," GQ (June 1, 2011).

10. Ibid.

11. Holly Kruse, "Local Identity and Independent Music Scenes, Online and Off," *Popular Music and Society* 33, no. 5 (2010): 625–39.

12. Nathaniel Penn, "Nirvana and Kurt Cobain."

13. Ibid.

14. Timothy Stenovec, 2011 (June 29), "Myspace History: A Timeline of the Social Network's Biggest Moments," *The Huffington Post.*

15. Norbert Michel, "The Impact of Digital File Sharing on the Music Industry: An Empirical Analysis," *Topics in Economic Analysis & Policy* 6 (1) (2006).

16. M. William Krasilovsky and Sidney Shemel, *This Business of Music: The Definitive Guide to the Music Industry,* 10th ed. (New York: Watson-Guptill Books, 2007).

17. François Moreau, "The Disruptive Nature of Digitization: The Case of the Recorded Music Industry," *International Journal of Arts Management* 15, no. 2 (2013).

18. Krasilovsky and Shemel, *This Business of Music*.

19. Martin Peitz and Patrick Waelbroeck, "Why the Music Industry May Gain from Free Downloading: The Role of Sampling," *International Journal of Industrial Organization* 24, no. 5 (2006): 907–13.

20. Ibid.

21. Rajiv Sinha, Fernando S. Machado, and Colin Sellman, "Don't Think Twice, It's All Right: Music Piracy and Pricing in a DRM-Free Environment," *Journal of Marketing* 74 (2010): 40–54.

22. Ibid.

23. Ibid.

24. Browne, David, *Goodbye 20th Century: A Biography of Sonic Youth*. 1st ed. (Da Capo Press ed.) (New York: Da Capo, 2008).

25. Michel, Norbert, "The Impact of Digital File Sharing on the Music Industry: An Empirical Analysis." *Topics in Economic Analysis & Policy* 6(1) (New Orleans: University of New Orleans, 2006).

26. David Browne, *Goodbye 20th Century*.

27. Ibid.

28. Ibid.

29. Ibid.

30. Joe La Puma, "A$AP Rocky: Rocky Road," *Complex* (January 2013).

31. Ibid.

32. Chris Franco, "A$AP Rocky's Transformation from YouTube Rap Star to International Touring Artist," *Hypebot* (April 5, 2013): 1.

33. Ibid.

34. Ibid.

35. Ibid.

36. "Radiohead: A Brief History," *NME News* (2007).

37. Ibid.

38. Josh Tyrangiel, "Radiohead Says: Pay What You Want," *Time* (October 1, 2007): 1.

39. Ibid.

40. Paul Thompson, "Radiohead's 'In Rainbows' Successes Revealed," *Pitchfork* (October 15, 2008):1.

41. Ibid.

42. "The Horrors Reveal Debut Album Title," *ClickMusic* (January 11, 2007).

BIBLIOGRAPHY

Browne, David. 2008. *Goodbye 20th Century: A Biography of Sonic Youth*. 1st ed. (Da Capo Press ed.) New York: Da Capo.

Eakin, Marah. 2012 (November 20). "Kathleen Hanna on Bikini Kill, Growing Up, and Being a Feminist Icon." *The A.V. Club*.

Franco, Chris. 2013 (April 5). "ASAP Rocky's Transformation from YouTube Rap Star to International Touring Artist." *Hypebot*, 1.

"The Horrors Reveal Debut Album Title." 2007 (January 11). *ClickMusic*.

Internet Society. 2014 (January 1). Brief History of the Internet. http://www .internetsociety.org/internet/what-internet/history-internet/brief-history-internet. Accessed January 2, 2015.

"Kathleen Hanna." Biography.com. http://www.biography.com/people/kathleen-hanna-17178854. Accessed January 3, 2015.

Krasilovsky, M. William, and Sidney Shemel. 2007. *This Business of Music: The Definitive Guide to the Music Industry*. 10th ed. New York: Watson-Guptill Books.

Kruse, Holly. 2010. "Local Identity and Independent Music Scenes, Online and Off." *Popular Music and Society* 33 (5): 625–39.

La Puma, Joe. 2013 (January). "A$AP Rocky: Rocky Road." *Complex*.

Michel, Norbert. 2006. "The Impact of Digital File Sharing on the Music Industry: An Empirical Analysis." In *Topics in Economic Analysis & Policy*. 1st ed. Vol. 6. New Orleans: University of New Orleans.

Moreau, François. 2013. "The Disruptive Nature of Digitization: The Case of the Recorded Music Industry." *International Journal of Arts Management* 15 (2).

Peitz, Martin, and Patrick Waelbroeck. 2006. "Why the Music Industry May Gain from Free Downloading: The Role of Sampling." *International Journal of Industrial Organization* 24 (5): 907–13.

Penn, Nathaniel. 2011 (June 1). "Nirvana and Kurt Cobain: An Oral History." *GQ*.

"Radiohead: A Brief History." 2007. *NME News*.

Sinha, Rajiv, Fernando S. Machado, and Colin Sellman. 2010. "Don't Think Twice, It's All Right: Music Piracy and Pricing in a DRM-Free Environment." *Journal of Marketing* 74: 40–54.

Stenovec, Timothy. 2011 (June 29). "Myspace History: A Timeline of the Social Network's Biggest Moments." *The Huffington Post*.

Thompson, Paul. 2008 (October 15). "Radiohead's in Rainbows Successes Revealed." *Pitchfork* 1.

Tyrangiel, Josh. 2007 (October 1). "Radiohead Says: Pay What You Want." *Time* 1.

4

Narcissism or Self-Actualization? An Evaluation of "Selfies" as a Communication Tool

Christina Best

INTRODUCTION

A "selfie" is "a photograph that one has taken of oneself, typically taken with a smartphone or webcam and often uploaded to a social media website."[1] Selfies have found a permanent home on social media, especially on image-dependent sites such as Facebook and Instagram. It seems that everyone is taking selfies, including the Pope,[2] President Barack Obama,[3] and even our pets.[4] In 2013, Oxford Dictionaries named "selfie" its word of the year.[5] The selfie revolution undoubtedly hit its saturation point when Ellen DeGeneres posted to Twitter photos of herself and numerous high-profile celebrities during the 2014 Oscars.[6]

Pew Research Center found that 46% of all Internet users polled "post original photos and videos online [that] they have created."[7] Furthermore, 55% of "Millennials" (18 to 33 year olds) "have shared a selfie on a photo sharing or social networking site such as Facebook, Instagram or Snapchat."[8] Not surprisingly, Millennials also comprise the greatest percentage of users on Facebook and Instagram, with 83% of 18- to 29-year-olds polled using Facebook and 27% using Instagram.[9] Although vital for understanding social media use demographically, these figures do not explain what motivates individuals to post selfies.

This research explores how selfies can be interpreted as self-actualizing and empowering as well as narcissistic and inauthentic, in addition to revealing the universality of seeing ourselves as others do. The author concludes that selfies transcend the concepts of "the good, the bad, and the ugly" because—despite being displayed as JPEGS and in pixels—selfies represent a timeless human desire: to capture one's essence despite life's impermanence.

A BRIEF HISTORY OF THE SELFIE

It's easy to write off the selfie as a product of a technological revolution, but "the selfie is a smartphone-produced version of the self-portrait, which has been a staple of art and photography history since artists first began examining their own images in the mirror."[10] Both painters and photographers have recreated images of themselves using their artistic mediums. This section briefly explores the history of self-portraiture to provide context for its current iteration: the selfie.

Self-Portraiture before the Camera Phone

People have been the subject of artistic works for centuries, and self-portraits have been created by many well-known artists, including Rembrandt van Rijn[11] and Frida Kahlo.[12] Although vanity cannot be discounted as a motive for creating a self-portrait, artists often paint themselves because they are a readily available[13] and inexpensive model[14] who will not "complain about the results when a painting session [is] over."[15] Self-portraits also allow an artist to develop his or her technique while stimulating self-discovery.[16]

Art historians hypothesize that Rembrandt experimented with self-portraiture "to study and practice [the] depiction of specific emotions"[17] by examining his face in a mirror.[18] Not only was Rembrandt sought after for his portraiture, but "throughout his life he produced an unprecedented number of drawn, etched, and painted self-portraits."[19] It is estimated that he painted more than 80 self-portraits,[20] with some historians placing that figure at 98,[21] making Rembrandt the most prolific self-portrait artist "of any major painter in history."[22] In his self-portraits, "Rembrandt frequently portrayed himself in period and other types of costumes"[23] with corresponding hats. Self-portraiture enabled Rembrandt to express unique versions of himself.[24] Each portrait conveys distinct and universally human emotions; Rembrandt is the instrument through which the meaning of the painting is created.[25]

Kahlo produced 143 paintings in her lifetime, 55 of which are self-portraits.[26] After contracting polio early in her childhood and "nearly [dying] in a bus accident as a teenager,"[27] Kahlo suffered from chronic pain.[28] She used art as a reprieve from her physical pain during her recovery after the bus accident, which initially confined her to a body cast.[29] In Kahlo's piece titled "The Broken Column," she "is depicted nearly naked, split down the middle, with her spine presented as a broken decorative column"[30] with nails piercing her skin.[31] Unlike Rembrandt's focus on portraying himself as various characters portraying specific emotions, Kahlo's self-portraits "capture her spiritual and physical suffering in anatomical, surgical detail."[32]

Kahlo and Rembrandt are only two of a myriad of artists who used (and continue to use) self-portraiture as a creative outlet to convey their life experiences. What makes Kahlo particularly of note "is her obsession with visual auto-biography, with defining her own image through the objects and items she accumulated around her, [which] connects her to this . . . heavily accessorized decade."[33] What Stephen Marche is referring to is the tendency for social media users to overshare various aspects of their lives, and it is his opinion that "the urge to share with the world a photo of your most recent meal, your sexy new handbag, or the sonogram of your unborn baby is an expression of the same fundamental desires that motivated Kahlo. She was an oversharer."[34] The desire to reveal to others fragments of our life's story, no matter how excessive or mundane, is what links Kahlo and her self-portrait predecessors to the selfie phenomenon.

The Rise of the Modern Selfie

The selfie would not be possible without the invention of the camera. Photography can be traced back as early as the fifth century BCE.[35] Although the camera obscura, a dark space "into which light is admitted through a tiny opening in one of the walls or windows,"[36] had existed for centuries prior, it remained undefined until it was most famously used by Leonardo da Vinci in the 13th century. In the centuries that followed, artists used this tool as a "compositional aid,"[37] sketching the projected images onto paper.[38] What this historical snapshot reveals is that "by 1800 the camera had long since been invented, but no one had created film for it."[39]

The first "film" was invented in the 1930s in France by chemist Joseph Nicéphore Niépce[40] with the assistance of inventor Louis Daguerre.[41] Niépce and Daguerre discovered a way to transfer images onto a copper plate.[42] This process, named the "daguerreotype,"[43] became a popular novelty among the middle class during the Industrial Revolution.[44] Just as artists of centuries past used self-portraiture to perfect their craft and explore, so, too, have photographers. One of the oldest daguerreotype self-portrait photographs in existence was taken in the 1830s by chemist Robert Cornelius in Philadelphia.[45] In the photo, Cornelius is described as "looking intently at a camera, making sure it's working."[46] Less about self-expression, Cornelius' photo represents a time when photography was focused on the process and not the outcome. But, by 1889, George Eastman created roll film,[47] shifting the focus to the outcome—the photo—of the photographic process. Roll film "made photography more portable"[48] which encouraged people to adopt photography as a hobby.[49]

Lee Friedlander, a prominent self-portrait and street photographer active from the 1950s through the 1980s,[50] focused his work on conceptualizing "the self as just another object in the world."[51] Friedlander deliberately broke standard photography practices, experimenting with "getting [his] own shadow in the picture, lining up a head with a tree."[52] He attempted to photograph himself and the world around him exactly as it appeared, including the use of "normal, reasonably well-shaped women marked by acne [and] bruises" as subjects in his nude photographs.[53]

In the 1970s, Cindy Sherman redefined the boundaries of self-portraiture.[54] Unlike Friedlander's meticulous attempts to convey life as realistically as possible,[55] Sherman used self-portraiture to create theatrical narratives.[56] She turned the still photograph into a piece of performance art. In 1976, she released a series titled "Bus Riders," in which she pretends to be passengers on a bus by wearing different outfits and using various props.[57] Contemporary self-portraiture combines Sherman's performance-art style and Friedlander's realism in ways that "depict the essence of a sitter's presence and . . . describe a narrative scene."[58]

It was during the lives of Sherman and Friedlander that cameras became "lighter and smaller,"[59] eventually leading to the creation of the digital camera in 1969,[60] though it was not until 1991 that the digital camera was introduced as a commercial product.[61] Digital photography has allowed the meaning of the photographed subject become more "fluid."[62] Instead of film cameras, which create static, "fixed image[s],"[63] digital cameras make capturing moments a more fluid process. Various settings can be manipulated in the field and further changes can be made afterward. Because digital images can be uploaded and shared in innumerable settings, "images' meanings morph, move, and can exist in multiple places and meanings at one time."[64] The advent of the built-in cell phone camera further expanded self-perception in the digital age.

With no way to easily document the birth of his first child, Philippe Kahn, a technology developer, saw an opportunity to combine digital photography with cell phone technology together.[65] The first adaptors of Kahn's technology were J-Phone, a Japanese company, that "introduced a mobile phone called the J-SH04, the first to have a built-in camera" in 2000.[66] By 2002, "American and European consumers"[67] could purchase camera phones. Cameras, along with Internet access, are a ubiquitous part of the modern cell phone. A 2013 report conducted by Pew found that "63% of cell phone owners now use their phone to go online."[68] Additionally, more than 90% of American adults own a cell phone,[69] and 56% of cell phone users own a smartphone.[70]

Shawn DuBravac's summary of recent technological trends can be applied to how photography has adjusted to the digital age: "The first digital era began when analog devices were replaced by their superior digital equivalents, but this second digital era is driven by the broader digitization of our physical space."[71] In photography's case, film cameras were replaced by digital cameras by consumers as a whole, only to become a camera phone substitute. Unlike physical digital cameras, people can quickly capture a moment "without the need for separate equipment and without the need to plan ahead."[72] As a result, "visual communication rules our technocentric lifestyles, and the selfie is a natural outgrowth of how we communicate and connect in a hyper-social-networked world."[73] Digital photography has created new ways for people to relate to one another and, perhaps more noticeably, it has allowed people to explore their identities in more detailed and immediate ways than ever before.

SELFIES AS COMMUNICATION TOOLS

From the camera obscura to the digital camera, documenting one's life has become easier to the point of being instantaneous, particularly with the addition of cameras to cell phones. Combined with the pervasiveness of social media sites, humans have redefined communication norms in the 21st century. In an interview with *Wired*, author and photographer Marvin Heiferman describes how photographs evolved from being static mementos to communication devices.

> In the past, it was more conventional; we had to have reason to make a picture and it was usually to document something specific. Whereas now people . . . take pictures because the camera is there [in their hand]. It has got to the point where sometimes if you ask people why they take pictures they can't even say. I think people are using images in a completely different way and as a communicative tool.[74]

Stephen Mayes, also interviewed for *Wired*, affirms Heiferman's observation, suggesting that "the way we relate to imagery is changing. Our new relationship is less about witness, evidence and document and much more about experience, sharing . . . and streaming."[75] Or, as Sherry Turkle puts it: "I share, therefore I am."[76] Sharing experiences through photos, no matter how mundane[77] they appear to an observer, provides people with "a sense of importance."[78] Furthermore, the rapid accessibility of social media has "fundamentally changed the way we read and watch: we think about how we'll share something, and whom we'll share it with, as we consume it."[79] Selfies often are taken with the intention of being shared via social media. Selfies have become a form of "self-announcement"[80]—a way for us to let others know through "electronic means"[81] that "mark our short existence and hold it up to others as proof that we were here."[82]

The increasing capabilities of camera phones to take higher-resolution images "mak[e] it easy to take and share self-made pictures online. It is also safe to say that the rise of social media, especially YouTube and newer services like Pinterest, Instagram, and Tumblr, has made curating activities easier because they are organized for easy image and videosharing."[83] Sharing remains the reason why we use social media, "but now our first focus is to have, to possess, a photograph of our experience." [84] Although "photos of friends enjoying a party, a newborn baby"[85] and selfies are "taken for the express purpose of sharing,"[86] the motivation to post these comes from a universal human need to "to stand outside of ourselves and look at ourselves."[87] Even President Obama "showed us how he, too, lives in our culture of documentation" when "he took a selfie at [Nelson] Mandela's memorial service."[88]

Selfies are ultimately a very personal extension of the person sharing them, which makes them reminiscent of the self-portraits in art and photography that came before them. Even though these "self-portraits are worlds—and decades—apart . . . they are threaded together by a timeless delight in our ability to document our lives and leave behind a trace for others to discover."[89] What differentiates selfies from earlier self-portraiture is their "level of self-conscious authenticity that is different from even a candid photograph— [selfies] are more raw and less perfect."[90] Selfies are also more about sharing an "experience at that moment"[91] than portraying a Kahlo-like allegory of the human condition.

In sum, selfies are a modern communication tool ushered in by the digital age, and they are a form of immediate self-expression shared with others as a way to justify our life's experiences. As a communication device, selfies are more complex than an individual simply snapping a photo of himself or herself. Selfies also can be seen as an extension of our "ongoing adolescence"[92] and an exercise in self-discovery.[93]

Much like performance-art self-portrait photographer Cindy Sherman's deliberate use of props and scene-setting discussed above, people use selfies to both explore their individuality and to create a narrative for their friends and followers.[94] This is defined as "identity as performance,"[95] a concept developed by sociologist Erving Goffman in the mid-20th century. It suggests that we assume various identities any time we exchange information with others.[96] Social networks enable a user to deliberately construct his or her self-identity using "tools and technologies to project, renegotiate, and continuously revise their consensual social hallucination. Users manipulate these communicative codes . . . to create not only online selves, but also to create the staging and setting in which these selves exist."[97] Alicia Eler affirms this point, stating that "we are ourselves, but we are also meta-selves and performances of our own perception of identities online."[98] What this means is that social networking sites enable us to create countless versions of ourselves, if we so choose. We are the curators of modern culture and with "each of us creating a world that is increasingly the fruit of our own experiences: our own videos, books, blogs, and images made from the materials of our own lives. Increasingly, we own our own worlds."[99] In our "global social networked age"[100] truth is a more fluid concept, defined by what information we choose to share online.

"Identity as performance"[101] also can be described, in more technical terms, as "self-presentation,"[102] which is "the process thorough which [social media] users attempt to control the impressions others may form of them."[103] In a study of Facebook users, researchers found that online users take special care to construct message expressions by revising or abandoning their ideas until they create satisfactory impressions."[104] People manage their online selves through the following "information control"[105] categories: "expressive information control, privacy information control, and image information control."[106] Expressive information refers to the user's opportunity to "'post,' 'comment,' and 'like,'"[107] other's content, privacy information is controlled using the privacy settings built into the social media site, and image information includes all information displayed visually, including profile pictures and banner images.[108] Social networking sites have continued to give users greater control over these categories,[109] which "means that most information about the virtual self and its place in the network is given through deliberate construction of signs."[110] We carefully construct and maintain our online identities using the features and tools of each social media site and join the "struggle to be recognized as a unique individual or as a member of a social group."[111]

Thus far, this chapter has focused on how "selfies represent 'the shift of the photograph [from] memorial function to a communication device,'"[112] and has considered how selfies are used as a performance, allowing the individuals posting the images to explore and define his or her self-identity. Although selfies have become "part of an ongoing virtual documentary of modern life,"[113] they have simultaneously come to represent a phenomenon "Buzz-Feed" columnist Amanda Petrusich defines as "the nostalgia of now."[114] Certain social media applications, such as Instagram, enable users to apply colored overlays, called "filters," to make images taken using their cell phones look old. In a time when a digital photo is just one "share" button away from being posted online, people are left "feeling nostalgic for a time [they] never actually knew,"[115] a time when a physical copy of a photo was "itself a precious object."[116] This "contemporary phenomenon"[117] is reflected in the use of Instagram filters and the resurgence of vinyl records.

Instagram filters become a part of an individual's online performance; it "is not just a filter, it's a meta-commentary—a deliberate aesthetic and narrative choice."[118] Instagram has given people greater control over the life stories they construct online. Although filters are not required for posting a photo on Instagram, people can use filters to slightly alter reality, thereby visually communicating to their friends and followers a certain tone or a feeling that is, in itself, a reflection of who they are.

WHY PEOPLE POST SELFIES

In the age of nonstop digital communication, billions of people all over the world send and receive messages every second of every day. Because selfies have become an online communication tool they have assumed multiple meanings, depending on who receives and interprets them. Taken as a symbol of the social-networking era, selfies largely are seen as either narcissistic or self-actualizing. This section explores both opinions on this subject, to help in understanding how selfies represent both "the good" and "the bad and the ugly" of digital communication.

Shameless self-promotion is a behavior commonly cited as evidence of the selfies' narcissistic nature. For those who hold this opinion, seeing a selfie can "trigger perceptions of self-indulgence or attention-seeking social dependence that"[119] indicate "either narcissism or low self-esteem."[120] Because selfies are used as a performance device, they can be interpreted as a less authentic view of reality; they are a "way to polish public-facing images of who we are, or who we'd like to appear to be."[121] They also inevitably "veer into scandalous or shameless territory . . . and, at their most egregious, raise all sorts of questions about vanity . . . and our obsession with beauty and body image."[122] Therefore, as a communication device, selfies have a selfish motivation: To seek approval of one's physical appearance from his or her peers in the form of likes and comments.[123]

Selfies also can be used to brag about our lives and achievements, a type of human behavior that "is practically coded into social media's DNA."[124] These messages of "self-enhancement—the human tendency to oversell ourselves,"[125] cause people to engage in "social comparison"[126]—a "social-psychology phenomenon"[127] that occurs when people compare themselves to "like-minded peers."[128] We engage in self-enhancement when we reflect on "the interestingness of our companions, the solidity of our relationships, [or] the fabulousness of our meals"[129] on social media. As a result, others could become envious of the lives we appear to live,[130] and although "we want to learn about other people and have others learn about us"[131] we could come to "resent both others' lives and the image of ourselves that we feel we need to continuously maintain."[132] In a study of Facebook use, researchers found that jealousy is more likely to occur in people who spend more time browsing others' content "as opposed to actively creating content and engaging with it."[133] For some people it can be difficult to resist the temptation to compare themselves to the achievements they read about or see on friends' social media pages.[134]

Oversharing personal information also is cited as a reason that selfies and the exchanging of information on social media are self-centered behaviors. The urge to overshare is likely a result of social media's pervasiveness, but it also is a reflection of how capitalism functions at an individual and personal level.[135] Images are created and curated by people—each with distinct personal brands[136]—and these images are mass-consumed.[137] It is Stephen Marche's opinion that social media users tend to "spend an inordinate amount of time photographing themselves with their stuff, defining themselves through things" and share these visual timelines in excess, thus "supply[ing] the world with constant images of themselves wearing designer shoes and clothes and bags and anything else they might be consuming or wanting to consume."[138]

Social networking sites are "owned by capitalist enterprises,"[139] and the content uploaded and shared is a reflection of the capitalist system. As a result, personal identity has been "commodified"[140]; identity has become a product we visually consume.[141] Roger Cohen, writing for *The New York Times*, likens social media to a "global high-school reunion at which"[142] people share "their numb faces at the dentist, their waffles and sausage, their appointments with their therapists, their personal hygiene, their pimples and pets, their late babysitters, their grumpy starts to the day . . . and all the rest."[143] This "ooze of status updates,"[144] Cohen says, is a result of peoples' "compulsion to share,"[145] no matter how mundane or off-putting the content might be.

Although there is validity to the perspective that selfies are a reflection of our propensity for narcissism, it is overly simplistic[146] and fails to acknowledge that we are, above all things, social beings and "visual communicators."[147] Communication and "interconnectedness"[148] are essential parts of the "human condition."[149] So it is not surprising that visual communication tools "are often more effective at conveying a feeling or reaction than text"[150] because seeing "the face of the person you're talking to brings back the human element of the interaction."[151] Additionally, although selfies seem to emphasize the importance of a person's physical appearance, "humans are hardwired to pay attention to looks and continually make both upward and downward comparisons. Social comparison is not a moral failing or an indication of misplaced values."[152] There are people who are insecure who post selfies "to get attention . . . and social validation,"[153] but selfies are not the root cause of this type of behavior.[154]

Sharing our experiences with others through digital communication can have positive effects on a person's well-being. Psychologist Matthew D. Lieberman, in an article about Facebook use published in *The New Yorker*, concludes that "the experience of successful sharing comes with a psychological and physiological rush that is often self-reinforcing. . . . The mere thought of successful sharing activates our reward-processing centers, even before we've actually shared a single thing."[155] A study conducted in 2010 about social well-being and social networking use[156] affirms the positive impact of sharing on social media, determining "that, when people engaged in direct interaction with others—that is, posting on walls, messaging, or 'liking' something—their feelings of bonding and general social capital increased, while their sense of loneliness decreased."[157] So although some people might either overshare or develop an unhealthy obsession with validation from their peers, selfies generally are a positive form of communication.

Selfies also can "be empowering and even normalizing and reaffirm the drive for authenticity that is the hallmark of social media." Social media affords people the "freedom of representation,"[158] making self-expression and self-truth more flexible concepts. In our increasingly digital world, expressing ourselves through selfies is an important exercise "in a modern world that bombards us with reasons to feel bad about ourselves. . . . Online, we're safe to note our achievements, our loves, our tiny daily triumphs in a bid for a little positive feedback." According to the "selfie king,"[159] James Franco, "attention seems to be the name of the game when it comes to social networking."[160] Franco states that "attention is power"[161] in the "struggle to be recognized as a unique individual"[162] in our information economy. Which begs the question: "Is that so wrong—to want to be noticed, to ask people to see us the way we see ourselves?"[163]

CONCLUSION

The following quote is credited to Frida Kahlo, "'I paint myself because I am often alone and I am the subject I know best.'"[164] Much like Kahlo, we "selfie ourselves" because our own lives are the subject that we know best. As much as we come to understand about ourselves over time, we never can physically see ourselves as others do. Selfies are the closest we can get to actually seeing ourselves. We also use selfies to document "a passing moment" and provide "an instant visual communication of where we are, what we're doing, who we think we are, and who we think is watching."[165] As with any social phenomenon, there always will be multiple viewpoints—both positive and negative. It is the conclusion of this author, however, that selfies—despite, at times, oversaturating the social media landscape—are a positive form of self-expression.

Ultimately, vanity and a preoccupation with our outward appearances are part of what makes us human. Taken at face value, selfies—and social media as a whole—seem to discourage face-to-face communication. The rapid development of digital technology has made humans adapt the way we communicate and, for better and for worse, a lot of conversation happens online. Although taking a selfie appears to be an act of solitude, it actually is a smaller part of the larger picture of how we project ourselves and connect with our friends and others online. Alicia Eler's summary of the selfie phenomenon encapsulates this concept quite well.

> If the selfie is the ultimate mirror in our internet house of mirrors, and we can frame our photos and curate ourselves as we want others to see us, then surely the selfie is an act of taking back the gaze. We look through the reversed mirror of the iPhone, into an actual mirror (camera flash reflection optional), or gaze longingly into a computer webcam. We self-consciously perform these moments from inside our private, domestic spaces, for ourselves and for our internet friends and "friends," who are also voyeurs. They are our voyeurs, and we willingly welcome them into our curated worlds.[166]

NOTES

1. "Oxford Dictionaries Word of the Year 2013," *Oxford Dictionaries*, last modified November 19, 2013, http://blog.oxforddictionaries.com/press-releases/oxford-dictionaries-word-of-the-year-2013/ (accessed January 20, 2015).

2. Sam Frizell, "Pope Francis Takes Selfies with Crowd after Palm Sunday Homily," *Time*, last modified April 13, 2014, http://time.com/60699/pope-francis-palm-sunday-selfie/ (accessed January 13, 2015).

3. Alyssa Newcomb, "President Obama Poses for Selfie at Nelson Mandela's Memorial Service," *ABC News*, last modified December 10, 2013, http://abcnews.go.com/Politics/president-obama-poses-selfie-nelson-mandelas-memorial-service/story?id=21162957 (accessed January 13, 2015).

4. The "These Animal 'Selfies' Prove that Cats, Dogs and Sloths Take the Best Self-Portraits," *The Huffington Post*, last modified July 29, 2013, http://www.huffingtonpost.com/2013/07/20/animal-selfies-pet-selfie-photos_n_3624985.html (accessed January 13, 2015).

5. *Oxford Dictionaries Word of the Year 2013* (*supra* note 1).

6. Kirthana Ramisetti, "The Most Headline-Making Celebrity Tweets of 2014: From Grumpy Kanye to Ellen's Selfie," *NY Daily News*, last modified December 31, 2014, http://www.nydailynews.com/entertainment/gossip/kanye-west-ellen'-degeneres-top-celebrity-tweets-2014-article-1.2062122 (accessed January 13, 2015).

7. Lee Rainie, Joanna Brenner, and Kristen Purcell, "Photos and Videos as Social Currency Online," *Pew Research Center*, last modified September 13, 2012, http://www.pewinternet.org/files/old-media//Files/Reports/2012/ PIP_OnlineLifeinPictures_PDF.pdf (accessed January 13, 2015).

8. Pew Research Center, "Millennials in Adulthood: Detached from Institutions, Networked with Friends," *Numbers, Facts and Trends Shaping the World*, last modified March 7, 2014, http://www.pewsocialtrends.org/files/2014/03/2014-03-07_generations-report-version-for-web.pdf (accessed January 13, 2015).

9. Lee Rainie, Joanna Brenner, and Kristen Purcell, "Photos and Videos As Social Currency Online," *Pew Research Center*, last modified September 13, 2012, http://

www.pewinternet.org/files/old-media//Files/Reports/2012/PIP_OnlineLifeinPictures
_PDF.pdf (accessed January 13, 2015).

10. Alicia Eler, "Before the Selfie, the Self-Portrait," *Hyperallergic*, last modified
August 5, 2013, http://hyperallergic.com/76218/before-the-selfie-the-self-portrait
/ (accessed January 10, 2015).

11. Albert Rothenberg, "Rembrandt's Creation of the Pictorial Metaphor of
Self," *Metaphor & Symbol* 23, no. 2 (2008): 108–29, accessed January 19, 2015, doi:
10.1080/10926480801944269.

12. Stephen Marche, "The Argument: Why Frida Kahlo Is the Patron Saint of
Internet-Enabled Narcissism," *Toronto Life*, last modified November 12, 2012, http://
www.torontolife.com/informer/features/2012/11/12/frida-kahlo-online-narcissism
/ (accessed January 19, 2015).

13. Albert Rothenberg, "Rembrandt's Creation of the Pictorial Metaphor of Self,"
Metaphor & Symbol 23, no. 2 (2008): 108–29, accessed January 19, 2015, doi:
10.1080/10926480801944269.

14. Ibid.

15. Marion Boddy-Evans, "Painting Self Portraits," *About.com*, accessed January
21, 2015, http://painting.about.com/cs/figurepainting/a/selfportraits.htm.

16. Ibid.

17. Albert Rothenberg, "Rembrandt's Creation of the Pictorial Metaphor of Self,"
Metaphor & Symbol 23, no. 2 (2008): 108–29, accessed January 19, 2015, doi:
10.1080/10926480801944269.

18. Ibid.

19. Ann Jensen Adams (2004). "Rembrandt van Rijn," *Europe, 1450 to 1789:
Encyclopedia of the Early Modern World*, Vol. 5, 174–77. Gale Virtual Reference
Library, Web.

20. Ibid.

21. Albert Rothenberg, "Rembrandt's Creation of the Pictorial Metaphor of Self,"
Metaphor & Symbol 23, no. 2 (2008): 108–29, accessed January 19, 2015, doi:
10.1080/10926480801944269.

22. Ibid.

23. Ibid.

24. Ibid.

25. Ibid.

26. "Frida Kahlo and Her Paintings," *Frida Kahlo*, accessed January 21, 2015,
http://www.fridakahlo.org/.

27. Ibid.

28. "Frida Kahlo Biography," *Frida Kahlo*, accessed January 21, 2015, http://www
.fridakahlo.org/frida-kahlo-biography.jsp.

29. "Frida Kahlo and Her Paintings," *Frida Kahlo*, accessed January 21, 2015,
http://www.fridakahlo.org/.

30. Ibid.

31. Ibid.

32. Mary Kay Vaughan, "Kahlo, Frida (1907–1954)," *Encyclopedia of Latin
American History and Culture* 2, vol. 4 (2008): 72–73, Gale Virtual Reference Library.

33. Stephen Marche, "The Argument: Why Frida Kahlo Is the Patron Saint of Internet-Enabled Narcissism," *Toronto Life*, last modified November 12, 2012, http://www.torontolife.com/informer/features/2012/11/12/frida-kahlo-online-narcissism/ (accessed January 19, 2015).

34. Ibid.

35. "Cameras," *Encyclopedia of Products & Industries—Manufacturing* (2008): 149–56. Gale Virtual Reference Library, Web.

36. Janice L. Neri, "Camera Obscura," *Europe, 1450–1789*, Vol. 1 (2004): 372–74. Gale Virtual Reference Library, Web.

37. Ibid.

38. "Cameras," *Encyclopedia of Products & Industries—Manufacturing* (2008): 149–56. Gale Virtual Reference Library, Web.

39. "Photography," *Gale Encyclopedia of Science*, Vol. 4, edited by K. Lee Lerner and Brenda Wilmoth Lerner (2008): 3308–13. Gale Virtual Reference Library, Web.

40. "Cameras," *Encyclopedia of Products & Industries—Manufacturing* (2008): 149–56. Gale Virtual Reference Library.

41. "Photography" (*supra* note 39).

42. Ibid.

43. "Cameras," *Encyclopedia of Products & Industries—Manufacturing* (2008): 149–56. Gale Virtual Reference Library.

44. "Photography," *Gale Encyclopedia of Science*, Vol. 4, edited by K. Lee Lerner and Brenda Wilmoth Lerner (2008): 3308–13. Gale Virtual Reference Library.

45. Tim Davis, "Portrait of the Artist," *Print* 67(4) (2013): 50–55. Academic Search Premier.

46. Ibid.

47. Gillian S. Holmes, "Camera," *How Products Are Made: An Illustrated Guide to Product Manufacturing*, Vol. 3 (1998): 67–71. Gale Virtual Reference Library.

48. Ibid.

49. Ibid.

50. Stewart Kampel, "Friedlander, Lee," *Encyclopedia Judaica*, Vol. 7 (2007): 277–78. Gale Virtual Reference Library.

51. Tim Davis, "Portrait of the Artist," *Print* 67 (4) (2013): 50–55. Academic Search Premier.

52. Ibid.

53. Stewart Kampel, "Friedlander, Lee," *Encyclopedia Judaica*, Vol. 7 (2007): 277–78. Gale Virtual Reference Library.

54. Tim Davis, "Portrait of the Artist," *Print* 67 (4) (2013): 50–55. Academic Search Premier.

55. Stewart Kampel, "Friedlander, Lee," *Encyclopedia Judaica*, Vol. 7 (2007): 277–78. Gale Virtual Reference Library.

56. Tim Davis, "Portrait of the Artist," *Print* 67 (4) (2013): 50–55. Academic Search Premier.

57. Ibid.

58. Ibid.

59. "Photography," *Gale Encyclopedia of Science*, Vol. 4, edited by K. Lee Lerner and Brenda Wilmoth Lerner (2008): 3308–13. Gale Virtual Reference Library.

60. "Cameras," *Encyclopedia of Products & Industries—Manufacturing* (2008): 149–56. Gale Virtual Reference Library.

61. Ibid.

62. Pete Brook, "Photographs Are No Longer Things, They're Experiences," *Wired*, last modified November 15, 2012, http://www.wired.com/2012/11/stephen-mayes -vii-photography/all/ (accessed January 7, 2015).

63. Ibid.

64. Ibid.

65. Andrew A. Kling, "Cameras, Video, and the Web," *Cell Phones* (2010): 64–79. Gale Virtual Reference Library.

66. Ibid.

67. Ibid.

68. Maeve Duggan and Aaron Smith, "Cell Internet Use 2013," *Pew Research Center*, last modified September 16, 2013, http://www.pewinternet.org/files/old-media //Files/Reports/2013/PIP_CellInternetUse2013.pdf (accessed January 22, 2015).

69. Lee Rainie, "Cell Phone Ownership Hits 91% of Adults," *Pew Research Center*, last modified June 6, 2013, http://www.pewresearch.org/fact-tank/2013/06/06/cell-phone-ownership-hits-91-of-adults/ (accessed January 22, 2015).

70. Aaron Smith, "Smartphone Ownership—2013 Update," *Pew Research Center*, last modified June 5, 2013, http://www.pewinternet.org/files/old-media//Files/Reports /2013/ PIP_Smartphone_adoption_2013_PDF.pdf (accessed January 22, 2015).

71. Shawn DuBravac, "Business Models in the Innovation Economy," *Consumer Electronics Association*, (2015): 25–28, http://content.ce.org/PDF/2014_5Tech_web .pdf (accessed January 29, 2015).

72. Andrew A. Kling, "Cameras, Video, and the Web," *Cell Phones* (2010): 64–79. *Gale Virtual Reference Library*, Web.

73. Alicia Eler, "All the Pretty Selfies Are Here to Stay," *Hyperallergic*, last modified December 23, 2013, http://hyperallergic.com/97441/all-the-pretty-selfies-are-here-to -stay/ (accessed January 10, 2015).

74. Pete Brook, "Photography Is the New Universal Language, and It's Changing Everything," *Wired*, last modified August 20, 2013, http://www.wired.com/2013/08 /raw-meet-marvin-heiferman/ (accessed January 13, 2015).

75. Pete Brook, "Photographs Are No Longer Things, They're Experiences," *Wired*, last modified November 15, 2012, http://www.wired.com/2012/11/stephen -mayes-vii-photography/all/ (accessed January 7, 2015).

76. Sherry Turkle, "The Documented Life," *New York Times*, last modified December 15, 2013, http://www.nytimes.com/2013/12/16/opinion/the-documented-life .html (accessed January 10, 2015).

77. Andrew A. Kling, "Cameras, Video, and the Web," *Cell Phones* (2010): 64–79. Gale Virtual Reference Library, Web.

78. Ken Eisold, "Why Selfies?" *Psychology Today*, last modified December 21, 2013, https://www.psychologytoday.com/blog/hidden-motives/201312/why-selfies (accessed January 11, 2015).

79. Maria Konnikova, "How Facebook Makes Us Unhappy," *New Yorker,* last modified September 10, 2013, http://www.newyorker.com/tech/elements/how-facebook-makes-us-unhappy (accessed January 9, 2015).

80. Ben Agger, "Hegel's Internet," *disClosure* (2014): 47–64. Academic Search Premier, EBSCOhost.

81. Ibid.

82. Jenna Wortham, "My Selfie, Myself," *New York Times,* last modified October 19, 2013, http://www.nytimes.com/2013/10/20/sunday-review/my-selfie-myself.html?pagewanted=all&_r=1& (accessed January 12, 2015).

83. Lee Rainie, Joanna Brenner, and Kristen Purcell, "Photos and Videos as Social Currency Online," *Pew Research Center,* last modified September 13, 2012, http://www.pewinternet.org/files/old-media/ /Files/Reports/2012/PIP_OnlineLifeinPictures _PDF.pdf (accessed January 13, 2015).

84. Ibid.

85. Andrew A. Kling, "Cameras, Video, and the Web," *Cell Phones* (2010): 64–79. Gale Virtual Reference Library, Web.

86. Pamela B. Rutledge, "Making Sense of Selfies," *Psychology Today,* last modified July 6, 2013, https://www.psychologytoday.com/blog/positively-media/201307/making-sense-selfies (accessed January 10, 2015).

87. Jenna Wortham, "My Selfie, Myself," *New York Times,* last modified October 19, 2013, http://www.nytimes.com/2013/10/20/sunday-review/my-selfie-myself.html?pagewanted=all&_r=1& (accessed January 12, 2015).

88. Sherry Turkle, "The Documented Life," *New York Times,* last modified December 15, 2013, http://www.nytimes.com/2013/12/16/opinion/the-documented-life.html (accessed January 10, 2015).

89. Jenna Wortham, "My Selfie, Myself," *New York Times,* last modified October 19, 2013, http://www.nytimes.com/2013/10/20/sunday-review/my-selfie-myself.html?pagewanted=all&_r=1& (accessed January 12, 2015).

90. Pamela B. Rutledge, "Making Sense of Selfies," *Psychology Today,* last modified July 6, 2013, https://www.psychologytoday.com/blog/positively-media/201307/making-sense-selfies (accessed January 10, 2015).

91. Chris Ziegler and Dieter Bohn, "Instagram Is the Best, Instagram Is the Worst," *The Verge,* last modified April 9, 2012, http://www.theverge.com/2012/4/9/2928975/instagram-filters-ping-counterping/ (accessed January 7, 2015).

92. Alicia Eler, "I, Selfie: Saying Yes to Selfies," *Hyperallergic,* last modified June 24, 2013, http://hyperallergic.com/73362/saying-yes-to-selfies/ (accessed January 10, 2015).

93. Ibid.

94. Tim Davis, "Portrait of the Artist," *Print* 67, No.4 (2013): 50–55. Academic Search Premier, Web.

95. Erika Pearson, "All the World Wide Web's a Stage: The Performance of Identity in Online Social Networks," *First Monday,* last modified March 2, 2009, http://journals.uic.edu/ojs/index.php/ fm/article/view/2162/2127 (accessed January 7, 2015).

96. Ibid.

97. Ibid.

98. Alicia Eler, "On the Origin of Selfies," *Hyperallergic*, last modified August 26, 2013, http://hyperallergic.com/79448/the-origin-of-selfies/ (accessed January 10, 2015).

99. Maria Bustillos, "Everyone Shoots First: Reality in the Age of Instagram," *The Verge*, last modified September 18, 2013, http://www.theverge.com/2012/9/18 /3317324/hall-of-mirrors-remaking-reality-camera-obsessed-world (accessed January 13, 2015).

100. Alicia Eler, "On the Origin of Selfies," *Hyperallergic*, last modified August 26, 2013, http://hyperallergic.com/79448/the-origin-of-selfies/ (accessed January 10, 2015).

101. Erika Pearson, "All the World Wide Web's a Stage: The Performance of Identity in Online Social Networks," *First Monday*, last modified March 2, 2009, http://journals.uic.edu/ojs/index.php/fm/article/view/2162/2127 (accessed January 7, 2015).

102. Feng-Yang Kuo, Chih-Yi Tseng, Fan-Chuan Tseng, and Cathy S. Lin, "A Study of Social Information Control Affordances and Gender Difference in Facebook Self-Presentation," *Cyberpsychology, Behavior, and Social Networking* 9, Vol. 16 (2013): 635–644. Academia.edu, Web.

103. Ibid.

104. Ibid.

105. Ibid.

106. Ibid.

107. Ibid.

108. Ibid.

109. Ibid.

110. Erika Pearson, "All the World Wide Web's a Stage: The Performance of Identity in Online Social Networks," *First Monday*, last modified March 2, 2009, http://journals.uic.edu/ojs/index.php/fm/article/view/2162/2127 (accessed January 7, 2015).

111. Gina Schlesselman-Tarango, "Searchable Signatures: Context and the Struggle for Recognition," *Information Technology & Libraries* 32, no. 3 (2013): 5–19, Academic Search Premier, EBSCOhost.

112. Jerry Saltz, "Art at Arm's Length: A History of the Selfie," *Vulture*, last modified January 26, 2014, http://www.vulture.com/2014/01/history-of-the-selfie.html (accessed January 10, 2015).

113. Richard Brody, "Status Update," *New Yorker*, last modified December 12, 2012, http://www.newyorker.com/culture/richard-brody/status-update (accessed January 19, 2015).

114. Amanda Petrusich, "Instagram, The Nostalgia of Now and Reckoning the Future," *Buzzfeed*, last modified April 24, 2012, http://www.buzzfeed.com/petrusich /instagram-the-nostalgia-of-now-and-reckoning-the#.hfLX8Qg3La (accessed January 9, 2015).

115. Ibid.

116. Christopher Bonanos, "Instantly Old," *New York Magazine*, last modified April 13, 2012, http://nymag.com/news/intelligencer/instagram-2012-4/ (accessed January 10, 2015).

117. Amanda Petrusich, "Instagram, The Nostalgia of Now and Reckoning the Future," *Buzzfeed*, last modified April 24, 2012, http://www.buzzfeed.com/petrusich /instagram-the-nostalgia-of-now-and-reckoning-the#.hfLX8Qg3La (accessed January 9, 2015).

118. Ibid.

119. Pamela B. Rutledge, "Making Sense of Selfies," *Psychology Today*, last modified July 6, 2013, https://www.psychologytoday.com/blog/positively-media/201307 /making-sense-selfies (accessed January 10, 2015).

120. Ibid.

121. Jenna Wortham, "My Selfie, Myself," *New York Times*, last modified October 19, 2013, http://www.nytimes.com/2013/10/20/sunday-review/my-selfie-myself.html ?pagewanted=all&_r=1& (accessed January 12, 2015).

122. Ibid.

123. Feng-Yang Kuo, Chih-Yi Tseng, Fan-Chuan Tseng, and Cathy S. Lin, "A Study of Social Information Control Affordances and Gender Difference in Facebook Self-Presentation," *Cyberpsychology, Behavior, and Social Networking* 9, vol. 16 (2013): 635–44. Academia.edu, Web.

124. Evan Ratliff, "Self-Service: The Delicate Dance of Online Bragging," *Wired*, last modified June 22, 2010, http://www.wired.com/2010/06/st_essay_tweet/ (accessed January 7, 2015).

125. Ibid.

126. Maria Konnikova, "How Facebook Makes Us Unhappy," *New Yorker*, last modified September 10, 2013, http://www.newyorker.com/tech/elements/how -facebook-makes-us-unhappy (accessed January 9, 2015).

127. Ibid.

128. Ibid.

129. Evan Ratliff, "Self-Service: The Delicate Dance of Online Bragging," *Wired*, last modified June 22, 2010, http://www.wired.com/2010/06/st_essay_tweet/ (accessed January 7, 2015).

130. Maria Konnikova, "How Facebook Makes Us Unhappy," *New Yorker*, last modified September 10, 2013, http://www.newyorker.com/tech/elements/how -facebook-makes-us-unhappy (accessed January 9, 2015).

131. Ibid.

132. Ibid.

133. Ibid.

134. Ibid.

135. Gina Schlesselman-Tarango, "Searchable Signatures: Context and the Struggle for Recognition," *Information Technology & Libraries* 32, no. 3 (2013): 5–19, Academic Search Premier, EBSCO host.

136. Stephen Marche, "The Argument: Why Frida Kahlo Is the Patron Saint of Internet-Enabled Narcissism," *Toronto Life*, last modified November 12, 2012, http://

www.torontolife.com/informer/features/2012/11/12/frida-kahlo-online-narcissism / (accessed January 19, 2015).

137. Pete Brook, "Photography Is the New Universal Language, and It's Changing Everything," *Wired*, last modified August 20, 2013, http://www.wired.com/2013/08 /raw-meet-marvin-heiferman/ (accessed January 13, 2015).

138. Stephen Marche, "The Argument: Why Frida Kahlo Is the Patron Saint of Internet-Enabled Narcissism," *Toronto Life*, last modified November 12, 2012, http:// www.torontolife.com/informer/features/2012/11/12/frida-kahlo-online-narcissism / (accessed January 19, 2015).

139. Gina Schlesselman-Tarango, "Searchable Signatures: Context and the Struggle for Recognition," *Information Technology & Libraries* 32, no. 3 (2013): 519. Academic Search Premier, EBSCO host.

140. Ibid.

141. Pete Brook, "Photography Is the New Universal Language, and It's Changing Everything," *Wired*, last modified August 20, 2013, http://www.wired.com/2013/08 /raw-meet-marvin-heiferman/ (accessed January 13, 2015).

142. Roger Cohen, "Thanks for Not Sharing," *New York Times*, last modified December 6, 2012, http://www.nytimes.com/2012/12/07/opinion/roger-cohen-thanks -for-not-sharing.html?_r=2& (accessed January 10, 2015).

143. Ibid.

144. Ibid.

145. Ibid.

146. Jenna Wortham, "My Selfie, Myself," *New York Times*, last modified October 19, 2013, http://www.nytimes.com/2013/10/20/sunday-review/my-selfie-myself.html ?pagewanted=all&_r=1& (accessed January 12, 2015).

147. Alicia Eler, "On the Origin of Selfies," *Hyperallergic*, last modified August 26, 2013, http://hyperallergic.com/79448/the-origin-of-selfies/ (accessed January 10, 2015).

148. Richard Brody, "Status Update," *New Yorker*, last modified December 12, 2012, http://www.newyorker.com/culture/richard-brody/status-update (accessed January 19, 2015).

149. Ibid.

150. Pamela B. Rutledge, "Making Sense of Selfies," *Psychology Today*, last modified July 6, 2013, https://www.psychologytoday.com/blog/positively-media/201307 /making-sense-selfies (accessed January 10, 2015).

151. Ibid.

152. Pamela Rutledge, "Selfie Use: Abuse or Balance?," *Psychology Today*, last modified July 8, 2013, http://www.psychologytoday.com/blog/positively-media /201307/selfie-use-abuse-or-balancese-or-balance (accessed January 10, 2015).

153. Ibid.

154. Ibid.

155. Maria Konnikova, "How Facebook Makes Us Unhappy," *New Yorker*, last modified September 10, 2013, http://www.newyorker.com/tech/elements/how -facebook-makes-us-unhappy (accessed January 9, 2015).

156. Moira Burke, Cameron Marlow, and Thomas Lento, "Social Network Activity and Social Well-Being," *Association for Computing Machinery* (2010): 1909–1912. ACM Digital Library, Web, doi: 10.1145/1753326.1753613.

157. Konnikova, "How Facebook Makes Us Unhappy" (see footnote 155).

158. Pamela Rutledge, "Selfie Use: Abuse or Balance?" *Psychology Today*, last modified July 8, 2013, http://www.psychologytoday.com/blog/positively-media/201307 /selfie-use-abuse-or-balancese-or-balance (accessed January 10, 2015).

159. James Franco, "The Meanings of Selfies," *New York Times*, last modified December 26, 2013, http://www.nytimes.com/2013/12/29/arts/the-meanings-of-the-selfie .html?_r=0 (accessed January 26, 2015).

160. Ibid.

161. Ibid.

162. Gina Schlesselman-Tarango, "Searchable Signatures: Context and the Struggle for Recognition," *Information Technology & Libraries* 32, no. 3 (2013): 5–19. *Academic Search Premier*, EBSCO host.

163. Alicia Eler, "Validating Me and My Selfie," *Hyperallergic*, last modified July 22, 2013, http://hyperallergic.com/74996/validating-me-and-myselfie/ (accessed January 10, 2015).

164. "Frida Kahlo Biography," *Frida Kahlo*, accessed January 21, 2015, http://www .fridakahlo.org/frida-kahlo-biography.jsp.

165. Jerry Saltz, "Art at Arm's Length: A History of the Selfie," *Vulture*, last modified January 26, 2014, http://www.vulture.com/2014/01/history-of-the-selfie .html (accessed January 10, 2015).

166. Alicia Eler, "I, Selfie: Saying Yes to Selfies," *Hyperallergic*, last modified June 24, 2013, http://hyperallergic.com/73362/saying-yes-to-selfies/ (accessed January 10, 2015).

BIBLIOGRAPHY

Adams, A. J. (2004). "Rembrandt van Rijn (1606–1669)." In J. Dewald (ed.), *Europe, 1450 to 1789: Encyclopedia of the Early Modern World* Vol. 5, 174–77). New York: Charles Scribner's Sons.

Boddy-Evans, M. (2015). Painting Self Portraits. About.com. http://painting.about .com/cs/figurepainting/a/selfportraits.htm. Retrieved January 21, 2015.

Bonanos, C. (2012, April 13). "Instantly Old." *New York Magazine*. http://nymag.com /news/intelligencer/instagram-2012-4/. Retrieved January 10, 2015.

Brody, R. (2012, December 6). "Status Update." *New Yorker*. http://www.newyorker .com/culture/richard-brody/status-update. Retrieved January 9, 2015.

Brook, P. (2013, August 20). Photography Is the New Universal Language, and It's Changing Everything. *Wired*. http://www.wired.com/2013/08/raw-meet-marvin -heiferman/. Accessed January 13, 2015.

Brook, P. (2012, November 15). Photographs Are No Longer Things, They're Experiences. *Wired*. http://www.wired.com/2012/11/stephen-mayes-vii-photography/all/. Accessed January 7, 2015.

Burke, M., C. Marlow, and T. Lento. (2010). "Social Network Activity and Social Well-Being." ACM *Digital Library*, 1909–1912.

Cameras. In P. J. Bungert and A. J. Darnay (eds.). (2008). *Encyclopedia of Products & Industries—Manufacturing* Vol. 1, 149–56. Detroit: Gale.

Cohen, R. 2012 (December 6). "Thanks for Not Sharing." *New York Times*. http://www.nytimes.com/2012/12/07/opinion/roger-cohen-thanks-for-not-sharing.html?_r=2&. Accessed January 10, 2015.

Davis, T. 2013 (August). "Portrait of the Artist." *Print*, 50–55.

DuBravac, S. (2015). *Business Models in the Innovation Economy*. Arlington, VA: Consumer Electronics Association.

Duggan, M., and A. Smith. (2013). *Cell Internet Use 2013*. Pew Research Center's Internet and American Life Project. Washington, DC: Pew Research Center.

Eisold, K. 2013 (December 21). "Why Selfies?" *Psychology Today*. https://www.psychologytoday.com/blog/hidden-motives/201312/why-selfies. Accessed January 11, 2015.

Eler, A. 2013a (August 26). "On the Origin of Selfies." *Hyperallergic*. http://hyperallergic.com/79448/the-origin-of-selfies/. Accessed January 10, 2015.

Eler, A. 2013b (August 5). "Before the Selfie, the Self-Portrait." *Hyperallergic*. http://hyperallergic.com/76218/before-the-selfie-the-self-portrait/. Accessed January 10, 2015.

Eler, A. 2013c (July 22). "Validating Me and Myselfie." *Hyperallergic*. http://hyperallergic.com/74996/validating-me-and-myselfie/. Accessed January 10, 2015.

Eler, A. 2013d (June 24). "I, Selfie: Saying Yes to Selfies." *Hyperallergic*. http://hyperallergic.com/73362/saying-yes-to-selfies/. Accessed January 10, 2015.

Frano, J. 2013 (December 26). "The Meanings of the Selfie." *New York Times*. http://www.nytimes.com/2013/12/29/arts/the-meanings-of-the-selfie.html?_r=0. Accessed January 26, 2015.

Frida Kahlo and Her Paintings. (n.d.). Frida Kahlo. https://owl.english.purdue.edu/owl/resource/717/05/. Accessed January 21, 2015.

Frizell, S. 2014 (April 13). *Time*. http://time.com/60699/pope-francis-palm-sunday-selfie/. Accessed January 13, 2015.

Holmes, G. S. 1998. "Camera." In *How Products Are Made: An Illustrated Guide to Product Manufacturing* Vol. 3, 67–71. Edited by K. M. Krapp and J. L. Longe. Detroit: Gale.

Hubbard, G. 1992. "Artists by Artists." *Arts & Activities*, 26–29.

Huffington Post. 2013 (July 29). "These Animal 'Selfies' Prove that Cats, Dogs and Sloths Take the Best Self-Portraits." *The Huffington Post*. http://www.huffingtonpost.com/2013/07/20/animal-selfies-pet-selfie-photos_n_3624985.html. Accessed January 13, 2015.

Kampel, S. 2007. "Friedlander, Lee." In *Encyclopedia Judaica*, 2nd ed., Vol. 7, 277–78. Edited by M. Berenbaum and F. Skolnik. Detroit: Macmillan Reference USA.

Kling, A. A. 2010. "Cameras, Video, and the Web." In *Cell Phones*, 64–79. Detroit: Lucent Books.

Konnikova, M. 2013 (September 10). "How Facebook Makes Us Unhappy." *New Yorker*. http://www.newyorker.com/tech/elements/how-facebook-makes-us-unhappy. Retrieved January 9, 2015.

Kuo, F.-Y., C.-Y. Tseng, F.-C. Tseng, and C. S. Lin. 2013. "A Study of Social Information Control Affordances and Gender Difference in Facebook Self-Presentation."
Cyberpsychology, Behavior, and Social Networking 16 (9), 635–44.

Lerner, K. L., and B. W. Lerner (eds.). 2008. "Photography." In *The Gale Encyclopedia
of Science*, 4th ed., Vol. 4, 3308–13. Detroit: Gale.

Marche, S. 2012 (November 12). The Argument: Why Frida Kahlo Is the Patron
Saint of Internet-Enabled Narcissism. *Toronto Life*. http://www.torontolife.com
/informer/features/2012/11/12/frida-kahlo-online-narcissism/. Retrieved January
19, 2015.

Neri, J. L. 2004. "Camera Obscura." In J. Dewald (Ed.), *Europe, 1450 to 1789:
Encyclopedia of the Early Modern World* Vol. 1, 372. New York: Charles Scribner's
Sons.

Newcomb, A. 2013 (December 10). *ABC News*. http://abcnews.go.com/Politics
/president-obama-poses-selfie-nelson-mandelas-memorial-service/story?id
=21162957. Retrieved January 13, 2015.

Oxford Dictionaries. 2013 (November 19). http://blog.oxforddictionaries.com/press
-releases/oxford-dictionaries-word-of-the-year-2013/. Retrieved January 20, 2015.

Pearson, E. 2009 (March 2). "All the World Wide Web's a Stage: The Performance of
Identity in Online Social Networks." *First Monday*. http://journals.uic.edu/ojs
/index.php/fm/article/view/2162/2127. Retrieved January 7, 2015.

Petrusich, A. 2012 (April 24). "Instagram, The Nostalgia of Now and Reckoning the
Future." *BuzzFeed* http://www.buzzfeed.com/petrusich/instagram-the-nostalgia-of
-now-and-reckoning-the#.hfLX8Qg3La. Retrieved January 9, 2015.

Pew Research Center. (2014). *Millennials in Adulthood: Detached from Institutions,
Networked with Friends*. Washington, D.C.: Pew Research Center.

Rainie, L., J. Brenner, and K. Purcell. (2012). *Photos and Videos as Social Currency
Online*. Washington, DC: Pew Research Center.

Ramisetti, K. 2014 (December 31). "The Most Headline-Making Celebrity Tweets of
2014: From Grumpy Kanye to Ellen's Selfie." *NY Daily News*. http://www.nydaily
news.com/entertainment/gossip/kanye-west-ellen-degeneres-top-celebrity-tweets
-2014-article-1.2062122. Retrieved January 13, 2015.

Ratliff, E. 2010 (June 22). "Self-Service: The Delicate Dance of Online Bragging."
Wired. http://www.wired.com/2010/06/st_essay_tweet/. Retrieved January 7, 2015.

Rothenberg, A. (2008). "Rembrandt's Creation of the Pictorial Metaphor of Self."
Metaphor & Symbol 2 (23), 108–29.

Rutledge, P. B. 2013 (July 6). "Making Sense of Selfies." *Psychology Today*. https://
www.psychologytoday.com/blog/positively-media/201307/making-sense-selfies.
Accessed January 10, 2015.

Saltz, J. 2014 (January 26). "Art at Arm's Length: A History of the Selfie." *Vulture*. http://
www.vulture.com/2014/01/history-of-the-selfie.html. Accessed January 10, 2015.

Schlesselman-Tarango, G. 2013. "Searchable Signatures: Context and the Struggle
for Recognition." *Information Technology & Libraries* 32 (3), 5–19.

Turkle, S. 2013 (December 15). *The Documented Life. New York Times*. http://www
.nytimes.com/2013/12/16/opinion/the-documented-life.html. Accessed January
10, 2015.

Vaughan, M. K. (2008). "Kahlo, Frida (1907–1954)." In *Encyclopedia of Latin American History and Culture*, 2nd ed., Vol. 4, 72–73. Edited by J. Kinsbruner and E. D. Langer. Detroit: Charles Scribner's Sons.

Ziegler, C., and D. Bohn. 2012 (April 9). "Instagram is the Best, Instagram Is the Worst." *The Verge.* http://www.theverge.com/2012/4/9/2928975/instagram-filters -ping-counterping. Accessed January 7, 2015.

5

Everyday Expertise: Instructional Videos on YouTube

Jörgen Skågeby and Lina Rahm

This chapter explores how videos on YouTube have become an everyday form of producing and consuming instructions, "how-to" guides, and tutorials. Consider, for example, the following everyday use scenarios.

Dan is a "tween" (preteen) interested in Asian (visual) culture. He is particularly fascinated by a variety of K-pop (Korean pop music) bands—to the extent that he wants to learn the Korean language. His immediate thought is to consult YouTube and he swiftly finds a series of videos providing him with the first basic steps of conversational Korean. He watches the first installment in the series approximately 20 times over the following days and quickly demonstrates new language skills.

Angie just bought a digital drum kit. Although the sales representative gave her some basic tips, Angie immediately consults YouTube for additional guidance through the complexities of drumming (and setting up the kit properly). She finds continuous challenges and inspiration through online videos, and a drumming session soon becomes part of her daily routine.

Helen plays a game on her console. She is an accomplished player and so she suspects that there are hidden levels and bonus items to be unlocked. To explore the "full reality" of the game, Helen goes on YouTube to find a visual walkthrough.

The use cases provided above illustrate how YouTube has become the primary go-to source for those who desire to learn about something unfamiliar. This is—of course—remarkable, but also is not very surprising. Easily accessible instructional videos have a clear attraction—they are visual, they are relevant, and they are up to date. Many videos also are very detailed and often are well edited. This reminds us that YouTube is not only a platform to consume knowledge, but also is a platform for sharing tips, tricks, and skills. Although the examples provided here might seem diverse, that is the point. All of the cases represent everyday situations in which a desire to learn something can be immediately met by a preceding desire to share knowledge. As such, YouTube has become a primary outlet for what can be called "everyday expertise."

This chapter presents of a number of use scenarios as examples for readers to think about. These scenarios are grounded partly in interviews with young adults as well as in more anecdotal empirical experiences drawn from the authors' everyday lives. What the scenarios have in common is that YouTube always was discussed as an arena for the sharing of everyday expertise. The chapter begins by presenting a brief genealogy of the instructional video. The chapter then argues for a developed analysis of the instructional video on YouTube that is not restricted to efficiency. The instructional video then is discussed in terms of YouTube as an all-purpose archive of everyday expertise; YouTube as a non-neutral platform; and the political economy of expertise on YouTube. Drawing on contemporary media theory, the chapter concludes that instructional videos might work as both a risk and an opportunity.

BACKGROUND: FREQUENTLY ASKED QUESTIONS AND WALKTHROUGHS

The sharing of everyday expertise through various media is not, of course, something that emerged with the advent of YouTube. The history of everyday expertise sharing can be traced through many genres and spanning far back in history. This chapter, however, focuses on two significant and fairly recent instances—"frequently asked questions" (FAQ) and the "walkthrough."

The FAQ (frequently asked questions) forms a genre of everyday instruction widely adopted across forums, e-mail lists, and Web communities in the early days of the Internet. Put simply, an FAQ is a file or Web page that compiles common questions (and the answers) relating to a particular topic or activity. In practice, it often has been used as a way to separate those new to a particular forum ("newbies" or "n00bs") from the users who are more experienced with the culture of that same forum. Newbies often are referred to the FAQ as an initial starting point upon joining a forum or when posting a novice question. As such, the FAQ became part of an emerging "netiquette"—a continuously developing set of rules that facilitate social interaction in online contexts. Although certain netiquette rules can be applied more widely across digital cultures (e.g., avoiding typing in all caps) there also are rules specific to certain communities and genres.

The FAQ (and netiquette in general) primarily is a way to explicate rules that have emerged over time with the underpinning reason being to save individual, social, and technical resources. As such, an FAQ also became a way to effectively share everyday expertise and skills. Another interesting method for sharing everyday expertise relates more specifically to gaming—the walkthrough.

Today, for popular games, official strategy guides published in print by game producers themselves are quite common, and presumably also are a viable source of income. Before official strategy guides were produced, however, there emerged a genre of user-generated guides, termed "walkthroughs." Walkthroughs are similar to strategy guides in that they provide instructions for most "efficiently" playing a particular game—with the distinction being that they often are user generated, in text form, and published online.

> Once inside, run around a bit to encounter enemy troops, and make use of this opportunity to heal your characters as well as to stock up on curative items and spells. Take note of the ' "Blind" ' Draw Point to the left. Once you're ready, step onto the elevator and take it to the top (1st option).
>
> <!> Tip: Be sure to have saved [your game] first, and ensure that one of your party members has the ' "Draw" ' command; you'll need to draw a G.F. [Guardian Force] in an upcoming battle.
>
> The scene switches to Wedge (the blue guy) and Biggs (the guy in red). Biggs is trying to repair the satellite when Wedge comments about [a monster in the vicinity]. The former pays no attention, much to the chagrin of Wedge, who decides to look around first.
>
> Before long, Squall arrives, and in the commotion, Biggs starts the satellite up. He is elated, but not after the party engages him in battle. . . . (http://www.gamefaqs.com/ps/197343-final-fantasy-viii/faqs/4718)
>
> SUB-BOSSES:
>
> - Biggs HP: 610 Weak: Poison; Draw List: Fire, Thunder, Blizzard, Esuna
> - Wedge HP: 608 Weak: Poison; Draw List: Fire, Thunder, Blizzard, Cure
> - Total AP: 8[1]

As shown, a walkthrough basically is a step-by-step guide to exploring as many aspects of a game as is possible. Taking the FAQ and the walkthrough as two example genres of how sharing of everyday expertise has been performed previously (and continues to be), it could be argued that today the FAQ has expanded from a specific online context to a more general situation called, "life hacking." Life hacking refers to the general application of smart solutions to everyday problems. Having begun as a term specific to computer programmers, the life hacking now refers to "any trick, shortcut, skill, or nov-elty method that increases productivity and efficiency, in all walks of life."[2] This can in itself be seen as a sign of how digital information is now effec-tively interlaced with everyday situations of every kind.

The walkthrough, in turn, also has developed through the use of screen-recording software—such as Fraps (derived from frames per second), which is a benchmarking, screen capture, and screen-recording utility for Windows—to the "video walkthrough," through which the audience can visually follow gameplay. Looking at these developments in parallel, they illustrate how the sharing of everyday expertise today has very much become a use of instruc-tional videos. As mentioned, such instructional videos—reusing elements from both FAQs and walkthroughs—now pertain to virtually all aspects of life, from kayaking techniques to cutting a pomegranate properly to applying for Canadian citizenship.

ANALYZING THE INSTRUCTIONAL VIDEO ON YOUTUBE

It can be argued that YouTube is a source of everyday expertise simply by referring to its popularity. Combined, users of YouTube watch approximately 4 billion hours of video each month.[3] As YouTube becomes more popular, so do user-generated instructional videos.[4] One underpinning reason for this growth seems to be the perceived lack of efficiency in text-based manuals in

Table 1 Assessment Model for Instructional Videos

Design	Objective	Goal
Physical	Accessibility	Focus on areas relevant to instruction
	Viewability	High production quality
	Timing	Appropriate pacing of instruction
Cognitive	Accuracy	Error-free
	Completeness	Well-structured and detailed enough
	Pertinence	Content is relevant for instructional goal
Affective	Confidence	Establish trust
	Self-Efficacy	Persuade viewers they can complete task on their own
	Engagement	Spur interest and motivation

Source: M. Morain and J. Swarts (2012). "YouTutorial: A Framework for Assessing Instructional Online Video." *Technical Communication Quarterly* 21 (1), 6–24.

delivering instruction.[5] In a recent study of usability and perceived usefulness of instruction videos on YouTube,[6] the authors conclude that users regard instructional videos as helpful if they are relevant, timely, and sufficient for the users' needs. Likewise, to assess the effectiveness of instructional videos, Morain and Swarts[7] provide an analytical model consisting of nine points of analysis relating to physical, cognitive, and affective design.

Although this model and similar efforts used to assess effectiveness of learning certainly are practical and helpful, they also focus exclusively on the efficiency of instruction. This chapter instead acknowledges that efficiency (and its underlying definition of expertise) can be seen as a wider concept built on a more complex—and political—interplay between humans and technologies.

YOUTUBE AS AN FAQ FOR LIFE

As this chapter argues, YouTube has become the primary go-to source when desiring to learn (about) new things. Drawing upon interviews with YouTube users, the researchers identified that the important difference from the FAQ genre—as it was instantiated in early Internet culture—is that it now emerges from an everyday situation, which is mobile in space and time and can be pertaining to virtually any activity.

- It is Friday night and Ben and his partner are having friends over for dinner. Ben is preparing a sauce and consults YouTube for the proper way to cook it.
- Aini bought a race bike last summer. When preparing for the upcoming season, she notices that the bike's gears act a bit quirky. Fortunately, YouTube provides a range of videos for how to fine-tune the specific brand of gears her bike uses.
- Bingwen saw a documentary on beekeeping and searches YouTube to find out more. Bingwen is presented with approximately 94,000 results on various aspects of beekeeping and, after some viewing, decides to pursue this interest. While working at the hives in the subsequent months, Bingwen frequently returns to YouTube for topical tips.
- Having never tiled a shower wall, but looking to save some money (and learn a thing or two), Naeva looks on YouTube and quickly finds many tips on the specific procedure.

Drawing from these scenarios (and the scenarios introduced in the beginning of this chapter), it seems that instructional videos on YouTube effectively blur the boundaries between formerly separated notions. This chapter focuses on two dimensions for which the immediate access to instructional videos makes strict separations more difficult—namely between digital and material realities, and between formal and informal learning.

First, instructional videos on YouTube are a clear example of how digital and material realities are woven together. The idea that the virtual, digital, and material represent separated worlds is increasingly being challenged through ubiquitous computational devices.[8] Andy Clark, for example, states that learning can no longer be separated from the tools we use (if it ever could)—going so far as to say that, even at societal level, we have adopted what can be called "post-digital" values. Clark states, "As we move towards an era of wearable computing and ubiquitous information access, the robust, reliable information fields to which our brains delicately adapt their routines will become increasingly dense and powerful, further blurring the distinction between the cognitive agent and her best tools, props, and artifacts."[9]

The term "post-digital" does not mean something that will succeed or replace the digital. Rather, post-digital refers to the gradual acceptance of ubiquitous technologies that now are completely embedded in almost all everyday situations. This process of normalization and pervasiveness of the digital has been going on for decades, but it is only now that we can see how certain parts of cultures are accomplishing the full potential of this transformation. As such, the post-digital is also a phase where the lingering "newness" of digital media begins to evaporate. The disruptive qualities of digitization become mundane and other paradigms take their places as objects of futuristic desires. Another important aspect of the post-digital is that it acknowledges that information always comes with a physical counterpart. For long, the digital was conceptually closer to the virtual rather than to the material. Digitization was seen as a process where something was *de-materialized* and made into intangible information. Screens, wires, and chips, however, all are material necessities that are anything but ethereal when our information needs increase.

Returning to the instructional video, it includes—in production and consumption—both "material handling" as well as discursive (written or verbal) instruction. When producing an instructional video, material handling must be performed to be captured by video. At the same time, the material procedure is many times accompanied by a verbal instruction, making it an interesting multimodal combination of materiality and discourse. Similarly, in consumption contexts, the viewer often is engaged in the physical activity while consulting the video, making it an interesting form of consumption where instruction becomes highly situated and mobile—instruction can be delivered and consumed at any place and at any time. This interweaving of the material and the digital also works to challenge the distinctions between formal learning, informal learning, and nonformal learning; the production and consumption of instructional videos contain elements of all three types.

Formal learning typically is described as being a systematic and intentional delivery of learning conducted within the framework of an institution of some type (e.g., school, academy, college, institute) by trained teachers and instructors. Formal learning is thereby structured according to, for example, learning objectives, specific scheduling, and various measurements of learning outcomes.

> Formalization as the constitutive variable of the continuum manifests itself in a variety of ways. As mentioned, the number of possibilities of a learning process for its enactment is its defining characteristic. The manifestations of formalization are, among others, reduction of complexity, linearization, trivialization, standardization, and in special cases abstraction and quantification/mathematization.[10]

Informal learning commonly is described as the learning that is a part of everyday activities related to work, family, or recreation. These activities often are learner-initiated, but nonintentional improvements of skills and knowledge ensue. As such, these usually take place in more or less organized forms outside of the formal education systems. Nonformal learning can be said to be located in between formal and informal learning,[11] meaning that it is not as structured as typical formal learning (in terms of regulated curricula, syllabi, accreditations, or certifications), but also is more coordinated (as a learning context) than informal learning. Nonformal learning can include, for example, sports activities, organized hobbies, workshops, and study circles provided by voluntary or commercial adult educational associations and nongovernmental organizations (NGOs).

So, how do instructional videos on YouTube fit into this classification? As mentioned they touch upon aspects of all three types. For example, there are certainly formal aspects, in the shape of segmenting and linearization of content. There are also informal aspects, where much of it is user generated and intended to be consumed in situations where the use of an instructional video may only be a small part of a larger endeavor (which is clearly outside any formal educational institution). Likewise, it has nonformal aspects to it, as emerging cultural norms and networks, adding a clear element of organization to their, otherwise informal, utilization. The question is, could we go as far as to say that YouTube is a new institution for learning? According to the depth and width of instructional videos available, as well as the many situations where viewers call upon instruction from these, it is clearly emerging as a new learning practice. But, as people continue to learn all kinds of skills and understandings from YouTube videos, the more pressing question becomes what kind of platform is YouTube and how does its design implicate the delivery of instruction videos?

THE NON-NEUTRALITY OF YOUTUBE AND THE DEVELOPMENT OF CRITICAL AMBIVALENCE

When it comes to the Internet, public debate unfortunately seems to end up in entrenched positions where extreme drawbacks are put up against extreme benefits. In this polemic debate some argue that the Internet creates a form of pedagogical anarchy where traditional processes such as selection, authority, truth, and consensus are refuted. Others claim that the very same Internet supports tremendous possibilities for pedagogical self-actualization, collaborative learning, and democratization of education. The tension between these positions is interesting, but instead of taking a definitive stance this chapter argues that researchers should examine how various groups in fact reason about quality, selection, authority, and truth online in order to develop a critical ambivalence toward the media used by these groups.

Researchers must ask themselves (or the groups they want to study) about what criteria for quality guide choices online. Are users really experiencing a lack of authority in this context? What is users' view of truth online? What is self-realization to them? A developed understanding of such questions will make us better equipped to understand the digital world. At the same time, we must remain critical toward the preconditions under which everyday expertise is shared.

YouTube is not a neutral technology. It is by and large a commercial platform, which thereby also becomes a channel for advertisements and targeted commercials. Continuing this argument, we can also question the ways in which social activities are exploited in a type of voluntary production of commercials (many of the activities are concerned with the consumption of products). The learning conveyed through instructional videos many times has clear commercial aspects to it. In practically all the scenarios discussed in this chapter (and grounded in empirical data) there are commercial products lingering in the background. There is research that suggests that K-pop, for example, as a profit-making industry, is made possible through the sharing of user-generated material in social media.[12] Although learning to speak and read Korean is not a commercial endeavor per se, taken together with the traceable consumption and sharing of K-pop videos, further targeting and segmenting of users is made possible. In the gaming example discussed in this chapter, learning is built on the fact that a viewer or reader must have a copy of the game, and details subsequently are shared with friends, acquaintances, and others in a form of advertisement for the game itself. Race bikes, shower walls, and beekeeping all revolve around Western middle-class hobbies that are as much a question of everyday expertise as everyday consumption.

Of course, there are cases where everyday expertise is shared to subvert or to show up DIY (do-it-yourself) alternatives to market mechanisms; but, as always, there are ambivalent outcomes. Arguably, one risk is that the border between genres such as the instructional video and the commercial becomes blurred. Drawing on research that emphasizes the commodification of social relations, production and consumption of instructional videos could be seen as a form of voluntary labor.[13] These double-edged consequences of our increasingly media-saturated society are typical. Public debates, however, too often are entrenched in polemics. The most interesting issue is not whether these activities in fact *are* free labor, but that there always is a tension between creative outlets online and the exploitation of the very same. Our everyday media ecology lacks alternatives that take personal integrity, anonymity, and net neutrality seriously (and not only as marketing terms for half-baked substitutes). As such, we need to develop a critical ambivalence for the media that saturate our routine day.

> Scholars of communication technology need to begin attending critically to questions of ownership, a topic we have generally avoided. While once we socialized online through public sites such as newsgroups, increasingly people are conducting their online social activities within proprietary systems such as social networking sites, virtual worlds, and massively multiplayer games in which the users have few rights and limited, if any, ownership of their contributions.[14]

YouTube is a sociotechnical platform that carries certain affordances (i.e., revealing certain agency, and at the same time obscuring other); therefore we must take an increasingly political view of how instructional videos are being "prosumed" (a term that is the fusion of "producer" and "consumer"—and means that consumers, through digital media, increasingly act as producers of various culturally circulated media objects) and appropriated.

THE POLITICAL ECONOMY OF EXPERTISE AND PEER-LEARNING

"[T]echnologies are empathetically marked by their histories, which is to say that a series of processes carry—in bodies and in code, in hardware and in software, in machine archives and in social memory—various kinds of legacies and are conditioned by them. . . ."[15]

The political economy of expertise on YouTube contains both social and technical elements. On the social end, it is interesting to note how the range of available expertise on YouTube spans from "someone who knows just a little bit more than me" to "the leading authority in the world." Such a hierarchy not only is a result of the representation that YouTube displays (though this certainly plays a role), but many times also is imported from various other media sources. That is, YouTube is not an isolated arena removed from other social arenas, but co-performs expertise in an intricate mesh of humans, policies, and technologies. For example, Kruse and Veblen researching instructional videos pertaining to folk music teaching, found that the instructors of this subgenre tended to be white middle-aged males.[16] Although there is a lack of research on other areas of instruction, it seems likely that there also would be specific socio-cultural biases present in performing expertise in other topics. At the same time, YouTube can be used to present alternative perspectives and give voice to marginalized groups.[17, 18] Again, this illustrates how a critical ambivalence toward digital culture is necessary to shed light upon all of its aspects.

On a larger scale, and as has been hinted at previously, instructional videos on YouTube are dependent on a specific technical infrastructure—a technical infrastructure that is not distributed equally across the globe. In the ITU report on ICT facts and figures it states that "in developed countries, mobile-broadband penetration will reach 84 percent, a level four times as high as in developing countries (21 percent)."[19] This raises the question of how and where instructional videos find their users.

It also is important to remember that those who access YouTube regularly receive a *designed* form of decision support. It enhances decision making by making certain options more explicit, but it also obscures other choices (or even makes them for users). For a person who uses YouTube as a source of everyday expertise the results are already sorted—but the sorting *mechanism* is at the same time hidden from the user. It thereby could be argued that the political economy of choices is displaced from the actual context to a pre-made matrix in which potential options become presented facts and results. YouTube presents a seemingly diverse range of options, but they are at the same time shaped by computational operations, which in turn are designed with certain intentions.

CONCLUSION

Having demonstrated and discussed the emergence of the instructional video, the question now becomes how we should understand and interpret this phenomenon as a cultural expression. At a surface level, the idea of accessible and democratized learning in the shape of technology-mediated sharing of everyday expertise is tantalizing. A more critical analysis, however, could nuance the often self-explanatory and taken-for-granted notions of learning, expertise, and (neutral) technology. Technologies always will retain a certain degree of opacity, and their operations will carry both foreseen and unforeseen consequences.[20] It is within these conditions that we must develop a critical ambivalence toward media technologies—concurrently as "stabs in the back" as well as "great gifts."[21] "Thus, thinking politics *with* technology becomes part of the process of the reinvention of the political in our technicized globalized world."[22]

On a final note, YouTube is just one platform to which the arguments put forth here are applicable. An increasing part of our everyday media are today digital, computational, or "coded assemblages."[23] In fact, it could be argued that they can be reduced to code—code that is increasingly superimposed on more and more things (i.e., a combination of informational and physical spaces). As such, our interaction with media is changing shape. Instructional videos emphasize how consumption and production—and thereby learning—increasingly are a question of "produsage,"[24] where people, fairly effortlessly, can combine informational elements from various sources to form a new object. These sources can be other "prodused" objects, but also the outputs of traditional media houses as well as established and emerging tech companies (most notably Apple, Google, Microsoft, and Facebook). As such, studies of digital culture should focus more directly on how media technologies in themselves are capable of shaping information, skills, and knowledge. This requires a variety of frameworks that take informational, political, and material aspects into consideration so that we can pursue to develop the critical ambivalence that is necessary for dealing with the incongruity of contemporary digital culture.

NOTES

1. Caroline Bassett, "Feminism, Expertise and the Computational Turn," in *Renewing Feminisms: Radical Narratives, Fantasies and Futures in Media Studies*, edited by Helen Thornham and Elke Weissmann (London: I. B. Tauris, 2013), 199–214.

2. Cephiroth, "An Unofficial Final Fantasy VIII Walkthrough," http://www.gamefaqs .com/ps/197343-final-fantasy-viii/faqs/4897, accessed May 31, 2015.

3. Wikipedia, "Life Hacking." http://en.wikipedia.org/wiki/Life_hacking, accessed May 25, 2015.

4. Alias Norlidah, Siti Hajar Abd Razak, Ghada elHadad, Nurul Rabihah Mat Noh Kokila Kunjambu, and Parimaladevi Muniandy, "A Content Analysis in the Studies of Youtube in Selected Journals," *Procedia—Social and Behavioral Sciences* 103, no. 26 (2013): 10–18.

5. J. Swarts, "New Modes of Help: Best Practices for Instructional Video," *Technical Communication* 59, no. 3 (2012): 195–206.

6. E. Eiriksdottir and R. Catrambone, "Procedural Instructions, Principles, and Examples: How to Structure Instructions for Procedural Tasks to Enhance Performance, Learning and Transfer," *Human Factors: The Journal of the Human Factors and Ergonomics Society* 53, no. 6 (2011): 749–70.

7. D. Y. Lee and M. R. Lehto, "User Acceptance of Youtube for Procedural Learning: An Extension of the Technology Acceptance Model," *Computers & Education* 61 (2013): 193–208.

8. M. Morain and J. Swarts, "Yoututorial: A Framework for Assessing Instructional Online Video," *Technical Communication Quarterly* 21, no. 1 (2012): 6–24.

9. Sue Thomas, "The End of Cyberspace and Other Surprises," *Convergence* 12, no. 4 (2006): 383–91.

10. Andy Clark, "Re-Inventing Ourselves. The Plasticy of Embodiment, Sensing, and Mind," *The Journal of Medicine and Philosophy* 32, no. 3 (2007): 263–89.

11. Reinhard Zürcher, "A Sociomaterial Model of the Teaching-Learning Continuum," *European Journal for Research on the Education and Learning of Adults* 6, no. 1 (2015): 73–90.

12. Organisation for Economic Co-operation and Development (OECD), "Recognition of Non-Formal and Informal Learning," http://www.oecd.org/education/skills -beyond-school/recognitionofnon-formalandinformallearning-home.htm, accessed May 31, 2015.

13. Sun Jung, "K-Pop, Indonesian Fandom, and Social Media," *Transformative Works and Cultures* 8 (2011).

14. Tiziana Terranova, "Free Labor," in *Digital Labor: Internet as Playground and Factory*, edited by Trebor Scholz (New York: Routledge), 2013.

15. Nancy K. Baym, "A Call for Grounding in the Face of Blurred Boundaries," *Journal of Computer-Mediated Communication* 14, no. 3 (2009): 720–23.

16. Nathan Kruse and Kari K. Veblen, "Music Teaching and Learning Online: Considering Youtube Instructional Videos," *Journal of Music, Technology & Education* 5, no. 1 (2012): 77–87.

17. Gooyong Kim, "Online Videos, Everyday Pedagogy, and Female Political Agency: 'Learning from Youtube' Revisited," *Global Media Journal* 11, no. 18 (2011).

18. Kathy Nakagawa and Angela E. Arzubiaga, "The Use of Social Media in Teaching Race," *Adult Learning* 25, no. 3 (2014): 103–10.

19. International Telecommunication Union, *The World in 2014: ICT Facts and Figures* (Geneva: United Nations, 2014).

20. Don Ihde, "The Designer Fallacy and Technological Imagination," Chapter 4 in *Philosophy and Design* (Netherlands: Springer, 2008), 51–59.

21. Jörgen Skågeby, "The Performative Gift: A Feminist Materialist Conceptual Model," *Communication +1* 2, no. 1 (2013).

22. Federica Frabetti, *Software Theory: A Cultural and Philosophical Study* (London: Rowman & Littlefield 2015).

23. Rob Kitchin and Matthew Dodge, *Code/Space: Software and Everyday Life* (Cambridge, MA: MIT Press, 2011).

24. Axel Bruns, "Towards Produsage: Futures for User-Led Content Production," Paper presented at the Cultural Attitudes towards Communication and Technology, Tartu, Estonia (2006).

BIBLIOGRAPHY

Alias, Norlidah, Siti Hajar Abd Razak, Ghada elHadad, Nurul Rabihah Mat Noh Kokila Kunjambu, and Parimaladevi Muniandy. "A Content Analysis in the

Studies of YouTube in Selected Journals." *Procedia—Social and Behavioral Sciences* 103, no. 26 (2013): 10–18.

Bassett, Caroline. "Feminism, Expertise and the Computational Turn." In *Renewing Feminisms: Radical Narratives, Fantasies and Futures in Media Studies*. Edited by Helen Thornham and Elke Weissmann, 199–214. London: I. B. Tauris, 2013.

Baym, Nancy K. "A Call for Grounding in the Face of Blurred Boundaries." *Journal of Computer-Mediated Communication* 14 (3) (2009): 720–23.

Bruns, Axel (2006). "Towards Produsage: Futures for User-Led Content Production." In Fay Sudweeks, Herbert Hrachovec, & Charles Ess (eds.), *Cultural Attitudes Towards Communication and Technology 2006, 28 June – 1 July, Tartu, Estonia*.

Cephiroth. "An Unofficial Final Fantasy VIII Walkthrough." http://www.gamefaqs .com/ps/197343-final-fantasy-viii/faqs/4897. Accessed May 31, 2015.

Clark, Andy. "Re-Inventing Ourselves. The Plasticy of Embodiment, Sensing, and Mind." *The Journal of Medicine and Philosophy* 32, no. 3 (2007): 263–89.

Eiriksdottir, E., and R. Catrambone. "Procedural Instructions, Principles, and Examples: How to Structure Instructions for Procedural Tasks to Enhance Performance, Learning and Transfer." *Human Factors: The Journal of the Human Factors and Ergonomics Society* 53, no. 6 (2011): 749–70.

Frabetti, Federica. *Software Theory: A Cultural and Philosophical Study* (Media Philosophy Series). London: Rowman & Littlefield, 2015.

Ihde, Don. "The Designer Fallacy and Technological Imagination." In *Philosophy and Design*, Peter Kroes, Pieter E. Vermaas, Andrew Light, Steven A. Moore, Chapter 4, 51–59. Springer Netherlands, 2008.

International Telecommunication Union. "The World in 2014: ICT Facts and Figures." Geneva: United Nations, 2014.

Jung, Sun. "K-Pop, Indonesian Fandom, and Social Media." *Transformative Works and Cultures* 8 (2011).

Kim, Gooyong. "Online Videos, Everyday Pedagogy, and Female Political Agency: 'Learning from Youtube' Revisited." *Global Media Journal* 11, no. 18 (2011).

Kitchin, Rob, and Matthew Dodge. *Code/Space: Software and Everyday Life*. Cambridge, MA: MIT Press, 2011.

Kruse, Nathan, and Kari K. Veblen. "Music Teaching and Learning Online: Considering Youtube Instructional Videos." *Journal of Music, Technology & Education* 5, no. 1 (2012): 77–87.

Lee, D. Y., and M. R. Lehto. "User Acceptance of Youtube for Procedural Learning: An Extension of the Technology Acceptance Model." *Computers & Education* 61 (2013): 193–208.

Morain, M., and J. Swarts. "Yoututorial: A Framework for Assessing Instructional Online Video." *Technical Communication Quarterly* 21, no. 1 (2012): 6–24.

Nakagawa, Kathy, and Angela E. Arzubiaga. "The Use of Social Media in Teaching Race." *Adult Learning* 25, no. 3 (2014): 103–10.

Organisation for Economic Co-operation and Development (OECD). "Recognition of Non-Formal and Informal Learning." http://www.oecd.org/education/skills -beyond-school/recognitionofnon-formalandinformallearning-home.htm. Accessed May 31, 2015.

Skågeby, Jörgen. "The Performative Gift: A Feminist Materialist Conceptual Model." *Communication +1* 2, no. 1 (2013).

Swarts, J. "New Modes of Help: Best Practices for Instructional Video." *Technical Communication* 59, no. 3 (2012): 195–206.

Terranova, Tiziana. "Free Labor." In *Digital Labor: Internet as Playground and Factory*. Edited by Trebor Scholz. New York: Routledge, 2013.

Thomas, Sue. "The End of Cyberspace and Other Surprises." *Convergence* 12, no. 4 (2006): 383–91.

Wikipedia. "Life Hacking." http://en.wikipedia.org/wiki/Life_hacking. Accessed May 25, 2015.

Zürcher, Reinhard. "A Sociomaterial Model of the Teaching-Learning Continuum." *European Journal for Research on the Education and Learning of Adults* 6, no. 1 (2015): 73–90.

6

Online Education, Massive Open Online Courses, and the Accessibility of Higher Education

Kristen Chorba and R. Benjamin Hollis

INTRODUCTION

Take a minute right now to think about your access to this book. You're reading this text either in print or online—which means that you either had access to purchase this book from a bookstore (brick and mortar or online) or borrow it from a person or a library or a school. If you are reading it online, you have access to a computer, tablet, or phone; the Internet; and a database or some other type of search engine that enables you to access this information. You also have the knowledge and skills to use these tools in a way that will accomplish this goal. If you are able to read and understand this text, then you have access to a version of this book that is in a language that you can read, as well as to the resources that have enabled you to learn how to interpret this language into words and phrases that makes sense.

What type of information do you have access to? Why or how do you have that access? What is access? Or, even, what is information? These all are questions that are vitally important to education—from very informal learning about topics such as basic home repairs, to an Ivy League college education. Such issues, however, seldom are considered in the course of daily life. This chapter discusses how online education—in numerous forms—has transformed the ways in which we access and use information, as well as how it has transformed those who are able to access and use information.

Technology is everywhere. We see it in our cars, in our homes, at work, and in our entertainment. Arguably, one of the most exciting places that we can see the impact of technology is in education. Our classrooms, our schools, our homes, and our communities have numerous resources and tools that provide us with access to information of all kinds. We can learn anything we choose! Or, at least, almost anything we want . . . if we have the skills, the knowledge, and a way to access technology that lets us do so.

Distance education has been around for a long time. In fact, "[d]istance education in various forms—written correspondence, radio, television, computer, Internet—has existed for well over a century in this country."[1] Online education in all of its current forms, however, has opened the doors for many people to access information and knowledge that just a couple of decades ago was not possible obtain. An almost infinite number of resources, fueled by the power of the Internet, now can be used as learning tools. Platforms such as YouTube host videos that range from very short to very long; amateur to professionally filmed and edited; and from basic "how-to" guides on various topics (crocheting, fixing a garbage disposal, using word-processing software) to elaborate educational presentations and lectures that feature experts in various fields (e.g., psychology, astronomy, anatomy, physiology). Videos on YouTube are free to access: All that is needed is a computer or "smart" device and an Internet connection. Companies such as Coursera provide actual classes—similar to those taught at a university—free of charge and delivered fully online. These classes can be accessed at any time, from any place that has offers Internet connection. Colleges and universities more and more commonly are offering fully online courses. Although these are not free services, the ability to participate in these courses completely online enables individuals who normally might not be able to take a face-to-face course (for a variety of reasons) to now do so.

Online education, in all its forms, has provided a level of access to information that is more attainable now than ever before. In the following pages, we will take an in-depth look at some of the ways that people are using technology to learn and explore some of the benefits of this type of accessible education.

SO WHAT IS "EDUCATION" . . . AND WHERE CAN I GET SOME?

What does accessibility really mean in terms of education? According to the United Nations, "education is a right, like the right to have proper food or a roof over your head."[2] This right extends to all people, of all ages, from all societies, with all ability levels, from around the globe. It includes those seeking specialized training as well as those seeking basic education on a specific topic. This is the foundation of the U.N. Education for All program. Education, then, is broadly defined and can range from more traditional, formalized learning, housed inside of academic institutions, to self-paced, informal, task- or skill-based training. Although education for all is a goal to strive for, it does not mean that this is currently the situation; there is, for some, a digital divide.

The term "digital divide" refers to inequities between those who have regular, reliable access to the Internet and digital technologies and those who do not."[3] There has been substantial research asserting that a digital divide indeed does persist, even though steps have been taken to help alleviate barriers to access. Libraries are one example of organizations that seek to bridge this gap. Libraries do so by providing free access to computers for education, job searching, and creating and maintaining social connections.[4] According to a report by the U.S. Department of Commerce, however, although more households used broadband Internet services (68 percent in 2011), the

> demographic and geographic disparities demonstrate a persistent digital divide among certain groups. . . . [L]ower income families, people with less education, those with disabilities, Blacks, Hispanics, and rural residents generally lagged the national average in both broadband adoption and computer use [which] stands in sharp contrast to the digital access enjoyed by well-educated middle- and upper-class White households.[5]

These types of statistics are not new, nor are they surprising: access often requires money or the ability to find someone who can provide you with free equipment and Internet connection . . . and always requires at least some base level of understanding. These issues, however, although important, are beyond the scope of this chapter and have been written about and debated extensively.[6] Instead, this chapter explores the current situation of online education for what it has to offer, examines how various organizations are working to increase access and online educational opportunities, and discusses some of the tools that are available to help create successful online learning experiences.

In summary, the good news is that there are many positive things happening. Education of all types is becoming more accessible, even if it presently is not always and completely such. Individuals are finding more and more ways to access information. In fact, in 2014, research by the Pew Center found that "the vast majority of Americans believe their use of the web helps them learn new things, stay better informed on topics that matter to them, and increases their capacity to share ideas and creations with others."[7] And access itself is becoming easier. The widespread use and availability of the Internet, a variety of ways to connect, and organizations dedicated to helping bridge the gap between what people know about technology and what they *need to know* about technology to be successful, is making education more accessible than ever.

WHO'S ONLINE, ANYWAY?

So, how many people, broadly speaking, have access to some form of on-line education? And who are these people? Although this chapter does not specifically address all areas or all populations, the following statistics from the Pew Research Center provide a broad idea of how much access people have and what type of access is available to various populations.

First, it's important to talk about some general statistics. These are based on adults (i.e., age 18 or older) in the United States. These numbers are also in aggregate and address overall population trends: due to space limitations and the focus of this chapter, we do not specifically break down populations further by ethnicity, sex, income, education, household size, or other demographic factors, with the exception of general age ranges. Much of that information, however, can be found using the sources we have cited here.

Generally speaking, according to data about adult users in the United States, collected and published in 2014:

- 87 percent use the Internet[8];
- 68 percent use smartphones and/or tablets to get online[9];
- 58 percent use a smartphone[10]; and
- 42 percent own a tablet.[11]

According to 2013 statistics, the following groups have either high-speed Internet (e.g., broadband) or a smartphone at home:

- 95 percent of adults aged 18 to 29;
- 89 percent of adults aged 30 to 49;
- 77 percent of adults aged 50 to 64; and
- 46 percent of adults aged 65 and older.[12]

The use of cell phones and smartphones has become far more prominent in the past several years. Cell phones are basic mobile phones, which might or might not have a full QWERTY keyboard (similar to a typical computer keyboard), and can make calls; send and receive text messages; and take, send, and receive pictures.[13] Smartphones are cell phones that have "built-in applications and Internet access, such as an iPhone or an Android."[14] According to data collected and published in 2014:

- 90 percent of adults own cell phones, up from 53 percent in 2000; and
- 58 percent of adults own smartphones, as compared to only 35 percent in 2011.[15]

The most recent data available (May 2013) state that "34% of cell Internet users go online *mostly* using their phones, and not using some other device, such as a desktop or laptop computer."[16] The Pew Center also notes that smartphones help to increase online access—even if it is not always full access to all Web sites and all browser functions—by offering a way to connect from home without having to own a computer or pay for a subscription to an Internet service.[17] Additionally, the Pew Center points out that, among certain groups, cell phone and smartphone access to the Internet lessens the gap in online access for many populations. According to Pew Center's statistics, "while blacks [sic] and Latinos are less likely to have access to home broadband than Whites, their use of smartphones nearly eliminates the difference."[18] That said, the opposite is true of older adult populations: The gap in access actually widens when Internet access is considered as both broadband connection and smartphone access.[19]

To be sure, many people access the Internet from home or via their smartphones. Although phones do not always provide full access to all sites or all applications on the Web (they have smaller screens and can be more difficult to fully access or operate certain sites) the way that a computer does, for some users phones can be a cheaper more accessible way to get online. Specifically, considerations about cost of one technology versus the other include the cost of the computer itself; the cost of Internet access (often a monthly subscription); the cost of the phone and service (often a monthly fee); and the cost of a home or landline, which might not be necessary for those with a cell phone or smartphone, but which often still is needed for a computer to be used to access the Internet. Additionally, although landlines only can be used to make standard voice calls, cell phones and smartphones can be used to send text messages as well as to make regular calls, and often can be used to take and send pictures and videos. Cell phones and smartphones are absolutely key access points, however, which can help people find and use information and provide a way to access online education.

Teen Users

Teens (individuals who are between the ages of 12 and 17 years of age) also account for a large proportion of those who access the Internet. This population is particularly important to note, as they fall outside of the age range of studies that target "adults" (again, those people older than age 18). According to 2013 statistics from the Pew Research Center:

- 93 percent of teens have at home access to a computer, whether it is their own or shared (71 percent) with others in their home;
- 78 percent of teens have access to their own cellphone; 47 percent of those phones are smartphones;

- 74 percent of teens use their cell or smartphones to go online; of the teens that use their cell or smartphones to go online, about 25 percent use the phone for most of their Internet access; and
- 23 percent of teens have a tablet.[20]

Older Adult Users

Older adults (that is, those aged 65 and older) also have become a group of Internet users. Data from a 2014 study by the Pew Research Center found that 59 percent of seniors use the Internet. Additionally, about 77 percent of seniors own a cell phone; however, only about 18 percent of these cell phones are smartphones. Finally, about 18 percent of this population owns a tablet. The data also indicate that this older generation of adults falls into two categories, each with differing attitudes toward and use of the Internet. Whereas younger seniors with more education and wealth are associated with a positive attitude toward computing and technology use, older seniors with health issues or disabilities are more likely to disconnect from technologies altogether. Seniors older than age 75 demonstrate a steep decline in technology and Internet adoption and use.[21]

THEN WHO IS NOT ONLINE?

A discussion about who is online is incomplete without briefly mentioning who is *not* online. According to 2013 statistics gathered by the Pew Research Center, 15 percent of adults are not online; additionally, 9 percent more do go online, but not from home.[22] For these users, the data indicate that:

- 19 percent do not use the Internet because of the cost;
- 7 percent do not use the Internet because they do not have access to it;
- 33 percent said that the Internet was too difficult for them to use, for a variety of reasons (including age, fear of hackers and viruses, and physical abilities);
- 34 percent said that they were simply not interested in using the Internet (i.e., they did not have time, were too busy, or thought it was a waste of time); and
- Only 8 percent of those adults who stated they were offline expressed a desire to be online; the other 92 percent reported that they did not care to access the Internet.[23]

Generally speaking, lots of people have access to the Internet and, as such, to online education in its various forms. Whether people access it from home, work, the library, or school; via a phone, tablet, or computer, users across a wide range of ages are interacting with tools and resources online. The next section explores some of the ways in which this online education is happening.

HOW WE'RE LEARNING ONLINE

Higher Education

Online education in colleges and universities (i.e., education after high school or after grade 12) has grown significantly over the last several years. Online courses in this realm typically are for credit, with a fee structure similar to that of traditional face-to-face classes, and require admission to a program of study at the specific school.

According to the Allen and Seaman's Babson Survey Research Group, 7.1 million higher education students (33.5 percent of all higher education students) are taking at least one online course.[24] Some argue that the numbers associated with Babson Survey Research Group's report are possibly inflated due to the reporting scope (e.g., only counting degree-granting institutions)[25] and difficulties collecting consistent enrollment data from universities (e.g. U.S. Department of Education only started asking universities to report distance learning enrollments in 2012, and possible inconsistencies reporting and quantifying online student enrollment).[26] The number of students taking online courses for college credit somewhere, however, estimates from 5.5 million[27] to the most recent estimate of 7.1 million[28] students taking at least one online course, thus the quantity of students turning to online learning in higher education is significant.

As institutions consider ways to make learning more accessible, relevant, and far-reaching, online course options have taken two main forms: fully online and hybrid. In addition, many teaching traditional, face-to-face courses use some kind of online platform—such as Blackboard or Moodle—to help supplement and organize their course. This sort of online course supplementation can take the form of posting notes, outlines, slides, or syllabi online, or allowing students to submit certain assignments online. Hybrid courses are more of a mix of face-to-face and online components. Hybrid courses could meet face to face some or most of the sessions, but also can meet online for some sessions, either synchronously or asynchronously. Another application of a hybrid course is a "flipped classroom"[29] approach, in which lectures and other typically in-class, direct instruction approaches instead are completed outside of the classroom, usually via an online platform. Class time then is reserved for group work, collaboration, discussion, and deep thought about the content, guided by the instructor. Lastly, fully online courses are just that, courses that are held only online, often asynchronously. All course content, assignments, and activities, including lectures, quizzes and tests, discussions, grading, and feedback, are accessed from and submitted to an online learning management system. This fully online format allows students who otherwise might not be able to attend that school (for example, due to full-time work requirements, location, or time constraints; distance; or myriad other reasons), to have the opportunity to complete classes and work toward degree completion.

Kindergarten through Twelfth Grade

Schools teaching kindergarten through twelfth grade (K–12) across the United States are also taking advantage of the Internet in various ways, to provide an online alternative to the traditional classroom grade-school experience. Just like the colleges and universities described above, grade schools are utilizing fully online, hybrid, and face-to-face classes plus online resources models to help students achieve. Depending on the state and the program, some of these are public and some are private. Some have tuition fees associated with them and others do not. Some also provide books, access to instructors, and online chat sessions for asking questions. Although it is not the authors' goal to describe these or nationwide trends in detail, they highlight a few initiatives that are happening within one specific area: Los Angeles County (United States).[30]

Some high schools also are partnering with local colleges to help students gain access to experiences and information that will help them in seriously considering or enrolling in college. Through some programs, students can register for free at a partner higher education institution and take classes online, if they cannot go to the school in person. Another way that some teachers are helping their student access higher education is through a classroom "account" to an online, college or college-level class (such as a specific college class that has an agreement with a particular high school, or even a MOOC, which is discussed elsewhere in this chapter). Although individual users typically must have a minimum grade point average to sign up for such classes, high school teachers can sign up for an account and then use it in their classes as a tool for learning. This also is a good strategy when there are not enough computers or laptops available in a high school classroom, and students individually cannot be expected to access the supplemental college course online.

Lastly, having these types of activities available to high school students allows them to have actual experience, in varying forms, with a college-level class, as well as with the types of technology they will encounter. This gives students the opportunity to see what a college course is "really" made up of, including what it is like and how difficult it is. For high-achieving students this can reassure them that college is a possibility. It also can serve as a wake-up call, however, for those students who might not know other people who have attended college, and can open these students' eyes to the level of difficulty of college classes and college-level work. In either case, this experience provides a way for students to contextualize the reasons behind why teachers push them to work hard to be prepared for college.

High school teachers also are using online courses and resources as a way to support students who need additional help with some aspect of their academic lives. Remediation is one way in which the Internet is being used for this purpose. If a student has fallen behind, or is struggling in a specific course or discipline, schools can use online learning to help the student master that content before moving on. This is beneficial because it does not require a full classroom or constant supervision; allows flexibility in scheduling, for the student and the school; and enables users to access the content from places outside of the school.

Online education can also help to provide alternative routes to schooling for students experiencing difficulties related to face-to-face education settings, including situations such as bullying, Attention Deficit and Hyperactivity Disorder (ADHD), anxiety, and varying types and degrees of disabilities. Large class sizes, for example, which can be necessary due to budget restrictions, school closings, or the need for specialized courses that serve large populations of students, can make some students feel anxious about having to interact within and adapt to that setting. In such cases, online courses that students can take from the comfort and safety of their own homes could create a more positive learning environment than a face-to-face environment provides. These types of challenges are a reality for many students, and online education is one way to help them deal with such challenges and alleviate some of the setbacks to having a quality learning environment and experience.

Finally, online education gives schools a way to help large numbers of students without having to find the physical space to hold students and accommodate class schedules. Using online learning tools is one way to support students who need help meeting graduation requirements and those who are interested in or need help with college preparation.

MASSIVE ONLINE OPEN COURSES

What Is a Massive Online Open Course?

A massive online open course or "MOOC" (mo͞ok) can take different forms (see xMOOCs and cMOOCs, explained below), but at its core, a MOOC is an online course made available to high student enrollments, scalable to serve tens of thousands of students in a single offering.[31] MOOC providers (e.g., edX, Coursera, Udacity) partner with the best universities in the country and abroad (e.g., Princeton, MIT, Harvard, National Taiwan University, Peking University) to deliver courses on topics ranging from biology to statistics to literature. MOOCs typically are free and offer certificates of course completion; in some cases, a nominal fee is required. MOOC completion certificates have a varying range of value. The American Council on Education (ACE) approved five MOOCs for university credit; however, individual institutions determine whether they will allow the transfer of MOOC completion credits for university credits.[32] Only a small handful of universities currently recognize MOOCs for university credits. Duke and UC Irvine—institutions with ACE-approved MOOCs—do not allow those ACE-approved MOOCs for credit. Recently, Udacity, Georgia Tech, and AT&T collaborated to offer an online master's degree in computer science on the Udacity-MOOC platform with a total tuition of less than $7,000—far less than the cost of competing online degree programs.[33]

Massive online open courses typically are divided into two categories of construction and philosophies: xMOOCs and cMOOCs. The xMOOCs closely resemble a traditional instructional paradigm, one where a professor disseminates content in the form of readings and recorded lectures and students demonstrate understanding by completing quizzes/tests and participating in online discussion forums[34]; these xMOOCs are commonly associated with university-partnered or for-profit platforms such as Coursera.[35] The "x" in xMOOC stands for "eXtended" and was intended to identify these MOOCs as extensions of core offering, just as MITx is the MOOC extension of the prestigious Massachusetts Institute of Technology (MIT).[36] The xMOOC instructors typically can be contacted by e-mail or during virtual office hours, and their role is more traditional in the sense of being the subject-matter expert and tutor for learners.[37]

The cMOOCs are driven by a connectivism theory of learning, hence the "c" for "connectivism," and the participants explore a topic with other co-learners;[38] in a cMOOC environment, "each attendee act(s) as both a participant learning and expert."[39] The cMOOC facilitators provide an outline or framework of the course, and the learners make decisions on which ideas and materials to pursue with the goal of developing a deeper understanding of the subject matter collectively.[40] Just as the role of learners changes, so does the role of the instructor/facilitator. Instead of assuming the traditional role of instructor/tutor for learners, the cMOOC facilitator shares information, challenges ideas, amplifies concepts, curates to arrange helpful readings and resources, filters useful versus nonuseful, models behaviors, and supervises course activities.[41] Fundamentally, for learners the difference between xMOOCs and cMOOCs is that "in one, they will be part of a process (cMOOC); in the other they will be part of a product (xMOOC)."[42]

Rodriguez (2012) noted the conceptual differences in openness, that xMOOCs are open in terms of public access and cMOOCs have an open curriculum.[43] Further research is needed to highlight the differences between xMOOCs and cMOOCs in terms of popularity, completion rates, and student achievement across formats, as most reports generalize MOOCs as a single identity.

Why Make a Massive Online Open Course?

Ferdig (2013) noted three important factors that led to the creation of MOOCs that speak directly to the theme of this chapter. First, MOOCs began with an effort to make access to content open to the public. Second, MOOCs operate alongside the movement of social media and technological advances in creating and sharing information; a single point of expertise is replaced by multiple sources of expertise. The third factor is the emergence of blended (online and in-person) learning in K–12 schools that can utilize these free education resources made available to the public in multiple formats.[44] Yan and Powell (2013) noted that MOOCs were designed with the "ideals of openness in education" in mind, that "knowledge should be shared freely, and the desire to learn should be met without demographic, economic, and geographical constraints."[45] MOOC advocacy increases university philanthropy and marketing to students, faculty, alumni, and donors; it also permits faculty to experiment with teaching and learning in online environments with high student enrollments.[46]

MOOC enrollments typically consist of advanced high school students, college students, faculty (for peer-based professional development), and hobbyists.[47] Both Harvardx and MITx provide access to data on the more than 1.5 million users in their combined 17 MOOC offerings. Across both institutions, approximately 67 percent of all users have at least a bachelor's degree or greater: 33 percent have a bachelor's degree; 23 percent have a master's degree; and 5 percent have a doctoral degree.[48] Due to the digital literacy skills needed to register and navigate MOOCs,[49] some argue that MOOCs are not uniformly equitable among all potential participants.[50]

2012, the Year of the Massive Online Open Course

As the first wave of MOOCs and this radical concept of open access to university content surfaced, most articles covering MOOCs or massive online open courses noted that 2012 was the "year of the MOOC."[51] The movement to provide free online courses from prestigious universities was a revolutionary change at higher education. Some people believed this revolution would become a disruptive innovation—one that changed the landscape of higher education and pressured institutions to restructure the traditional online learning and tuition-based frameworks to consider open access.[52]

2013, the Year of the Anti–Massive Online Open Course

One year later, others deemed 2013 the year of the *anti*-MOOC.[53] Criticisms of MOOCs centered on the pedagogical challenges of delivering and assessing an authentic learning experience among thousands of other students and low completion rates with only 2 percent to 10 percent of students typically completing a MOOC.[54] Students drop MOOCs for a variety of reasons, including lack of incentive; insufficient prior knowledge; unfocused discussion forums; unclear assignments; difficultly with learning the materials and receiving instructor support; lack of time due to other commitments and priorities; personal interests; wanting to complete specific modules of interest versus an entire course[55]; and a feeling of isolation with a lack of interactivity in the MOOC environment.[56]

Additionally, MOOC considerations at universities are modest with 53 percent of institutions undecided about MOOCs, 33 percent with no plans for MOOC development, and 5 percent of universities currently delivering a MOOC. Although not expressly correlated, the modest consideration for MOOC planning and development is likely linked to the lack of perceived sustainability and long-term future of MOOCs by university administrators, with only 23 percent viewing MOOCs as sustainable and 22 percent believing in the long-term future of MOOCs. Both of these values declined from percentages in 2012. Greater enrollment (more than 15,000 students) doctoral/research institutions with resources for development are more likely to offer and plan MOOCs than are smaller institutions (those having fewer than 3,000 students) with a baccalaureate-degree focus. Leaders at institutions with existing online programs noted that a primary objective for offering a MOOC was marketing to increase institutional visibility; whereas universities without online programs planned to offer MOOCs for student recruitment. Institutions were less motivated by other MOOC-related objectives, such as reducing costs and generating income.[57]

IF LESS THAN 10 PERCENT COMPLETE A MOOC, AND UNIVERSITIES AREN'T SO SURE ABOUT DEVELOPING MOOCS, THEN WHY BOTHER?

With 2013 being the year of the "anti-MOOC" and such low MOOC completion rates, why then are MOOCs still in the conversation of online learning in the year 2015? In a summary of MOOC literature, Hew and Cheung (2014) reported four categories motivating learners to enroll in MOOCs.

- *Category One—Free, timely learning*: To increase or refresh knowledge about a topic at no cost or to learn something in a timely fashion that could help with work without seeking course credit or completion;
- *Category Two—Curiosity*: To experience a MOOC and engage in a learning experience with thousands of other learners;
- *Category Three—Personal Challenge*: To see whether the student can complete a course from a prestigious university; and
- *Category Four—Hobby*: To earn as many certificates or badges as possible.[58]

Note that badges are achievements recognized in a digital " badge"-style format. The term and concept are derived from the merit-badge achievement system commonly associated with the Boy Scouts and Girl Scouts. Learners earn badges for achievements in online courses and often display them on Web sites or digital backpacks.[59] In his article, *MOOCs—Completion Is Not Important*, Matthew LeBar (2014) noted that a majority of students taking a MOOC already held a college degree and were using the MOOC to bolster job-market readiness or professional development; therefore, the ability to gain or supplement knowledge was not dependent on completing a MOOC.[60]

Fredig (2013) noted the practical and conceptual values of MOOCs that also have little to do with MOOC completion rates. From a practical standpoint, MOOCs can serve as supplemental instruction for K–12 schools, where students can get a start on college-level coursework, potentially meet state graduation requirements for online learning experiences, and explore content areas outside of the scope of school funding or faculty expertise. Given the diversity of MOOC enrollments, learners can experience multiple perspectives from other learners with similar or dissimilar perspectives[61]; "motivated learners who are willing to capitalize on this opportunity can begin to see the content in new ways."[62] MOOCs also can be used for continuing professional development and strengthening communities of practice.

Fredig (2013) also considered the conceptual benefits of MOOCs, such as furthering advancements in connected learning; utilizing digital badge assessment for student engagement and motivation; reusing, revising, remixing, and redistributing open content; promoting learner artifact creation in an open-access environment, not limited to the standard instructor-student viewership.[63] As evidence of these practical and conceptual values in practice, Vivian Falkner and Nickolas Falkner (2014) reported success at applying a MOOC to solve an instructional need at a national scale. A MOOC was used to deploy new computing curriculum training to in-service teachers across Australia.[64] In this case, the scalability of the MOOC helped reach uses across the nation to deliver a flexible learning framework with ad hoc interactions that promoted a successful community of practice and professional development.

WHAT WILL THE FUTURE HOLD FOR MOOCS?

The year 2015 might not be the "resurgent year of the MOOC." Due to the low adoption and development rates, MOOCs might not be the disruptive innovation that pressures reform in higher education costs and course delivery. Pro-MOOC arguments presented in this chapter, however, urge us to look beyond the numbers of completion rates and pedagogical challenges to at least move us beyond the years of the anti-MOOC and acknowledge that the "anti-MOOC is so 2013."[65] In a recent interview, MOOC pioneer and innovator, George Siemens stated:

> I think we are starting to see a maturing of our understanding of what MOOCs are. . . . The hype is finally starting to die. We're now starting to realize that MOOCs don't do everything. But they do serve a particular need, and they are an important research opportunity for universities to help transition their activities and offerings in those spaces. . . . I'm hoping that we will start to talk more about digital and online learning and blended learning, because I think that's what MOOCs really reflect. MOOC is still a term that allows us to encapsulate a movement. I'm reminded of what happened with Web 2.0. In 2004 and 2005 we were inundated with Web 2.0 nonsense. Web 2.0 was supposed to do everything, make our coffee, raise our kids, change society. We all hated the term by the time it was done. MOOCs, I think, will do exactly the same thing. Going forward, I hope we will be able to do away with the MOOC acronym. If you overhype something, you eventually learn to hate it. The legacy of MOOCs in the higher education system, I think, will be valuable. The acronym, though, not so much.[66]

Given the opportunities for supplemental instruction, professional development, and communities of inquiry, MOOCs still stand to contribute value to K–12 classrooms, higher education, professional development, and hobbyists alike seeking to broaden existing perspectives on content, others, and self.

INTERESTED IN EXPLORING A MOOC?

The following is a list of the top MOOC providers.

edX | edX.org

Harvard and MIT created the nonprofit platform, "edX." Global university schools and partners from UC Berkley to Seoul National University have authored 408 courses (including 44 high school courses), covering topics ranging from biology to literature. Students using edX have earned more than 100,000 course certificates.

Coursera—Coursera.org

Coursera is a for-profit platform but offers free student enrollments. It has a catalog of nearly 2,000 courses, ranging from the arts to teacher professional development, and includes specializations (sequences of courses) in various topics, including data mining and systems biology—authored by faculty representing universities located around the world.

UDACITY—Udacity.com

Udacity is a for-profit platform that offers programming and technology courses and currently features four "nanodegree programs": Front-end web developer; full stack web developer; data analyst; and iOS developer. The nanodegrees cost $200 per month and typically take 6 to 12 months to complete.

NONPROFIT ORGANIZATIONS

Nonprofit organizations also are major players in the online education world. Some organizations, such as the Fulfillment Fund, have a specific mission and work with high school–aged students to make higher education a reality for them. Organizations such as this work hard to help students get the access and knowledge that they need. This includes things such as using e-mail and obtaining access to or learning how to use a computer. These types of tasks—which sometimes are basic computer and computing functions—often are unfamiliar to students who are under-resourced. Students who are under-resourced are not necessarily "poor." Many students and their families who are in this situation have enough money; they just do not have enough *access*. Nonprofit organizations often work hard to bridge gaps in accessibility, and could be working with a relatively small staff to try to prepare a growing number of students for the realities of higher education, including the technical skills they will need to be successful. Preparing students in this way is essential: "If you are preparing students for college, you're only doing half your job if you are not also preparing them for the technology part."[67]

Other nonprofit organizations are online education providers, such as Wikipedia, CODE.org, and the Khan Academy. Each of these sites provides information on a specific topic or range of topics and makes them freely available to anyone who can access the Internet. For example, CODE.org is a nonprofit organization that is dedicated to "expanding participation in computer science by making it available in more schools, and increasing participation by women and underrepresented students of color."[68] They provide free tutorials and lesson plans on a variety of topics and for a variety of age groups. Some programs are short (one hour) and some are long (up to 20 hours) and can be used by individual students at home, or by teachers to integrate into their classes. Depending on the program of interest, options for accessing the interactive tutorials from a browser, a tablet, a smartphone, and even without using an Internet connection.

Technology, Entertainment, and Design (TED), is a nonprofit organization that produces short videos (usually 10 to 60 minutes long) on a range of topics. These TED Talks often focus on important, powerful, and provocative topics, and help users of all types have access to some of "the world's most inspired thinkers" to "make great ideas accessible and spark conversation."[69] TED also conducts face-to-face events that frequently are filmed.

Wikipedia is an open-access online encyclopedia. Anyone can edit it and it is monitored regularly. At the time of this writing, Wikipedia contains almost five million entries and is accessed by more than a half billion people from around the globe. Although critics do not see Wikipedia as a sound educational tool—as it is not peer-reviewed and any topic can be edited by anyone (including laypersons)—it is a prominent tool for looking up information quickly. The amount of freely accessible information it contains is enormous. The user-centered, collaborative nature ensures that entries can be updated as information changes and becomes available. The ability for experts and nonexperts alike to interact with the content helps to keep entries and updates in perspective: If information is biased or is not backed up with evidence, users can include those notes, as well. Regardless of the scholarly nature of Wikipedia, it can be a good learning tool and starting point—especially for those who do not have easy access to more traditional forms of education.

Nonprofit organizations, such as the ones discussed above as well as many others not discussed here, have much that they can contribute to online education. Providing access to learning materials, as well as helping users to learn how to use these materials and powerful learning tools are important functions of nonprofit organizations.

NONPROFIT RESOURCES

TED—Ted.com

TED is a nonprofit, global community of speakers who share ideas and expertise in a series of 18-minute (or shorter) talks on various topics, such as, "Why you should care about whale poo" and "The art of stillness."

Khan Academy—Khanacademy.org

The Khan Academy is a nonprofit that offers free access to instructional videos and exercises for curriculums from kindergarten to higher education, along with test prep and coaching resources to help anyone, anywhere.

Wikipedia—Wikipedia.org

Wikipedia is the free encyclopedia that anyone can edit. Articles are user created, monitored, and published. Worldwide usage is so prominent and synonymous with Internet information that Wikipedia currently contains nearly five million articles.

Additionally, there are many free, online resources that are run by for-profit companies. The following are some examples of free online learning resources that are home to a vast amount of content.

OTHER FREE AND FOR-PROFIT RESOURCES

iTunesU—iTunes.apple.com

iTunesU is an Apple application (app) that is designed to distribute more than 750,000 pieces of free educational media (lectures, videos, books) from universities all over the world to mobile devices. iTunesU interacts with other mobile device apps to help students take notes and contribute content. Instructors can author, distribute, and conduct full courses in iTunesU.

YouTube Edu—Youtube.com/edu

YouTube is ubiquitous with Internet video. The exponential growth in contributors, subscribers, and video content available define this generation of authoring and sharing videos on the Web. The YouTube Edu channel is a space dedicated to topics ranging from business to social sciences.

NONTRADITIONAL ONLINE LEARNING: NEWSFEEDS, EXPERTS, AND AMAZON.COM

This chapter discussed systems dedicated to providing education to the masses, such as MOOCs and the Khan Academy. But often the interactions in settings without a direct educational focus can provide additional outlets for learning that influence our information intake and decision making on a daily basis. Facebook and Twitter gather popular topics tagged or referenced by other users of the platform. Trends provide instant access to top stories in the media. Social media users can also follow experts and organizations to customize this stream of information (newsfeed). Expert-finding also takes place in online discussion forums, in videos, in game environments, and when shopping.

Online forums, such as Stack Overflow (stackoverflow.com), utilize a community-driven, question-and-answer system to vote-up useful responses to questions until a correct or suitable response is identified. YouTube videos are ranked and ordered by viewership and a simple count of thumbs-up or thumbs-down votes. Such rankings help users easily gauge the popularity or usefulness of a resource and identify expertise in the subject matter. Expert gamers often are associated with higher-level rankings within the gaming environment—set apart from beginners and novices with lower-level indicators. Shoppers on Amazon.com can aggregate searches and inform decisions based on a star-based ranking system and comments generated by other shoppers.

The platforms and scenarios mentioned are certainly different than traditionally structured learning environments. The undeniable influence of social media and user-driven data, however, informs learning and decision making in daily interactions with the online world.

ONLINE EDUCATION: WHAT'S GOOD ABOUT IT?

This chapter highlights a number of exciting initiatives for and avenues to online education, in its many forms. We maintain that the landscape of higher education—including online education—is very positive. Although there are disparities in access for some populations, the authors believe that the number of people who do have access, as well as the decreasing divide in that access, shows promise. Additionally, people do not always realize that they have the potential and opportunity to be able to access all of this information or that these avenues to online education exist. Through the work of nonprofit organizations, educational institutions, and others, however, the gap in access is beginning to be addressed.

Online education brings a vast amount of information within reach of people's fingertips: What can be found, learned, explored, and studied is limitless. In a recent interview, Google executive chairman, Eric Schmidt said,

> The Internet will disappear. . . . It will be part of your presence all the time. Imagine you walk into a room, and the room is dynamic. And with your permission and all of that, you are interacting with the things going on in the room. . . . A highly personalized, highly interactive and very, very interesting world emerges.[70]

Learners no longer are limited to what they can find at the public library or through the card catalog, or from the daily newspaper or television news.

Technology can be challenging, can be stimulating, and can promote better learning. Technology also fills a gap when brick-and-mortar schools cannot accommodate students (i.e., college prep, graduation requirements, physical space, classrooms, teachers, remediation, special needs). Ultimately, "technology and online education are a necessity. You have to see it as that and you have to want to learn about it, because it is not going away. It is only growing. There is a huge potential for what is to come and for what students can learn."[71]

ACKNOWLEDGMENTS

The authors thank Elizabeth Kalfas for her insights and contributions to parts of this chapter.

NOTES

1. Campaign for the Future of Higher Education, 2013, "The Promises of Online Higher Education: Access—Campaign for the Future of Higher Education," 44.

2. UN.org, 2015, "United Nations Resources for Speakers on Global Issues—Education for All (EFA)," http://www.un.org/en/globalissues/briefingpapers/efa/index.shtml, accessed June 1, 2015.

3. "The Promises of Online Higher Education," 20.

4. Samantha Becker, Michael D. Crandall, Karen E. Fisher, Bo Kinney, Carol Landry, and Anita Rocha, 2010, *Opportunity for All: How the American Public Benefits from Internet Access at U.S. Libraries*. Washington, DC: Institute of Museum and Library Services. http://www.imls.gov/assets/1/assetmanager/opportunityforall.pdf, accessed June 1, 2015.

5. "The Promises of Online Higher Education," 21.

6. Sharon Strover, 2014, "The US Digital Divide: A Call for a New Philosophy." *Critical Studies in Media Communication* 31 (2): 114–22. doi:10.1080/15295036.2014.922207.

7. Kristen Purcell and Lee Rainie, 2014, "Americans Feel Better Informed Thanks to the Internet." *Pew Research Center's Internet & American Life Project*, http://www.pewinternet.org/2014/12/08/better-informed/, accessed June 1, 2015.

8. Susannah Fox and Lee Rainie, 2014, "The Web at 25 in the U.S.," *Pew Research Center's Internet & American Life Project*, http://www.pewinternet.org/2014/02/27/the-web-at-25-in-the-u-s/, accessed June 1, 2015.

9. Ibid.

10. Pew Research Center's Internet & American Life Project, 2013, "Mobile Technology Fact Sheet." http://www.pewinternet.org/fact-sheets/mobile-technology-fact-sheet/, accessed June 1, 2015.

11. Ibid.

12. Kathryn Zickuhr and Aaron Smith, 2013, "Home Broadband 2013," *Pew Research Center's Internet & American Life Project*, http://www.pewinternet.org/2013/08/26/home-broadband-2013/, accessed June 1, 2015.

13. Adam Fendelman, "How Are Cell Phones Different From Smartphones?," *About Tech*, http://cellphones.about.com/od/coveringthebasics/qt/cellphonesvssmart phones.htm, accessed June 1, 2015.

14. Bruce Drake, 2014, "Americans with Just Basic Cell Phones Are a Dwindling Breed," *Pew Research Center*. http://www.pewresearch.org/fact-tank/2014/01/09 /americans-with-just-basic-cell-phones-are-a-dwindling-breed/, accessed June 1, 2015.

15. Fox and Rainie, "The Web at 25 In the U.S."

16. Pew, "Mobile Technology Fact Sheet."

17. Zickuhr and Smith, "Home Broadband 2013."

18. Ibid.

19. Ibid.

20. Mary Madden, Amanda Lenhart, Maeve Duggan, Sandra Cortesi, and Urs Gasser, 2013, "Teens and Technology 2013," *Pew Research Center's Internet & American Life Project*, http://www.pewinternet.org/2013/03/13/teens-and-technology-2013 /, accessed June 1, 2015.

21. Aaron Smith, 2014, "Older Adults and Technology Use," *Pew Research Center's Internet & American Life Project*, http://www.pewinternet.org/2014/04/03/older -adults-and-technology-use/, accessed June 1, 2015.

22. Kathryn Zickuhr, 2013, "Who's Not Online and Why," *Pew Research Center's Internet & American Life Project*, http://www.pewinternet.org/2013/09/25/whos-not -online-and-why/, accessed June 1, 2015.

23. Ibid.

24. Elaine I. Allen and Jeff Seaman, 2014, *Grade Change: Tracking Online Education in the United States*. Babson Survey Research Group and Quahog Research Group, LLC. http://www.onlinelearningsurvey.com/reports/gradechange.pdf, accessed June 1, 2015.

25. Phil Hill, 2014, "Clarification: No, There Aren't 7.1 Million Students In US Taking at Least One Online Class," *E-Literate*, http://mfeldstein.com /clarification-arent-7-1-million-students-us-taking-least-one-online-class/, accessed June 1, 2015.

26. Steve Kolowich, 2014, "Exactly How Many Students Take Online Courses?" *The Chronicle of Higher Education*, http://chronicle.com/blogs/wiredcampus/exactly -how-many-students-take-online-courses/49455, accessed June 1, 2015.

27. Hill, "Clarification."

28. Allen and Seaman, "Grade Change."

29. Jonathan Bergmann and Aaron Sams, 2012, *Flip Your Classroom*. Eugene, OR: International Society for Technology in Education.

30. Elizabeth Kalfas, telephone conversation, November 20, 2014.

31. Rick E. Ferdig, 2013, *What Massive Open Online Courses Have to Offer K-12 Teachers and Students*, Lansing: Michigan Virtual Learning Research Institute, http:// media.mivu.org/institute/pdf/mooc_report.pdf, accessed June 1, 2015.

32. Steve Kolowich, 2013, "The Professors Who Make the MOOCs," *The Chronicle of Higher Education*, http://chronicle.com/article/The-Professors-Behind -the-MOOC/137905/≡=overview, accessed June 1, 2015.

33. Audrey Watters, 2012, "Top Ed-Tech Trends Of 2012," *Insidehighered.Com*. https://www.insidehighered.com/blogs/hack-higher-education/top-ed-tech-trends -2012-moocs, accessed June 1, 2015.

34. Khe Hew and Wing Sum Cheung, 2014, "Students' and Instructors' Use of Massive Open Online Courses (MOOCs): Motivations and Challenges," *Educational Research Review* 12: 45–58, doi:10.1016/j.edurev.2014.05.001.

35. Ferdig, "What Massive Open Online Courses Have to Offer."

36. Stephen Downes, 2013, "What the 'x' in 'xMOOC' Stands for." *Google+*, https:// plus.google.com/+StephenDownes/posts/LEwaKxL2MaM, accessed June 1, 2015.

37. Hew and Cheung, "Students' and Instructors' Use."

38. George Siemens. 2005. "Connectivism: A Learning Theory for the Digital Age," *International Journal of Instructional Technology And Distance Learning* 2 (1): 3–10.

39. Ferdig, "What Massive Open Online Courses Have to Offer," 4.

40. Hew and Cheung, "Students' and Instructors' Use."

41. Dave Cormier and George Siemens, 2010, "Through the Open Door: Open Courses as Research, Learning, and Engagement," *Educause Review* 45 (4): 31–38, https://net.educause.edu/ir/library/pdf/ERM1042.pdf, accessed June 1, 2015.

42. Ferdig, "What Massive Open Online Courses Have to Offer."

43. Osvaldo C. Rodriguez, 2012, "MOOCs and the AI-Stanford Like Courses: Two Successful and Distinct Course Formats for Massive Open Online Courses," *European Journal of Open, Distance and E-Learning*, 11, http://www.eurodl.org/?p=arch ives&year=2012&halfyear=2&article=516, accessed June 1, 2015.

44. Ferdig, "What Massive Open Online Courses Have to Offer."

45. Li Yuan and Stephen Powell, 2013, *MOOCs and Open Education: Implications for Higher Education*, Centre for Educational Technology and Interoperability Standards, http://publications.cetis.ac.uk/2013/667, accessed June 1, 2015.

46. Yvonne Belanger and Jessica Thornton, 2013, "Bioelectricity: A Quantitative Approach Duke University's First MOOC," *Hdl.Handle.Net*. http://hdl.handle .net/10161/6216. Accessed June 1, 2015.

47. Ry Rivard. 2013. "Measuring the MOOC Dropout Rate," *Inside Higher Ed*, https://www.insidehighered.com/news/2013/03/08/researchers-explore-who-taking -moocs-and-why-so-many-drop-out, accessed June 1, 2015.

48. Jimmy Daily, 2014, "HarvardX's and MITx's MOOC Data Visualized and Mapped," *EdTech Magazine*, http://www.edtechmagazine.com/higher/article/2014/02 /harvardxs-and-mitxs-mooc-data-visualized-and-mapped, accessed June 1, 2015.

49. Li and Powell, "MOOCs and Open Education."

50. Katy Jordan, 2014, "Initial Trends in Enrolment and Completion of Massive Open Online Courses," *The International Review of Research in Open and Distributed Learning* 15:1, http://www.irrodl.org/index.php/irrodl/rt/printerFriendly/1651/2774, accessed June 1, 2015. *See also* Yuan and Powell, "MOOCs and Open Education."

51. Laura Pappano, 2012, "Massive Open Online Courses are Multiplying at a Rapid Pace," *Nytimes.Com*, http://www.nytimes.com/2012/11/04/education/edlife /massive-open-online-courses-are-multiplying-at-a-rapid-pace.html?pagewanted=all& _r=2&, accessed June 1, 2015.

52. Li and Powell, "MOOCs and Open Education."

53. Audrey Watters, 2013, "Top Ed-Tech Trends of 2013: MOOCs and Anti-MOOCs," *Hack Education*, http://hackeducation.com/2013/11/29/top-ed-tech-trends-2013-moocs/, accessed June 1, 2015.

54. Li and Powell, "MOOCs and Open Education."

55. Hew and Cheung, "Students' and Instructors' Use."

56. Hanan Khalil and Martin Ebner, 2014, "MOOCs Completion Rates and Possible Methods to Improve Retention—A Literature Review," In *Proceedings of World Conference on Educational Multimedia, Hypermedia and Telecommunications* 1236–44. Chesapeake, VA: AACE.

57. Allen and Seaman, "Grade Change."

58. Hew and Cheung, "Students' and Instructors' Use."

59. Mozilla, "Open Badges," http://openbadges.org/, accessed June 1, 2015.

60. Matthew LeBar, 2014, "MOOCs—Completion Is Not Important," *Forbes*, http://www.forbes.com/sites/ccap/2014/09/16/moocs-finishing-is-not-the-important-part/, accessed June 1, 2015.

61. Ferdig, "What Massive Open Online Courses Have to Offer."

62. Ibid., at 8.

63. Ibid.

64. Rebecca Vivian, Katrina Falkner, and Nickolas Falkner, 2014, "Addressing the Challenges of a New Digital Technologies Curriculum: MOOCs As a Scalable Solution for Teacher Professional Development," *Research in Learning Technology* 22 (0). doi:10.3402/rlt.v22.24691.

65. Jonathan Ress, 2014, "I Am No Longer Anti-MOOC," https://moreorless bunk.wordpress.com/2014/06/06/i-am-no-longer-anti-mooc/, accessed June 1, 2015.

66. Rosanna Tamburri, 2014, "An Interview with Canadian MOOC Pioneer George Siemens," *University Affairs*, http://www.universityaffairs.ca/features/feature-article/an-interview-with-canadian-mooc-pioneer-george-siemens/, accessed June 1, 2015.

67. Kalfas, telephone conversation.

68. Code.org, 2014, "About Us," http://code.org/about, accessed December 14.

69. Ted.com, 2014, "Our Organization | About | TED," http://www.ted.com /about/our-organization, accessed December 14.

70. Georg Szalai, 2015, "Google Chairman Eric Schmidt: 'The Internet Will Disappear'," *The Hollywood Reporter*, http://www.hollywoodreporter.com/news/google -chairman-eric-schmidt-internet-765989, accessed June 1, 2015.

71. Kalfas, telephone conversation.

BIBLIOGRAPHY

Allen, I. Elaine, and Jeff Seaman. 2014. *Grade Change: Tracking Online Education in the United States*. Babson Survey Research Group and Quahog Research Group, LLC. http://www.onlinelearningsurvey.com/reports/gradechange.pdf. Accessed June 1, 2015.

Becker, Samantha, Michael D. Crandall, Karen E Fisher, Bo Kinney, Carol Landry, and Anita Rocha. 2010. *Opportunity for All: How The American Public Benefits from*

Internet Access at U.S. Libraries. Washington, DC: Institute of Museum and Library Services. http://www.imls.gov/assets/1/assetmanager/opportunityforall.pdf. Accessed June 1, 2015.

Belanger, Yvonne, and Jessica Thornton. 2013. "Bioelectricity: A Quantitative Approach Duke University's First MOOC." *Hdl.Handle.Net*. http://hdl.handle.net/10161/6216. Accessed June 1, 2015.

Bergmann, Jonathan, and Aaron Sams. 2012. *Flip Your Classroom*. Eugene, OR: International Society for Technology in Education.

Campaign for the Future of Higher Education. 2013. "The Promises of Online Higher Education: Access—Campaign for the Future of Higher Education." http://futureofhighered.org/promises-online-higher-education-access-2/. Accessed June 1, 2015.

Code.org. 2014. "About Us." http://code.org/about. Accessed December 14.

Cormier, Dave, and George Siemens. 2010. "Through the Open Door: Open Courses As Research, Learning, and Engagement." *Educause Review* 45 (4): 31–38. https://net.educause.edu/ir/library/pdf/ERM1042.pdf. Accessed June 1, 2015.

Daily, Jimmy. 2014. "HarvardX's and MITx's MOOC Data Visualized and Mapped." *EdTech Magazine*. http://www.edtechmagazine.com/higher/article/2014/02/harvardxs-and-mitxs-mooc-data-visualized-and-mapped. Accessed June 1, 2015.

Downes, Stephen. 2013. "What the 'x' in 'xMOOC' Stands for." *Google+*. https://plus.google.com/+StephenDownes/posts/LEwaKxL2MaM. Accessed June 1, 2015.

Drake, Bruce. 2014. "Americans with Just Basic Cell Phones Are a Dwindling Breed." *Pew Research Center*. http://www.pewresearch.org/fact-tank/2014/01/09/americans-with-just-basic-cell-phones-are-a-dwindling-breed/. Accessed June 1, 2015.

Fendelman, Adam. "How Are Cell Phones Different from Smartphones?" *About Tech*. http://cellphones.about.com/od/coveringthebasics/qt/cellphonesvssmartphones.htm. Accessed June 1, 2015.

Ferdig, Rick E. 2013. *What Massive Open Online Courses Have to Offer K–12 Teachers and Students*. Lansing: Michigan Virtual Learning Research Institute. http://media.mivu.org/institute/pdf/mooc_report.pdf. Accessed June 1, 2015.

Fox, Susannah, and Lee Rainie. 2014. "The Web at 25 in the U.S." *Pew Research Center's Internet & American Life Project*. http://www.pewinternet.org/2014/02/27/the-web-at-25-in-the-u-s/. Accessed June 1, 2015.

Hew, Khe Foon, and Wing Sum Cheung. 2014. "Students' and Instructors' Use of Massive Open Online Courses (MOOCs): Motivations and Challenges." *Educational Research Review* 12: 45–58. doi:10.1016/j.edurev.2014.05.001.

Hill, Phil. 2014. "Clarification: No, There Aren't 7.1 Million Students in US Taking at Least One Online Class." *E-Literate*. http://mfeldstein.com/clarification-arent-7-1-million-students-us-taking-least-one-online-class/. Accessed June 1, 2015.

Jordan, Katy. 2014. "Initial Trends in Enrolment and Completion of Massive Open Online Courses." *The International Review of Research in Open and Distributed Learning* 15:1. http://www.irrodl.org/index.php/irrodl/rt/printerFriendly/1651/2774. Accessed December 14, 2014.

Kalfas, Elizabeth, telephone conversation, November 20, 2014.

Khalil, Hanan and Martin Ebner. 2014. "MOOCs Completion Rates and Possible Methods to Improve Retention—A Literature Review." In *Proceedings of World Conference on Educational Multimedia, Hypermedia and Telecommunications* 1236–44. Chesapeake, VA: AACE.

Kolowich, Steve. 2014. "Exactly How Many Students Take Online Courses?" *The Chronicle of Higher Education.* http://chronicle.com/blogs/wiredcampus/exactly-how-many-students-take-online-courses/49455. Accessed June 1, 2015.

Kolowich, Steve. 2013. "The Professors Who Make the MOOCs." *The Chronicle of Higher Education.* http://chronicle.com/article/The-Professors-Behind-the-MOOC/137905/==overview. Accessed June 1, 2015.

LeBar, Matthew. 2014."'MOOCs—Completion Is Not Important." *Forbes.* http://www.forbes.com/sites/ccap/2014/09/16/moocs-finishing-is-not-the-important-part/. Accessed June 1, 2015.

Madden, Mary, Amanda Lenhart, Maeve Duggan, Sandra Cortesi, and Urs Gasser. 2013. "Teens and Technology 2013." *Pew Research Center's Internet & American Life Project.* http://www.pewinternet.org/2013/03/13/teens-and-technology-2013/. Accessed June 1, 2015.

Mozilla. "Open Badges." http://openbadges.org/. Accessed June 1, 2015.

Pappano, Laura. 2012. "Massive Open Online Courses Are Multiplying at a Rapid Pace." *Nytimes.com.* http://www.nytimes.com/2012/11/04/education/edlife/massive-open-online-courses-are-multiplying-at-a-rapid-pace.html?pagewanted=all&_r=2&. Accessed June 1, 2015.

Pew Research Center's Internet & American Life Project. 2013. "Mobile Technology Fact Sheet." http://www.pewinternet.org/fact-sheets/mobile-technology-fact-sheet/. Accessed June 1, 2015.

Purcell, Kristen, and Lee Rainie. 2014. "Americans Feel Better Informed Thanks to the Internet." *Pew Research Center's Internet & American Life Project.* http://www.pewinternet.org/2014/12/08/better-informed/. Accessed June 1, 2015.

Ress, Jonathan. 2014. "I Am No Longer Anti-MOOC." https://moreorlessbunk.wordpress.com/2014/06/06/i-am-no-longer-anti-mooc/. Accessed June 1, 2015.

Rivard, Ry. 2013. "Measuring the MOOC Dropout Rate." *Inside Higher Ed.* https://www.insidehighered.com/news/2013/03/08/researchers-explore-who-taking-moocs-and-why-so-many-drop-out. Accessed June 1, 2015.

Rodriguez, Osvaldo C. 2012. "MOOCs and the AI-Stanford Like Courses: Two Successful and Distinct Course Formats for Massive Open Online Courses." *European Journal of Open, Distance and E-Learning,* 11. http://www.eurodl.org/ ?p=archives&year=2012&halfyear=2&article=516. Accessed June 1, 2015.

Siemens, George. 2005. "Connectivism: A Learning Theory for the Digital Age." *International Journal of Instructional Technology and Distance Learning* 2 (1): 3–10.

Smith, Aaron. 2014. "Older Adults and Technology Use." *Pew Research Center's Internet & American Life Project.* http://www.pewinternet.org/2014/04/03/older-adults-and-technology-use/. Accessed June 1, 2015.

Strong, Robert A. 2014. "Strong: Miics, Not MOOCs, at Washington and Lee." *Richmond.Com.* http://www.richmond.com/opinion/their-opinion/columnists-blogs

/guest-columnists/strong-miics-not-moocs-at-washington-and-lee/article_c9efafa6
-aafe-56a1-a67e-32473c30cb35.html. Accessed June 1, 2015.

Strover, Sharon. 2014. "The US Digital Divide: A Call for a New Philosophy." *Critical Studies in Media Communication* 31 (2): 114–22. doi:10.1080/15295036.2014
.922207.

Szalai, Georg. 2015. "Google Chairman Eric Schmidt: 'The Internet Will Disappear.'" *The Hollywood Reporter.* http://www.hollywoodreporter.com/news/google
-chairman-eric-schmidt-internet-765989. Accessed June 1, 2015.

Tamburri, Rosanna. 2014. "An Interview with Canadian MOOC Pioneer George
Siemens." *University Affairs.* http://www.universityaffairs.ca/features/feature-article
/an-interview-with-canadian-mooc-pioneer-george-siemens/. Accessed June 1,
2015.

Ted.com. 2014. "Our Organization | About | TED." http://www.ted.com/about/our
-organization. Accessed December 14

UN.org. 2015. "United Nations Resources for Speakers on Global Issues—Education
for All (EFA)." http://www.un.org/en/globalissues/briefingpapers/efa/index.shtml.
Accessed June 1, 2015.

Vivian, Rebecca, Katrina Falkner, and Nickolas Falkner. 2014. "Addressing the Challenges of a New Digital Technologies Curriculum: MOOCs as a Scalable Solution
for Teacher Professional Development." *Research in Learning Technology* 22 (0).
doi:10.3402/rlt.v22.24691.

Watters, Audrey. 2013. "Top Ed-Tech Trends of 2013: MOOCs and Anti-MOOCs."
Hack Education. http://hackeducation.com/2013/11/29/top-ed-tech-trends-2013
-moocs/. Accessed June 1, 2015.

Watters, Audrey. 2012. "Top Ed-Tech Trends of 2012." *Insidehighered.Com.* https://
www.insidehighered.com/blogs/hack-higher-education/top-ed-tech-trends-2012
-moocs. Accessed June 1, 2015.

Yuan, Li, and Stephen Powell. 2013. *MOOCs and Open Education: Implications for
Higher Education.* Centre for Educational Technology and Interoperability Standards. http://publications.cetis.ac.uk/2013/667. Accessed June 1, 2015.

Zickuhr, Kathryn. 2013. "Who's Not Online and Why." *Pew Research Center's Internet
& American Life Project.* http://www.pewinternet.org/2013/09/25/whos-not
-online-and-why/. Accessed June 1, 2015.

Zickuhr, Kathryn, and Aaron Smith. 2013. "Home Broadband 2013." *Pew Research
Center's Internet & American Life Project.* http://www.pewinternet.org/2013/08/26
/home-broadband-2013/. Accessed June 1, 2015.

Part II

Democratization

7

Leaks, Whistle-Blowers, and Radical Transparency: Government Accountability in the Internet Age

Rekha Sharma

The idea of privileged information is recognized under the law in a variety of contexts. Spousal relationships are protected, as are doctor-patient, attorney-client, and clergy-penitent information exchanges.[1] The philosophy underlying these bonds is trust, and forcing one party to testify against the other would be a critical violation of that intimacy. Under the Fifth Amendment to the U.S. Constitution, one does not have to provide incriminating testimony about oneself—a measure meant to ensure due process and a fair trial. The common thread in all these forms of privileged information is that some speech must be protected to promote meaningful communication within a society.

Individuals shape their societies via dialogues in which information, ideas, and meaning can be shared. The ideals of free speech and a free press are keystones of American democracy because they catalyze and enrich the debates necessary for effective communication. To bring issues of public interest to light and to foster discussion about the actions of government, journalists sometimes use information from confidential sources who provide information "on background," "off the record," or in some other form of agreement that guarantees anonymity. Attribution of sources and, by extension, accountability of news organizations are important for readers to be able to judge the credibility of the news they receive. A promise of anonymity, however, sometimes is necessary to protect whistle-blowers, witnesses, and others who divulge information to the media—thus endangering their careers, safety, or reputations. Without individuals taking such risks, it often would be difficult for media to serve their role as the fourth estate of government, a check on the powers of the executive, legislative, and judicial branches of government at local, state, and national levels.

Journalists continue to wrestle with inconsistent legal protections for those who grant anonymity, however, as well as the ethical quandaries of reporting information provided by confidential sources. New communication technologies have exacerbated these dilemmas in a variety of ways, prompting news organizations to reexamine their policies and principles on this topic. Nuances exist when it comes to protecting sources when much that is available online already is anonymous (e.g., pseudonyms for bloggers, unidentified commenters, dubious or unauthorized links, source authenticity).

Compounding credibility issues is the notion of a "fifth estate" that serves as a watchdog over both government officials and the media. Bloggers and citizen journalists have been responsible for propagating dubious information, but they also have unearthed major stories and have used technology to expose the machinations of the powerful in unprecedented ways. Controversies such as the WikiLeaks disclosures have raised questions about the legal and ethical parameters of content creators unfettered by technological limitations. In turn, government officials have employed legislation such as the Patriot Act and the Espionage Act to prosecute people who have disclosed confidential information. Advances in technology continue to upset the status quo of who has access to the means of production and dissemination of news and opinion. Internet users should understand the level of anonymity that the virtual world truly offers, and people should understand and influence the mechanisms in place for maintaining transparency of government and the allocation of power in society.

This chapter addresses legal, ethical, and practical issues associated with anonymous sourcing in journalism, and how technological advances have complicated these issues—including the protection of whistle-blowers as well as the strategic leaking of information by anonymous government sources. Additionally, it explores the fifth estate of journalism (i.e., bloggers, citizen journalists), which serves as watchdog both of the government and of traditional media. Finally, it bridges prevalent cases from the past with contemporary controversies, providing possible direction for future research regarding anonymity, transparency, and news in a digital environment.

ANONYMITY VERSUS TRANSPARENCY IN THE NEWSGATHERING PROCESS

In the 1960s, many journalists began to doubt their ability to serve merely as objective gatekeepers and instead adopted the professional model of journalistic advocacy.[2] Under this framework, the journalist's guarantee of confidentiality to a source was of prime importance because it was a necessary condition of being able to collect and reveal information whenever it would serve the public or give voice to marginalized or suppressed groups.[3, 4] Since then, the practice of pledging confidentiality to sources has been utilized in newsrooms, sometimes with discretion and sometimes gratuitously.[5]

When former *Washington Post* editor Ben Bradlee tried to ban unnamed sources, his reporting staff believed they were so disadvantaged that the policy only lasted two days.[6, 7] Of course, Bradlee's experiment was more than 40 years ago and was implemented just before the Watergate scandal, in which FBI deputy director Mark Felt provided confidential leads that enabled Bob Woodward and Carl Bernstein to unravel the conspiracy within the Nixon administration.[8, 9, 10]

Known for decades only by the pseudonym "Deep Throat," Felt has become one of journalism's most famous examples of the necessity of anonymous sources. Ironically, the Watergate scandal stemmed from Nixon's attempts to smear the reputation of another whistle-blower, Daniel Ellsberg, who gave the infamous "Pentagon Papers" to *The New York Times* and *The Washington Post* to protest the Vietnam War.[11] For a separate account of the final days of Nixon's presidency, Woodward and Bernstein along with their research assistants interviewed 394 people "'on background,' with the understanding that the information would be used only if the source's identity remained confidential."[12] A promise of anonymity was essential to obtaining the depth of information required for Woodward and Bernstein's book, therefore no footnotes were included.[13]

Aside from the need to guarantee anonymity to gather information, journalists also must consider ethical responsibilities of granting confidentiality and the legal consequences of failing to protect anonymous sources. If promises of confidentiality are broken, then victims of disclosure may sue to collect damages. In *Cohen v. Cowles Media Co.* (1991), the Supreme Court ruled in favor of Dan Cohen, a public relations executive and political spokesperson who was fired after two newspapers revealed him as the source of information about a gubernatorial candidate's shoplifting conviction.[14] The newspapers' editors decided that readers needed to know that the disclosure came from the opposing candidate's camp. But in a 5–4 decision, the Court decided that Minnesota's law was constitutional. This meant that the First Amendment would not preclude the state from remedying injustices created by broken promises.[15]

As Christians, Fackler, Rotzoll, and McKee (2001) pointed out, reporters must be able to approach sources with skepticism and cooperation, becoming adversarial or friendly when the situation calls for it, stating: "[i]f newspeople become too intimate with important men and women, they lose their professional distance or develop unhealthy biases protecting them. However, to the degree that powerful sources are not cultivated and reporters establish no personal connections, the inside nuance and perspective may be lost."[16] Berry (2009) argued that although the use of anonymous sources can be valuable, it also can undermine a story's credibility.[17] Berry advised that investigative journalists verify information provided by anonymous sources against multiple "on-the-record" sources or documents and also employ probing interviewing techniques to determine how anonymous sources might have obtained the information they shared with the media. Most journalism industry codes of ethics also address anonymity, cautioning reporters to identify sources fully and completely whenever possible but at the same time to honor all promises of confidentiality.

Although some news organizations allow reporters to keep the identities of confidential sources completely secret, most require editors to be involved to help judge the credibility of the information being provided.[18] Editorial oversight of reporters' use of anonymous sources was deemed especially important after a scandal involving former *Washington Post* reporter Janet Cooke. In what would become a cautionary tale for future generations of journalists, Cooke won a Pulitzer Prize in 1981 for a story about an eight-year-old heroin addict named "Jimmy." After 11 hours of interrogation by her editors, Cooke admitted that she had fabricated the entire account.[19] Financial journalist Dan Dorfman was fired from *Money* magazine in 1996 after he refused to reveal his confidential sources to his editors, who stated that they could not continue to employ him if they could not evaluate the quality of his reporting.[20]

Other problems at major newspapers have caused editors to reevaluate newsroom policies on the use of anonymous sources. For instance, the standards editor of *The New York Times* randomly checks stories containing information provided by anonymous sources each week to make sure another editor knows the name of the source.[21] This practice went into effect after a *Times* investigation revealed that reporter Jayson Blair had fabricated or plagiarized his coverage of several news events.[22] *USA Today* reporter Jack Kelley resigned after an investigation showed that he had fabricated and plagiarized stories for more than a decade. Investigators concluded that Kelley had exploited close relationships with editors, who did not inspect his work. Although rules limiting the use of anonymous sources were in place at the newspaper, editorial scrutiny of Kelley's stories was found to be lax.[23] Moreover, the paper's rules on anonymous sources contained a loophole stating that reporters did not always have to divulge their sources to an editor.[24]

Not only have some reporters abused the practice of anonymous sourcing, but sources themselves sometimes have employed strategic leaks to spin a story favorably or to damage an opponent.[25] Sources in government sometimes leak information anonymously to journalists to gauge public support for ideas without committing to them as set policies.[26, 27] This can occur because journalists must cover the story in a way that removes identifying information about the source, omitting the context of events, comments, or issues as well as failing to disclose the source's possible motive for speaking to the press.[28] Sometimes unwillingness to go on the record can make a source seem more provocative and his or her information seem more newsworthy. According to Bennett (2001), "At times, the political message is not important enough to be guaranteed coverage if released through normal press channels or presented as a pseudo-event. If the right reporter is given a scoop based on the information, however, the chances are pretty good that the story will receive special attention."[29] In this manner, sources can orchestrate news coverage and events insidiously, with reporters playing the patsy. This practice of shifting culpability onto the media might be one motivating factor for journalists to reveal sources' identities.

An example of a possible planned leak involved the identification of Valerie Plame as a CIA operative in a column by Robert Novak of the *Chicago Sun-Times*.[30] Novak revealed this information in the context of writing about Plame's husband, Joseph C. Wilson, who was a retired diplomat sent by the Central Intelligence Agency to Africa to investigate rumors that Iraq was attempting to buy yellowcake uranium from Niger. The veracity of this claim would become important in evaluating the decision of U.S. president George W. Bush to launch a military incursion into Iraq to topple Saddam Hussein's regime. In discussing Wilson's involvement, Novak said that Plame—who worked for the CIA—recommended her husband for the mission. Her name was given to Novak by "two senior administration officials," but Novak refused to divulge the sources' identities.[31] Novak defended his decision to reveal Plame's identity by saying that although he was asked to withhold her name he never was told that naming her could endanger anyone. In an interview with CNN, Novak explained, "According to a confidential source at the CIA, Mrs. Wilson was an analyst, not a spy, not a covert operative and not in charge of undercover operators."[32]

McCann (2004) argued that, along with the Novak case, a multitude of other recent lawsuits indicated that the Justice Department had been exploiting weaknesses in media law to put pressure on reporters having confidential sources. McCann cited the U.S. attorney general's "Guidelines for Subpoenaing Members of the News Media" as a measure of protection for journalists. Although the guidelines "do not carry the force of law, [they] require that the news media subpoenas identify particular relevant information that cannot be obtained in any other way" and require that negotiations occur between the Justice Department and the journalist before a subpoena is issued against a media outlet.[33]

I. Lewis "Scooter" Libby, a former aide to Vice President Dick Cheney, was later convicted of perjury, making false statements, and obstruction of justice for concealing his role in leaking Plame's identity to the media.[34] In a separate lawsuit against Libby as well as other government officials, Plame alleged that the disclosure occurred as retaliation against her husband, Joe Wilson, a former U. S. ambassador to Iraq who criticized the Bush administration's prewar intelligence.[35] During the trial, Special Prosecutor Patrick Fitzgerald issued grand jury subpoenas to reporters from *Time The Washington Post*, and *NBC News* to compel them to reveal the source of the leak.[36] *New York Times* reporter Judith Miller served 85 days in jail in 2005 because she chose not to divulge the name of the person who exposed Plame's affiliation with the CIA.[37, 38]

In addition to ethical codes in the industry, journalists should be aware of the ongoing legal debate over how much protection reporters can extend to sources. Section 403 of the Federal Rules of Evidence allows judges "to quash subpoenas if information sought from reporters would duplicate information already available."[39] In this way, the legal system attempts to protect journalists and their sources and maintain the fairness of trial procedure delineated in the Sixth Amendment. As a matter of law, First and Sixth Amendment freedoms are equally essential. However, reporters from mainstream media outlets often are sued for refusing to testify about the identities of anonymous sources, whereas bloggers rarely are held accountable for unattributed information.[40]

The landmark case for journalists' protection of sources is *Branzburg v. Hayes.*[41] The 5–4 majority ruled against reporters who refused to testify in front of grand juries. The opinions in the case, however, have been interpreted in divergent ways—both as a protection for journalists and as a rationale for requiring them to divulge their sources.[42] Justice Byron White wrote the plurality opinion that placed the public interest in law enforcement as paramount to an uncertain burden on news organizations. He said there was no reason to believe that many sources would disappear into silence if reporters did not guarantee anonymity. Rather, White argued that sources would always seek mass media attention to reach larger audiences. White did not want to saddle courts with the responsibility of having to decide on qualified privilege claims in case after case or to define who was eligible to be called a "journalist."[43, 44]

Justice Potter Stewart wrote an influential dissenting opinion favoring First Amendment protection for journalists, and created a three-part test for justifying the infringement of their rights by officials.[45] Stewart said that to override the rights of journalists to guard their sources, the government would have to demonstrate "(1) a probable cause to believe that a reporter has information 'clearly relevant' to a specific violation of law; (2) evidence that the information sought cannot be obtained by alternative means less destructive of First Amendment values; [and] (3) 'a compelling and overriding interest in the information.'"[46]

Given the possibility that journalists could be summoned to testify about their confidential sources or to provide evidence in such cases, news organizations need to rearticulate their policies about the proper conditions for granting anonymity. In a digital environment, even records of communication such as archived e-mails, cell phone records, video outtakes, and instant messages could be stored electronically by a news organization. Carter, Franklin, and Wright (2001) prompted journalists to consider whether they could be compelled to provide information to law enforcement. Reporters and editors should understand the length of time such information could be retrieved and the level of computer security their newsrooms possess. Such considerations could prove vital for making decisions as technology spawns ethical and legal ambiguities.[47]

THE ASCENDANCE OF THE FIFTH ESTATE: CHALLENGES OF NEW TECHNOLOGY

Inevitably, news organization policies must adapt to changing social, legal, and technological influences. These forces undoubtedly will impact the relationships between journalists and their sources on a practical level, and journalists must be able to anticipate potential challenges and defend their decisions to grant or withhold promises of confidentiality.

Most news organizations require verification of information by two or three sources before printing a story, as well as including specific attribution whenever possible.[48] Attribution often serves as an important media literacy tool for critical readers to determine the source's motives for supplying the information and possible outcomes of providing the information.[49, 50] Although media scholars, practitioners, and legal experts have dealt with situations in which a reporter might know the source but refuse to divulge it, they must also confront situations in which users contribute content anonymously to media outlets or in which journalists gather information from unnamed sources online. For instance, the addition of multimedia features on cellular telephones has created an army of self-proclaimed stringers who gather content as firsthand witnesses for mainstream media outlets.

Camera phones were first launched in 2000. Since then, sales have risen sharply and the quality of images has been greatly enhanced.[51] Because of the portability and often inconspicuous nature of camera phones, amateur photojournalists can use them to capture everyday events, celebrity sightings, and breaking news.[52] Aside from posting these images—including video—to personal blogs that are accessible from their cell phones,[53] cell phone users contribute footage and photos to mainstream media outlets. For example, "citizen reporters" equipped with cell phones provided pictures of the July 7, 2005, subway and bus bombings in the United Kingdom within minutes to ITV, the BBC, Sky News, and national newspapers when transporting journalists to the scene would have been physically impossible.[54]

The ubiquity of cell phones and the usefulness of tech-savvy eyewitnesses at breaking news events should prompt mainstream news organizations to consider their editorial policies to ensure credibility, accuracy, and consistency.[55, 56] Perhaps more importantly, though, mainstream journalists must look within their organizations to determine whether their policies and information-gathering routines remain adequate in the age of citizen journalists wielding cell phones. As Stephens (2007) explained, with the proliferation of raw images, accounts, and reports available on the Internet, journalists must provide insight and interpretation rather than a regurgitation of sound bites and information to remain viable in the coming years.

> We are still very early in the evolution of the form, but surely industrious bloggers won't always need reporters to package such materials before they commence picking them apart. Mainstream journalists are making a mistake if they believe their ability to collect and organize facts will continue to make them indispensable.[57]

News organizations also should consider the credibility of information gathered from online sources such as bulletin boards, chat rooms, and blogs. Because users can post information under one or many pseudonyms, it could be difficult to identify users, let alone hold them accountable for comments, inaccuracies, or misinformation.[58] Crain (2008) raised concerns that an anonymous blogger on a media outlet's site could make disparaging comments about individuals or companies simply because he or she had an axe to grind.[59] The editorial staff of some sites moderates the electronic responses to news stories because users are able to post comments without providing their names or affiliations.[60] Also, journalists who post messages, criticisms, or questions in virtual forums must understand their news organization's policies about making their identities known to other users. A blogger from the *Los Angeles Times* and a senior editor at the *New Republic* magazine were suspended for violating their publications' ethical guidelines about identifying themselves when dealing with the public after they invented online identities—called "sock puppets" or "sock-puppet accounts" in the Internet vernacular—to spar with critics and attack other media outlets in their own blogs and on other Web sites.[61, 62] Journalists sometimes act as individuals, but they sometimes represent their organization or the media industry.[63] The protection journalists can expect to enjoy for their online communication could vary depending on the writer's role and the context of the writing.

Anonymity in the computer world has exacerbated the challenges journalists face with regard to attribution of sources and verification of information. The sophistication of technological features enables users to cloak their identities more thoroughly, thus increasing the complexity of the social, legal, and ethical landscapes as they relate to news. This is compounded by technology such as anonymous remailer programs, which forward e-mail messages to different users after removing the information about the origin of the message. Remailers could preserve the anonymity of political dissidents, activists, and informers who fear retribution yet feel compelled to expose wrongdoings or injustices in the media. The same technology, however, can be used to perpetrate cybercrimes.[64] Cowan (2008) worried that technology has allowed people to wield power without consequence or accountability.[65]

There is no guarantee of absolute anonymity thanks to expanding and ever-evolving technologies. Reporters therefore have an ethical obligation to clarify the limits of protection they can provide to a source. Nissenbaum (1999) cautioned that stripping the name from a person would not necessarily mean that he or she would be anonymous on the Internet.[66] Rather, a person could be identified through a constellation of personal data, information-gathering habits, and official records from a myriad of sources. The profits to be gained from collecting and selling such information already have caused some companies to violate individuals' privacy and some individuals to relinquish details about their personal lives willingly.[67]

Even the best-intentioned of policies might not be able to keep up with and address new technologies that provide access to information considered confidential. Given the uncertainty about how users interact with technology, Allen (1999) argued that global anonymity regulations on the Internet would not be feasible, and that policies should be based on the context of engagement between users.[68] That is, whereas a corporate whistle-blower or human rights activist might be able to justify their need for anonymity when communicating with journalists, terrorist organizations or child pornographers would not deserve the same level of protection. Anonymity decentralizes authority on the Internet, but it also limits the ability of law enforcement to regulate unprotected forms of speech, such as fraud and threats.[69] Likewise, electronic communication as a form of intellectual property can range from diary entries to purchase orders, and therefore would be afforded different levels of legal protection.[70]

Clearly, creating context-based policies is easier said than done. Allen's (1999) suggestions—which included regulating online contexts in accordance with analogous offline scenarios, developing mechanisms for policy enforcement, and determining conditions for surveillance—appeared logical on paper, but would do little to resolve actual impasses.[71] After all, law enforcement officials, journalists, and individuals have been struggling to find a balance between privacy and open communication using these very steps. Likewise, the American Association for the Advancement of Science offered suggestions such as allowing online communities to determine their own rules about anonymity.[72] Online communities often are vaguely defined,

however, and allowing them to police themselves would not really help officials to stop unlawful activity. These examples highlight the need for more
thoughtful recommendations. The process of policy building would be better
served by more detailed proposals.

Of particular note, news organizations must understand the ethical responsibilities and legal obligations of providing content as compared to providing
venues for interaction with audiences. The Good Samaritan provisions of
the 1996 Telecommunications Act, also known as the Communications
Decency Act, might offer some protection to news organizations that allow
users to post messages in electronic forums such as bulletin boards, as the
courts seem to have recognized that multimedia capabilities have altered
the traditional dissemination functions of news outlets.[73] Even Internet
service providers (ISPs) that typically were treated as common carriers not
to be held responsible for the content of messages they transmitted, however,
have begun to reevaluate their roles.[74] Internet service providers have had
to protect the privacy of their customers while cooperating with law enforcement officials to unmask the identities of criminals.[75] Thus, even if news
organizations are not required to provide identifying information about users
posting anonymously on their Web sites, the editors and publishers would
do well to consider whether reactions to legal challenges will serve to foster
conditions for transparency and dialogue or create a chilling effect on free
speech.

OPEN SECRETS: DEBATING THE MORALITY AND
LEGALITY OF HACKING AND LEAKING

In response to the terrorist attacks of September 11, 2001, the U.S. Congress passed a law to expand the intelligence-gathering powers of government
officials investigating suspected terrorists.[76] The "Uniting and Strengthening
America by Providing Appropriate Tools Required to Intercept and Obstruct
Terrorism" law came to be known more popularly as the Patriot Act.[77] This
law—along with the establishment of closed deportation hearings and military tribunals, expansion of executive privilege, reclassification of public information, and refusal to detail the government's increased surveillance
activities—marked a distinct effort by the George W. Bush administration to
withhold information from the public, ostensibly to protect homeland security.[78] For example, the president, the attorney general, and the president's
chief of staff encouraged government agencies to interpret exemptions of the
Freedom of Information Act to withhold classified, sensitive, or even previously public information.[79] Although the executive branch spearheaded these
efforts, the administration received considerable support from the legislative
and judicial branches of government.[80]

Since 9/11, whistle-blowers working in the defense and intelligence areas of government have weighed in on the morality of government actions, bringing to light issues such as flawed military equipment, abuse of prisoners, and domestic surveillance programs.[81] Government workers who attempt to point out problems to their supervisors, however, often are disciplined or penalized. According to Taylor (2014), "[M]ore than 8,700 defense and intelligence employees and contractors have filed retaliation claims with the Pentagon inspector general since the 9/11 attacks, with the number increasing virtually every year."[82] Rather than facing lengthy legal battles, professional demotion, or character assassinations, some whistle-blowers opt to take their stories to the news media rather than voice their concerns via government channels.[83]

According to David Carr (2012) and other journalists, President Barack Obama promised that his administration would enhance legal protections for federal government whistle-blowers, but it actually has been investigating and prosecuting the sources of leaks of classified information more aggressively than all prior presidential administrations.[84] These prosecutions have taken place under the aegis of the Espionage Act, which was passed by the U.S. Congress in 1917 with the aim of protecting military secrets and preventing sabotage.[85] Carr (2012) noted that John Kiriakou was charged with leaking information about Central Intelligence Agency officers, some of whom had been involved with the agency's enhanced interrogation program.[86] Carr criticized the government for prosecuting Kiriakou—a former CIA officer and a staff member on the Senate Foreign Relations Committee—merely for talking to the press, when the government had not taken any legal action against people who had authorized or engaged in waterboarding suspected terrorists.

Carr also pointed out the administration's prosecution of Thomas A. Drake, who worked for the National Security Agency and told a reporter that a software program developed in-house would be more effective, less expensive, and less intrusive than a private-sector vendor's data-monitoring program. Carr questioned whether the charges had less to do with national security concerns and more to do with preventing scrutiny of government actions and mitigating public embarrassment. Conversely, Carr said that the administration has changed its attitude about government secrecy whenever releasing information was deemed advantageous, stating, "Reporters were immediately and endlessly briefed on the 'secret' operation that successfully found and killed Osama bin Laden. And the drone program in Pakistan and Afghanistan comes to light in a very organized and systematic way every time there is a successful mission."[87]

In 2010, the nonprofit, online whistle-blowing organization WikiLeaks posted a video of a U.S. helicopter gunship in Iraq killing civilians and journalists and then published 90,000 classified documents about U.S. military action in Afghanistan.[88, 89] The site also released 40,000 documents related to the Iraq War and 250,000 U.S. State Department files concerning correspondence with diplomats.[90] WikiLeaks and its founder, Julian Assange, vociferously advocate for government transparency; users can upload documents to the site anonymously, making the site a haven for whistle-blowers.[91] The U.S. government investigated the leaks, determining the source to be Private First Class Bradley Manning, a U.S. Army intelligence analyst.[92] Manning was convicted on multiple counts—including violating the Espionage Act—and was sentenced to 35 years in prison.[93] As the government worked to establish a definitive connection between Manning and Assange, the WikiLeaks founder was confronted with an arrest warrant for sexual assault and has attempted to avoid extradition to Sweden for years. Some people suspect that the assault charges actually were a retaliation measure for the leaks.[94, 95]

The government has paid private-sector whistle-blowers substantial rewards for information about corporate malfeasance, but blowing the whistle on the clandestine activities of government officials and agencies often is regarded less favorably.[96] Although some whistle-blowers are lauded for their motives and actions, other whistle-blowers are labeled as "traitors" or "snitches" and are harshly criticized.[97] Linda Tripp—who revealed President Bill Clinton's affair with White House intern Monica Lewinsky—considered herself to be a whistle-blower,[98] as did Edward Snowden, a former NSA contractor who disclosed classified information about the agency's extensive electronic surveillance activities.[99] Snowden explained that he was motivated by principle rather than profit, seeking to preserve privacy and opportunities for creativity and intellectual exploration. He added that he evaluated each document he released to refrain from harming people and only shared them with journalists who would use discretion in assessing each disclosure's contribution to the public interest.[100] Because Snowden expected to be demonized by the media and punished by the government for his actions, however, he fled the United States.[101]

News organizations, legislators, and the courts have taken steps to protect source anonymity in selected journalistic contexts. For example, 49 states and the District of Columbia have some form of common law, statutory, or rule-based protections for journalists to keep them from having to divulge the names of confidential sources.[102] To provide some legal continuity at the national level for journalists, the U.S. House of Representatives approved the Free Flow of Information Act of 2009.[103] The bill addressed the conditions under which federal courts could compel journalists to testify about anonymous sources. U.S. attorney general Michael B. Mukasey (2008) voiced concern that the bill would encourage leaks that would damage national security and hinder law enforcement.[104] Eventually, that bill was stalled after being assigned to the U.S. Senate Committee on the Judiciary.[105, 106, 107] The Senate Judiciary Committee approved it in 2009, but it never received a floor vote because of controversy over the WikiLeaks disclosures.[108]

The idea of implementing a federal shield law was resurrected, however, after federal investigators secretly subpoenaed phone and fax records from Associated Press reporters and bureaus to investigate who might have leaked information to the press about a CIA operation that thwarted a bombing plot by Al-Qaeda in Yemen.[109] In 2013, President Obama's Senate liaison asked Democratic Senator Charles Schumer to reintroduce a version of the Free Flow of Information Act, perhaps to mitigate negative news coverage about the Justice Department's failure to notify the Associated Press of the seizure of its phone records.[110] The White House request came hours before Attorney General Eric Holder was to go before a House Judiciary Committee hearing concerning the matter.[111] Holder voiced support for a shield law that would protect journalists, but not for individuals who leaked classified information. Holder said in his testimony that "[t]he focus should be on those people who break their oath and put the American people at risk, not reporters who gather this information."[112]

President Obama, along with U.K. prime minister David Cameron, announced their intentions to create a cybersecurity unit, allowing law enforcement and intelligence agencies to collaborate to combat threats from hackers.[113] The government initiative arose from a growing recognition of the vulnerability of technological information systems, pointedly publicized after the hacking of Sony Pictures Entertainment.[114] The effort also buttressed the governments' arguments that intelligence agencies required access to encrypted communication on the Internet and in social media to be able to detect and avert terrorist plots as well as to secure economic infrastructures.[115] Intensifying cybersecurity efforts, however, must be weighed against civil liberties, privacy rights, and the potential value of anonymity for dissenting voices in a democratic society.

CONCLUSION

The practice of granting anonymity to sources never has been a simple endeavor for journalists because it creates tensions between competing professional and personal values. Conferring anonymity to a source often has meant the sacrifice of transparency for the sake of gaining information. Therefore, the decision to withhold a source's identity must be considered carefully. New communication technologies have added a more challenging dimension to this assessment in that computer-mediated communication has altered the social, legal, and ethical environment as well as the roles of news organizations and audiences.

This chapter serves as an introduction to some broad and complex topics. Only some of the issues regarding the use of anonymous sources in news organizations are covered, and as users, providers, and newsgatherers develop new uses for new technologies, more questions and puzzles will emerge as worthy areas of study. Future research should include analysis of journalistic organizations, including policies specific to individual media outlets as well as codifications of professional subfields such as print journalism, broadcast journalism, public relations, and advertising. As industry associations revise their ethical guidelines, they must address the ways that technological convergence impacts their stances on anonymity, transparency, and other professional norms.

Practitioners should collaborate with researchers to develop more detailed, cogent recommendations for legislation. More distinct protections are needed to balance the right of the public to access information with the right of individuals to retain their personal privacy without tying the hands of law enforcement officials. Civil liberties and social order can be preserved equitably and consistently, perhaps through the passage of a national shield law that would supersede state-by-state regulations. By bridging the divides between journalism, political science, law, computer technology, sociology, and psychology, scholars can lead the way toward a new understanding of an information society. The ever-changing nature of the Internet and ever-expanding access to interactive communication technologies require vigilance on the part of ethicists and continued inquiry from communication researchers.

NOTES

1. T. Barton Carter, Mark A. Franklin, and Jay B. Wright (2001), *The First Amendment and the Fourth Estate: The Law of Mass Media* (8th ed.), New York: Foundation Press.

2. Morris Janowitz (1975), "Professional Models in Journalism: The Gatekeeper and the Advocate," *Journalism Quarterly* 52, 618–26.

3. Ibid.

4. Joseph Straubhaar and Robert LaRose (1997), *Communications Media in the Information Society*, Belmont, CA: Wadsworth.

5. Margaret Sullivan (2013, October 12), "The Disconnect on Anonymous Sources," *New York Times.* http://www.nytimes.com/2013/10/13/opinion/sunday/the-public-editor-the-disconnect-on-anonymous-sources.html?_r=0. Accessed January 19, 2015.

6. Ben Bagdikian (2005, August/September), "When the Post Banned Anonymous Sources," *American Journalism Review* 27, 33.

7. R. J. Haiman (1999), *Best Practices for Newspaper Journalists*, Arlington, VA: The Freedom Forum.

8. Ben Bagdikian (2005, August/September), "When the Post Banned Anonymous Sources," *American Journalism Review* 27, 33.

9. Matt Carlson (2010), "Embodying Deep Throat: Mark Felt and the Collective Memory of Watergate," *Critical Studies in Media Communication* 27, 235–50. doi: 10.1080/15295030903583564

10. Bob Woodward (2006), *"The Secret Man: The Story of Watergate's Deep Throat*, New York: Simon & Schuster.

11. Thomas L. Tedford and Dale A. Herbeck (2009), *Freedom of Speech in the United States* (6th ed.), State College, PA: Strata Publishing.

12. James W. Davidson and M. H. Lytle (1986), *After the Fact: The Art of Historical Detection* (2nd ed.), New York: Alfred A. Knopf.

13. Ibid.

14. *Cohen v. Cowles Media Company*, 501 U.S. 663 (1991).

15. Kent R. Middleton, Robert Trager, and Bill F. Chamberlin (2000), *The Law of Public Communication* (5th ed.), New York: Addison Wesley Longman.

16. Clifford G. Christians, Mark Fackler, Kim B. Rotzoll, and K. B. McKee (2001), *Media Ethics: Cases and Moral Reasoning* (6th ed.), New York: Addison Wesley Longman.

17. Stephen J. Berry (2009), *Watchdog Journalism: The Art of Investigative Reporting*, New York: Oxford University Press.

18. Christians, Fackler, Rotzoll, and McKee (2001), *Media Ethics*.

19. Howard Kurtz (1997b), "Janet Cooke's Untold Story: 15 Years after Her Famous Hoax the Disgraced Reporter Speaks," in *Messages 4: The Washington Post Media Companion*, edited by T. Beell, 204–6, Boston, MA: Allyn & Bacon.

20. Howard Kurtz (1997a), "Ethics and the Mass Media: A Tipster Takes a Tumble; Dan Dorfman, Embroiled in a Controversy over Sources, Is Fired by Money Magazine," in *Messages 4: The Washington Post Media Companion*, edited by T. Beell, 204–6, Boston, MA: Allyn & Bacon.

21. Ibid.

22. Byron Calame (2006, June 18), "Preventing a Second Jayson Blair," *New York Times*, 12.

23. Kevin McCoy (2004, April 24), "Report: Newsroom Culture Enabled Kelley," *USA Today*, 10A.

24. Ibid.

25. James Cowan (2008, December 31), "The Orwellian Power of Anonymity; the Web Has Changed the Rules of Politics and Journalism," *National Post*, A6.

26. Haiman, *Best Practices*.

27. Straubhaar and LaRose, *Communications Media*.

28. W. Lance Bennett (2001), *News: The Politics of Illusion* (4th ed.), New York: Addison Wesley.

29. Ibid., 131.

30. Robert Novak (2003, July 14), "The Mission to Niger" [Editorial]. *Chicago Sun-Times*, 31.

31. Ibid.

32. David Ensor (2003, October 1), "Novak: 'No Great Crime' with Leak," CNN.com. http://www.cnn.com

33. Tom McCann (2004, October), "Journalist Groups See Surge in Legal Attacks on Privilege," *Chicago Lawyer* 90.

34. Carol D. Leonnig and Michael D. Shear (2007, March 8), "Prison May Be Long Way Off for Libby; Appeals Could Keep Former Cheney Aide Free Until after 2008 Election," *Washington Post*, A4.

35. Eric M. Weiss and Charles Lane (2006, July 14), "Vice President Sued by Plame and Husband; Ex-CIA Officer Alleges Leak of Her Name Was Retaliatory," *Washington Post*, A3.

36. McCann, "Journalist Groups."

37. Carlson, "Embodying Deep Throat."

38. Jennifer Harper (2009, September 10), "Journalist 'Shield Law' Gains Steam in Senate," *Washington Times*. http://www.washingtontimes.com

39. Middleton, Trager, and Chamberlin, *The Law of Public Communication*.

40. Andrew Keen (2007), *The Cult of the Amateur: How Today's Internet Is Killing Our Culture*, New York: Doubleday.

41. *Branzburg v. Hayes*, 408 U.S. 665 (1972).

42. Middleton, Trager, and Chamberlin, *The Law of Public Communication*.

43. Carter, Franklin, and Wright, *The First Amendment and the Fourth Estate*.

44. Middleton, Trager and Chamberlin, *The Law of Public Communication*.

45. *Branzburg v. Hayes*, 408 U.S. 665 (1972).

46. Middleton, Trager, and Chamberlin, *The Law of Public Communication*.

47. Carter, Franklin, and Wright, *The First Amendment and the Fourth Estate*.

48. Christians, Fackler, Rotzoll, and McKee, *Media Ethics*.

49. Rance Crain (2008, March 10), "I'm Going on the Record When It Comes to Anonymous Bloggers," *Advertising Age*, 14.

50. Art Silverblatt (1995), *Media Literacy: Keys to Interpreting Media Messages*, Westport, CT: Praeger.

51. L. Srivastava (2005), "Mobile Phones and the Evolution of Social Behaviour," *Behaviour & Information Technology* 24, 111–129, doi: 10.1080/01449290512331321910.

52. Ibid.

53. Angelo Fernando (2006, May-June), "That Third Screen—In Your Pocket!," *Communication World* 23, 11–12.

54. Emma Hall (2005), "Technology Makes Storytellers of Citizens," *Advertising Age* 76 (6). Communication & Mass Media Complete database, February 18, 2007.

55. Ken Kerschbaumer (2005, January 24), "Dialing for News," *Broadcasting & Cable* 135, 43–44.

56. Barb Palser (2001, September), "I Want My Breaking News," *American Journalism Review* 23, 66.

57. Mitchell Stephens (2007, January/February), "Beyond News," *Columbia Journalism Review* 45, 34–39.

58. Jane B. Singer (1996), "Virtual Anonymity: Online Accountability and the Virtuous Virtual Journalist," *Journal of Mass Media Ethics* 11, 95–106.

59. Crain, "I'm Going on the Record."

60. Ibid.

61. *New York Times* Editorial. (2006, September 13), "Sock Puppet Bites Man," *New York Times*, http://www.nytimes.com

62. Andrew Keen, *The Cult of the Amateur.*

63. Jane B. Singer, *Virtual Anonymity.*

64. Yaman Akdeniz (2002), "Anonymity, Democracy, and Cyberspace," *Social Research* 69, 223–37.

65. James Cowan, *The Orwellian Power of Anonymity.*

66. Helen Nissenbaum (1999), "The Meaning of Anonymity in an Information Age," *The Information Society* 15, 141–44. doi: 10.1080/019722499128592

67. Michael Bugeja (2005), *Interpersonal Divide: The Search for Community in a Technological Age*, New York: Oxford University Press.

68. Christina Allen (1999), "Internet Anonymity in Contexts," *The Information Society* 15, 145–46, doi: 10.1080/019722499128600.

69. Peter Wayner (1999), "Technology for Anonymity: Names by Other Nyms, *The Information Society* 15, 91–97," doi: 10.1080/019722499128556.

70. A. Michael Froomkin (1999), "Legal Issues in Anonymity and Pseudonymity. *The Information Society* 15, 113–27, doi: 10.1080/019722499128574

71. Christina Allen, *Internet Anonymity in Contexts.*

72. Al Teich, Mark S. Frankel, Rob Kling, and Ya-Ching Lee (1999), "Anonymous Communication Policies for the Internet: Results and Recommendations of the AAAS Conference," *The Information Society* 15, 71–77, doi: 10.1080 /019722499128538.

73. Madeline Johnson and Betsy D. Gelb (2002), "Cyber-Libel: Policy Trade-Offs," *Journal of Public Policy and Marketing* 21, 152–59.

74. Victoria S. Ekstrand (2003), "Unmasking Jane and John Doe: Online Anonymity and the First Amendment," *Communication Law and Policy* 8, 405–27, doi: 10.1207/S15326926CLP0804_02

75. Kara Swisher and John Schwartz (1997), "Walking the Beat in Cyberspace: On-Line Services Struggle with How to Combat Smut and Protect Privacy," in *Messages 4: The Washington Post Media Companion*, edited by T. Beell, 315–17, Boston, MA: Allyn & Bacon.

76. Tedford and Herbeck, *Freedom of Speech in the United States.*

77. Ibid.

78. Paul Haridakis (2006), "Citizen Access and Government Secrecy," *Saint Louis University Public Law Review* 25, 3–32.

79. Ibid.

80. Ibid.

81. Marisa Taylor (2014, December 30), "Intelligence, Defense Whistleblowers Remain Mired in Broken System," *McClatchy DC*, http://www.mcclatchydc.com/static /features/Whistleblowers/Whistleblowers-remain-mired-in-broken-system.html, accessed January 18, 2015.

82. Ibid.

83. Ibid.

84. David Carr (2012, February 7), "Blurred Line between Espionage and Truth," *New York Times*, B1, B7.

85. Tedford and Herbeck, *Freedom of Speech in the United States*.

86. David Carr, "Blurred Line between Espionage and Truth."

87. Ibid.

88. Christian Fuchs (2011), "WikiLeaks: Power 2.0? Surveillance 2.0? Criticism 2.0? Alternative Media 2.0? A Political-Economic Analysis," *Global Media Journal—Australian Edition* 5, 1–17.

89. Dawn L. Rothe and Kevin F. Steinmetz (2013), "The Case of Bradley Manning: State Victimization, Realpolitik, and WikiLeaks," *Contemporary Justice Review* 16, 280–92. doi: http://dx.doi.org/10.1080/10282580.2013.798694.

90. Fuchs, "WikiLeaks: Power 2.0?"

91. Ibid.

92. Rothe and Steinmetz, "The Case of Bradley Manning."

93. Julie Tate (2013, August 21), "Bradley Manning Sentenced to 35 Years in WikiLeaks case," *Washington Post*. http://www.washingtonpost.com/world/national-security/judge-to-sentence-bradley-manning-today/2013/08/20/85bee184-09d0-11e3-b87c-476db8ac34cd_story.html. Accessed January 19, 2015.

94. Alan Cowell (2014, November 20), "Swedish Court Upholds Order for Arrest of Julian Assange," *New York Times*. http://www.nytimes.com/2014/11/21/world/europe/swedish-court-rejects-appeal-by-julian-assange.html. Accessed January 19, 2015.

95. Fuchs, "WikiLeaks: Power 2.0?"

96. Steven D. Solomon (2014, December 30), "Whistle-Blower Awards Lure Wrongdoers Looking to Score," *New York Times*. http://dealbook.nytimes.com/2014/12/30/whistle-blower-awards-lure-wrongdoers-looking-to-score/?_r=0. January 18, 2015.

97. Ibid.

98. Ibid.

99. Glenn Greenwald, Ewen MacAskill, and Laura Poitras (2013, June 11), "Edward Snowden: The Whistleblower Behind the NSA Surveillance Revelations," *The Guardian*, http://www.theguardian.com/world/2013/jun/09/edward-snowden-nsa-whistleblower-surveillance, accessed January 18, 2015.

100. Ibid.

101. Ibid.

102. Harper, "Journalist 'Shield Law'."

103. "Protecting Sources; A Federal Shield Law Is within Reach" (2009, April 6), *Washington Post*, A14.

104. Michael B. Mukasey (2008, June 28), "A Legislative Shield for the Press?," *Washington Times*, M12.

105. "Protecting Sources," 2009.

106. Charles Schumer (2008, July 2), "Defending Free Speech; a Vital Interest for All Americans," *Washington Times*, A25.

107. U.S. Library of Congress (2009, December 11), "Bill Summary and Status 111th Congress (2009–2010) S. 448, http://thomas.loc.gov/cgi-bin/bdquery/z?d111:SN00448:@@@X," accessed March 29, 2011.

108. Charlie Savage (2013, May 15), "Criticized on Seizure of Records, White House Pushes News Media Shield Law," *New York Times*. http://www.nytimes.com/2013/05/16/us/politics/under-fire-white-house-pushes-to-revive-media-shield-bill.html?_r=3&%20-%20h&, accessed January 18, 2015.

109. Erik Wemple (2013, May 13), "AP: Government Subpoenaed Journalists' Phone Records," *Washington Post*, http://www.washingtonpost.com/blogs/erik-wemple/wp/2013/05/13/ap-government-subpoenaed-journalists-phone-records/, accessed January 18, 2015.

110. Savage, "Criticized on Seizure of Records."

111. Ibid.

112. Ibid.

113. Chris Strohm, Angela G. Keane, and Robert Hutton (2015, January 16), "Obama, Cameron Vow to Bolster Cybersecurity after Sony hack," *Bloomberg*. http://www.bloomberg.com/news/2015-01-16/obama-cameron-cybersecurity-agenda-shaped-by-paris-sony-attacks.html, accessed January 18, 2015.

114. Ibid.

115. Ibid.

BIBLIOGRAPHY

Akdeniz, Yaman (2002). "Anonymity, Democracy, and Cyberspace." *Social Research* 69, 223–37.

Allen, Christina (1999). "Internet Anonymity in Contexts." *The Information Society* 15, 145–46. doi: 10.1080/019722499128600.

Bagdikian, Ben (2005, August/September). "When the Post Banned Anonymous Sources." *American Journalism Review* 27, 33.

Bennett, W. Lance (2001). *News: The Politics of Illusion* (4th ed.). New York: Addison Wesley.

Berry, Stephen J. (2009). *Watchdog Journalism: The Art of Investigative Reporting*. New York: Oxford University Press.

Branzburg v. Hayes, 408 U.S. 665 (1972).

Bugeja, Michael (2005). *Interpersonal Divide: The Search for Community in a Technological Age*. New York: Oxford University Press.

Carlson, Matt (2010). "Embodying Deep Throat: Mark Felt and the Collective Memory of Watergate." *Critical Studies in Media Communication* 27, 235–50. doi: 10.1080/15295030903583564

Carr, David (2012, February 7). "Blurred Line between Espionage and Truth." *New York Times*, B1, B7.

Carter, T. Barton, Mark A. Franklin, and Jay B. Wright, (2001). *The First Amendment and the Fourth Estate: The Law of Mass Media* (8th ed.). New York, NY: Foundation Press.

Christians, Clifford G., Mark Fackler, Kim B. Rotzoll, and K. B. McKee (2001). *Media Ethics: Cases and Moral Reasoning* (6th ed.). New York: Addison Wesley Longman.

Cohen v. Cowles Media Company, 501 U.S. 663 (1991).

Cowan, James (2008, December 31). "The Orwellian Power of Anonymity; the Web Has Changed the Rules of Politics and Journalism." *National Post*, A6.

Cowell, Alan (2014, November 20). "Swedish Court Upholds Order for Arrest of Julian Assange." *New York Times*. http://www.nytimes.com/2014/11/21/world/europe/swedish-court-rejects-appeal-by-julian-assange.html. Accessed January 19, 2015.

Crain, Rance (2008, March 10). "I'm Going on the Record When It Comes to Anonymous Bloggers." *Advertising Age*, 14.

Davidson, James W., and M. H. Lytle (1986). *After the Fact: The Art of Historical Detection* (2nd ed.). New York: Alfred A. Knopf.

Ekstrand, Victoria S. (2003). "Unmasking Jane and John Doe: Online Anonymity and the First Amendment." *Communication Law and Policy* 8, 405–27. doi: 10.1207/S15326926CLP0804_02

Ensor, David (2003, October 1). "Novak: 'No Great Crime' with Leak." CNN.com. http://www.cnn.com

Fernando, Angelo (2006, May–June). "That Third Screen—In Your Pocket!" *Communication World* 23, 11–12.

Froomkin, A. Michael (1999). "Legal Issues in Anonymity and Pseudonymity." *The Information Society* 15, 113–27. doi: 10.1080/019722499128574

Fuchs, Christian (2011). "WikiLeaks: Power 2.0? Surveillance 2.0? Criticism 2.0? Alternative Media 2.0? A Political-Economic Analysis." *Global Media Journal—Australian Edition* 5, 1–17.

Greenwald, Glenn, Ewen MacAskill, and Laura Poitras (2013, June 11). "Edward Snowden: The Whistleblower behind the NSA Surveillance Revelations." *The Guardian*. http://www.theguardian.com/world/2013/jun/09/edward-snowden-nsa -whistleblower-surveillance. Accessed January 18, 2015.

Haiman, R. J. (1999). *Best Practices for Newspaper Journalists*. Arlington, VA: The Freedom Forum.

Hall, Emma (2005). "Technology Makes Storytellers of Citizens." *Advertising Age* 76 (6). Communication & Mass Media Complete Database. Accessed February 18, 2007.

Haridakis, Paul (2006). "Citizen Access and Government Secrecy." *Saint Louis University Public Law Review* 25, 3–32.

Harper, Jennifer (2009, September 10). "Journalist 'Shield Law' Gains Steam in Senate." *Washington Times*. http://www.washingtontimes.com

Janowitz, Morris (1975). "Professional Models in Journalism: The Gatekeeper and the Advocate." *Journalism Quarterly*, 52, 618–26.

Johnson, Madeline, and Betsy D. Gelb (2002). "Cyber-Libel: Policy Trade-Offs." *Journal of Public Policy and Marketing* 21, 152–59.

Keen, Andrew (2007). *The Cult of the Amateur: How Today's Internet Is Killing Our Culture*. New York: Doubleday.

Kerschbaumer, Ken (2005, January 24). "Dialing for News." *Broadcasting & Cable* 135, 43–44.

Kurtz, Howard (1997a). Ethics and the Mass Media: A Tipster Takes a Tumble; Dan Dorfman, Embroiled in a Controversy over Sources, is Fired by Money Magazine.

In *Messages 4: The Washington Post Media Companion*, edited by T. Beell, 204–206. Boston, MA: Allyn & Bacon.

Kurtz, Howard (1997b). "Janet Cooke's Untold Story: 15 Years after Her Famous Hoax the Disgraced Reporter Speaks." In *Messages 4: The Washington Post Media Companion*, edited by T. Beell. Boston, MA: Allyn & Bacon, 204–6.

Leonnig, Carol D., and Michael D. Shear (2007, March 8). "Prison May Be Long Way Off for Libby; Appeals Could Keep Former Cheney Aide Free Until after 2008 Election." *Washington Post*, A4.

McCann, Tom (2004, October). "Journalist Groups See Surge in Legal Attacks on Privilege." *Chicago Lawyer*, 90.

Middleton, Kent R., Robert Trager, and Bill F. Chamberlin (2000). *The Law of Public Communication* (5th ed.). New York: Addison Wesley Longman.

Mukasey, Michael B. (2008, June 28). "A Legislative Shield for the Press?" *The Washington Times*, M12.

New York Times Editorial. "Sock Puppet Bites Man." (2006, September 13). *New York Times*. http://www.nytimes.com

Nissenbaum, Helen (1999). "The Meaning of Anonymity in an Information Age." *The Information Society* 15, 141–44. doi: 10.1080/019722499128592

Novak, Robert (2003, July 14). "The Mission to Niger" [Editorial]. *Chicago Sun-Times*, 31.

Palser, Barb (2001, September). "I Want My Breaking News." *American Journalism Review* 23, 66.

"Protecting Sources; a Federal Shield Law Is Within Reach." (2009, April 6). *Washington Post*, A14.

Rothe, Dawn L., and Kevin F. Steinmetz (2013). "The Case of Bradley Manning: State Victimization, Realpolitik, and WikiLeaks." *Contemporary Justice Review* 16, 280–92. doi: http://dx.doi.org/10.1080/10282580.2013.798694

Savage, Charlie (2013, May 15). "Criticized on Seizure of Records, White House Pushes News Media Shield Law" *New York Times*. http://www.nytimes.com/2013/05/16/us/politics/under-fire-white-house-pushes-to-revive-media-shield-bill.html?_r=3&%20-%20h&. Accessed January 18, 2015.

Schumer, Charles (2008, July 2). "Defending Free Speech; a Vital Interest for All Americans." *Washington Times*, A25.

Silverblatt, Art (1995). *Media Literacy: Keys to Interpreting Media Messages*. Westport, CT: Praeger.

Singer, Jane B. (1996). "Virtual Anonymity: Online Accountability and the Virtuous Virtual Journalist." *Journal of Mass Media Ethics* 11, 95–106.

Solomon, Steven D. (2014, December 30). "Whistle-Blower Awards Lure Wrongdoers Looking to Score." *New York Times*. http://dealbook.nytimes.com/2014/12/30/whistle-blower-awards-lure-wrongdoers-looking-to-score/?_r=0. Accessed January 18, 2015.

Srivastava, L. (2005). "Mobile Phones and the Evolution of Social Behaviour." *Behaviour & Information Technology* 24, 111–29. doi: 10.1080/01449290512331321910

Stephens, Mitchell (2007, January/February). Beyond News. *Columbia Journalism Review* 45, 34–39.

Straubhaar, Joseph, and Robert LaRose (1997). *Communications Media in the Information Society*. Belmont, CA: Wadsworth.

Strohm, Chris, Angela G. Keane, and Robert Hutton (2015, January 16). "Obama, Cameron Vow to Bolster Cybersecurity after Sony Hack." *Bloomberg*. http://www .bloomberg.com/news/2015-01-16/obama-cameron-cybersecurity-agenda-shaped -by-paris-sony-attacks.html. Accessed January 18, 2015.

Sullivan, Margaret (2013, October 12). "The Disconnect on Anonymous Sources." *New York Times*. http://www.nytimes.com/2013/10/13/opinion/sunday/the-public -editor-the-disconnect-on-anonymous-sources.html?_r=0. Accessed January 19, 2015.

Swisher, Kara, and John Schwartz (1997). "Walking the Beat in Cyberspace: On-line Services Struggle with How to Combat Smut and Protect Privacy." In *Messages 4: The Washington Post Media Companion*, edited by T. Beell, 315–17. Boston, MA: Allyn & Bacon.

Tate, Julie (2013, August 21). "Bradley Manning Sentenced to 35 Years in WikiLeaks Case." *Washington Post*. http://www.washingtonpost.com/world/national-security /judge-to-sentence-bradley-manning-today/2013/08/20/85bee184-09d0-11e3-b 87c-476db8ac34cd_story.html. Accessed January 19, 2015.

Taylor, Marisa (2014, December 30). "Intelligence, Defense Whistleblowers Remain Mired in Broken System." *McClatchy DC*. http://www.mcclatchydc.com/static /features/Whistleblowers/ Whistleblowers-remain-mired-in-broken-system.html. Accessed January 18, 2015.

Tedford, Thomas L., and Dale A. Herbeck (2009). *Freedom of Speech in the United States* (6th ed.). State College, PA: Strata Publishing.

Teich, Al, Mark S. Frankel, Rob Kling, and Ya-Ching Lee (1999). "Anonymous Communication Policies for the Internet: Results and Recommendations of the AAAS Conference." *The Information Society* 15, 71–77. doi: 10.1080/019722499 128538

U.S. Library of Congress (2009, December 11). Bill Summary and Status 111th Congress (2009–2010) S. 448. http://thomas.loc.gov/cgi-bin/ bdquery/z?d111 :SN00448:@@@X. Accessed March 29, 2011.

Wayner, Peter (1999). "Technology for Anonymity: Names by Other Nyms." *The Information Society* 15, 91–97. doi: 10.1080/019722499128556

Weiss, Eric M., and Charles Lane (2006, July 14). "Vice President Sued by Plame and Husband; Ex-CIA Officer Alleges Leak of Her Name Was Retaliatory." *Washington Post*, A3.

Wemple, Erik (2013, May 13). "AP: Government Subpoenaed Journalists' Phone Records." *Washington Post*. http://www.washingtonpost.com/blogs/erik-wemple /wp/2013/05/13/ap-government-subpoenaed-journalists-phone-records/. Accessed January 18, 2015.

Woodward, Bob (2006). *The Secret Man: The Story of Watergate's Deep Throat*. New York: Simon & Schuster.

8

Rethinking Digital Democracy in a Time of Crisis: The Case of Spain

Salomé Sola-Morales

INTRODUCTION

In recent decades, various political parties, associations, and social movements around the world have developed their political action online. In fact,
it is not possible to speak about political actions outside of the Internet today,
because social networks are ubiquitous[1] and cyberspace is established as the
new "public agora"[2] or the "global agora."[3] In this new paradigm, Internet use
is key in the creation of collective debates, and fundamental in the dissemination of political ideas or the spread of community organizing. Moreover,
the Internet is creating an alternative space of communication to the hegemonic order, in which citizens are beginning to acquire new roles and gradually
becoming the real protagonists of political communication.

In recent years, Spain has witnessed many initiatives that promote a "direct digital democracy." Different politicians, citizen associations, or movements such as the X Party, *Podemos*,[4] 15-M Movement, and the Platform
Affected by Mortgage (PAH) have begun using online tools to transcend
traditional boundaries of time and space and to promote transnational or
massive participatory debates. In a climate of political disaffection and distrust of institutions that seems to have a worldwide scope,[5] the main objective of these new initiatives is to encourage a horizontal relationship among
citizens and to communicate with greater transparency. In this context, significant changes—which certainly go far beyond voting in an election—are
affecting the classical concept of political participation. As the role of citizens evolves in politics, it is essential to inquire about their new ways of
participating in the political process.

Yet, there are still many pending questions about Internet's reliability, as it
regards the internal and external decision-making process of an organization,

and the advantages and disadvantages of virtual voting and virtual decision making. Can the Internet be used to create a truly integrating, inclusive participatory democracy? Or, conversely, will the Internet end up creating a more exclusive system? Currently, Internet access is taken to be one of the basic principles of a democratic system. But how does one account for the growing digital divide among citizens, in which many citizens still lack Internet access or the economic or symbolic resources to use these new tools with assurance or guaranteed security? Are we witnessing the beginning of a revolution or is it only an appearance?

To tackle these questions, the main objective of this chapter is to reflect upon and discuss whether the Internet and the new media can generate a more participatory democracy. Further, the specific objectives of this research include describing the Spanish political party *Podemos*'s use of social networks and online tools, and providing an analysis of how these tools and networks are used in politics.

The starting premise is that the Internet has increased the role of citizenship by democratizing the media and political systems.[6] The present research tries to find evidence of how *Podemos* makes use of online tools to generate new forms of participation. More precisely, this research tries to identify how the instruments of "techno-politics" (as Rodotá would say) structure the role of citizen participation in Spain in the context of the current "crisis."[7]

As for the structure of this chapter, it first presents a theoretical consideration on cyberactivism and new forms of online participation. Next the concept of "digital democracy" and its growing development in Spain is explored, especially since the emergence of the political party *Podemos*. Next is an explanation of what online tools and type of political outreach and participation are promoted by *Podemos*. Finally, the political functions of Facebook within the organization are analyzed.

CYBER AND NETWORK PARTICIPATION, BETWEEN LIGHT AND SHADOW

The Internet has a great ability to transform political life[8] and it has a major impact on democracy today.[9] Cyberspace is not only promoting deliberation or modifying the organizational structure of institutions, but also transfiguring and redefining political mobilization, outreach, and participation;[10] the meaning of the assemblies; political parties; and even citizenship itself.[11]

Many researchers have argued that the development of "democracy 2.0" has increased and enhanced citizen participation[12] due to its fledgling potential. In fact, in the words of Colombo Villarrasa, the "Internet will raise the political participation of citizens both qualitatively and quantitatively."[13]

In the current context—where the digital surrounds any social or communicative process—the domain of the public no longer corresponds to an elite, but rather to "a vast majority of people."[14] Theoretically, participants are no longer an elite, but rather multitudes[15]—capable of establishing dialogues, self-organizing, and making decisions independently.[16] At first glance, online participation has many advantages, as it allows an "organized group of people"[17] to communicate instantly through free and user-friendly applications.[18] Additionally, all participants end up being players in the process and interact in a horizontal relationship on a massive participatory level.[19] At the same time, "the participants do not only discuss issues of common interest, but also form action groups, subscribe to causes, sign petitions, fund proposals and support reforms,"[20] so that participation increases qualitatively. In this sense, it could be argued that the Internet primarily provides a more open and accessible space, and less prior control than traditional media.[21] The Internet also provides a space in which political outreach can be more transparent, direct, and democratic;[22] and new venues for deliberation can emerge.[23]

Yet, other researchers have warned of the limitations and contradictions that the use of the Internet involves,[24] and have demonstrated the processes of exclusion and marginalization that derive from Internet access.[25] In fact, some scholars have pointed out that this is not just a question of access in its triple perspective—technology infrastructure, skills for effective use, and access to network services[26]—but instead is a question of the quality of Internet use.[27]

Moreover, assuming that no digital divide existed, that all people could browse the Internet at no cost, and all had the cognitive and symbolic resources to use the Web, it also would be necessary to have civic involvement by individuals, a collective commitment, and most importantly an interest or motivation to participate, which many passive or marginalized citizens lack.[28] Precisely for these reasons, some researchers have firmly declared that the Internet widens the gap of inequality,[29] and that the possible increase in online participation does not imply a higher quality of participation[30] because the Internet does not increase the share capital of a community.[31]

From this same perspective, other researchers think that the type of participation taking place in virtual environments is pseudo-participation, as it only generates weak ties.[32] This type of participation ends up being rather passive and it has no significant impact.[33] Recent empirical research also has shown that those who participate in political discussions online remain a minority.[34]

It is clear that there are many people who are marginalized or excluded and that their virtual participation is minor or nonexistent today. Thus, what truly is at stake are the power relations, that is to say, the basis of politics. This idea should be the focus of the debate on the democratic quality of any system, regardless of whether it takes place online or offline.

Therefore, and with an awareness of the limitations and risks that Internet use involves, this chapter emphasizes the positive changes that cyberspace is causing or could cause on the way that policies are made and on how to participate in policy making. The current political context finds more and more people outside of the decision-making process, and apathy and discouragement seem to be a shared leitmotif. This is evident, for example, in the high rates of absenteeism worldwide.[35] This context is where the Internet has a great potential—even if it is underdeveloped—to increase confidence, enrich democracy, and revitalize civic life; therefore it should be reconsidered.[36]

It is worth noting that for years the Internet has been bringing about a transformation in political systems. Regardless of the size or scope, these changes—more than structural or essential—rather are proving to be historic and are subjected to different contexts, media, and formats. Although political participation and communication might change or take various forms, some continuity with previous practice remains. Thus, one would prefer to admit contradictory possibilities and move away from the classic confrontations between apocalyptic and integrated visions (to quote Umberto Eco), or from "cyber-optimists" and "cyber-skeptics."[37] Instead, this chapter aims to show how the Internet can increase and improve the democratic system. To explore this assumption and illustrate it with a current example, the focus is on Spain and the use of digital democracy by the political party "*Podemos*."

DIGITAL DEMOCRACY UNDER SUSPICION

Talking about digital democracy today is not a utopia. New technologies have opened an "opportunity" for direct or participatory democracy,[38] which now is "physically and technologically possible."[39] In fact, currently in Spain citizens can—and increasingly do—use technology to communicate with the administration and with politicians, to express their views by voting, and to vote on referendums online. As shown in the latest news summary published by the October 2014 Centro de Investigaciones Sociológicas (CIS)[40] Barometer (Study No. 3041), Spaniards increasingly used technologies to communicate in political activities, such as contacting or communicating with a politician or a political party (6.8 percent); subscribed to an electronic mailing list on current issues to receive documents, requests, and campaigns (15 percent); posted comments on current affairs and social or political topics in a forum, blog, or social network (19.1 percent); made a donation to an association or organization (10.8 percent); signed a petition or joined a campaign or manifesto (20.1 percent); or participated in the call for a demonstration or protest act (10.8 percent). Although the percentages still might seem small, they show that today's political participation occurs partly online—a fact that was completely unthinkable a few years ago.

As Josep M. Llauradó mentioned, regardless of the name used to refer to it (e.g., digital democracy, hyperdemocracy, teledemocracy, computercracy, direct democracy, e-democracy, cyberdemocracy, wired democracy) and the discussions centered on it, what matters is that "digital democracy poses a desire to deepen the democratic system."[41] Hence, it is evident that the implementation of online participatory mechanisms does not "aspire to completely replace the existing representative system structures inherited from the nineteenth century, but to supplement and enrich them with a major source of legitimacy."[42]

These online tools therefore should not be wasted,[43] especially when Spanish society in general is in a climate of mounting disaffection, and politicians lack legitimacy and trust in the eyes of Spanish citizens. This was evidenced in the news summary published by the CIS Barometer (October 2104, Study No. 3041), which stated that only 2.1 percent of respondents expressed high confidence for the prime minister, Mariano Rajoy; 9.6 percent was fairly confident; 25.5 percent expressed little confidence; and 61.1 percent showed no confidence. When asked about the general political situation, 60.9 percent of respondents considered the current political situation to be very bad, 26.9 percent considered it bad, and 9.4 percent considered it regular, according news summary of the November CIS Barometer (Study No. 3045).

It therefore is not constructive to adopt a negative academic viewpoint that only sees the limits or risks that the Internet brings about. In the authors' view, the implementation of new online technologies "gives the possibility of opening a direct dialogue between the administrative structures and a larger population, increases the openness of power, and makes it easier for citizens to access information."[44] In short, this is an opportunity to encourage a population that is discouraged and disgusted by the institutional, political, and economic crisis currently ravaging Spain.

PODEMOS AND THE DIGITAL REVOLUTION

It is in this current climate that the Spanish political party *Podemos* has seized its opportunity. Born as a citizens' initiative, *Podemos* lays its foundation on participation, transparency, and cyberactivism. This party has used a very powerful and comprehensive political marketing strategy, as it has combined a large presence of their leaders in traditional mass media communication, and provides innovative online tools for citizens and supporters.

Born just five months before the European elections of May 2014, *Podemos* won 1,245,948 votes and five members of the European Parliament (MEPs). It is evident that the party utilized political communication strategically, combining the advantages of mediation with ePolicy. Although the use of mass media—especially television—has been the key to publicizing its political ideas and leaders, *Podemos* simultaneously has been able to create a space for a massive transnational dialogue online, through Facebook, Twitter, and other online tools. This seems to have increased the quantity and the quality of digital participation and, undoubtedly, has shown a large group of online supporters.[45]

SOCIAL MEDIA AND DIGITAL TOOLS USED BY *PODEMOS*

The political party *Podemos* has made a clear commitment to digital democracy because it uses different user-generated social networking sites and online tools as part of its political strategy. The following is a discussion of these tools and how each one is used politically.

The party has an official Web site (www.podemos.info), which shares information concerning the party and its program, policy documents, financial accounts, various news, links to e-mail contacts, links to the citizens' assembly or constituent body, and links to official social network accounts (e.g., Facebook, Twitter, Demoblog). The Web site also offers the opportunity to join a portal for online voting (there are more than 269,584 individuals presently registered) and to take sides in internal voting (such as voting for their program components and lists of candidates). The online voting portal is powered by Agora voting (https://agoravoting.com/) and uses software that allows users to vote online safely. In short, the official Web site can work mainly as the party's online headquarters, a matrix, or as a letter of presentation, and the site's components mainly are descriptive and informative. As noted, however, this Web site enables users to participate and to be a part of the constituent processes and internal organization if they are registered and have become what traditionally are known as members or supporters of the party.

The party also has a Facebook page (Podemos/Facebook) with nearly one million "likes." On the profile page, political ideas are disseminated, events and activities are advertised, and relevant information about the party is provided. This Facebook page works primarily as informational media, as most of its posts link to news from traditional media (i.e., television, radio, press). Yet, its main function is to spread the party's ideology by defending the values and ideology of the party, focusing and denouncing the government's or other parties' errors, and linking to relevant news and discussing it. Additionally, its Facebook page includes information about upcoming interviews and discussions, and where and when leaders will be featured in traditional media. Thus, the page works as informational media, as a reminder, and as an agenda of activities for supporters.

The party also has a Twitter account (@ahorapodemos), with more than 450,000 followers. Its main strategy for this account is to redistribute information and create a collective conversation. Looking at Podemos' time line, shows that the main Twitter account basically re-tweets:

- Messages from the party's top leaders (@ Pablo-Iglesias_; LolaPodemos; @pnique; @TeresaRodr_; @rierrejon),
- Messages linking to *Podemos* circles (@EuroPodemos; @PodemosCarab-Lat; @PodemosSa; @BrasilPodemos; @PodemosLeganes; @PodemosC-Madrid; @PodemosLinares; @PodemosVilaseca; @PodemosLeganes; @ PodemosSanidad; @PodemosARG, to name just a few),
- Messages linking to party leaders' TV appearances (@publico_es; @SextaNocheTv), and
- Messages on news from alternative media, or from related organizations (democraciareal or @la_tuerka), or even from anonymous people who send messages of encouragement to leaders or who share similar views.

In this regard, this tool also works as an information provider in the same way that Facebook does, in that it establishes links with other information. At the same time, it is complemented by all the party leaders' personal pages, which also establish a dialogue with @ahorapodemos and "re-tweet" what is posted there. This tool not only facilitates communication with the common citizen, regardless of whether they are *Podemos*'s members or supporters, but also favors internal communication and creates continuity.

The party also has a YouTube channel that includes information on the constitutional process of the party, the structure of the organization, political proposals, speeches made by party leaders, and interviews from traditional media. The site also has videos relating to the use of online tools for citizen participation proposed by the party, and information on democratic models and political issues of other countries. The main purpose of the YouTube channel is informational and educational.

Another online tool is Plaza *Podemos* (http://plaza.podemos.info). Unlike the other tools, its purpose is not to disseminate information or to expose information from other media outlets. Plaza *Podemos*—a community and forum (on the same page)—is defined as "*Podemos*' official meeting and discussion place," where people can share materials, opinions, ideas, and projects. Through the online tool reddit—which works as informational media—users can perform activities including making comments, providing input, sharing information, initiating threads in the community, participating in real-time open discussions, and moderating discussions.

This tool also allows for group interviews (#RuedaDeMasas) with the party leaders in real time, during which people can pose questions, get answers, vote on questions and answers, and make comments. The purpose of this tool is to create a space for a live debate online in which all users can register and participate, so long as they meet certain ethical standards so as not to disturb or harass other participants.

The party also uses an online tool called "titanpad" (https://titanpad.com) to hold virtual meetings with followers and supporters. Because there is no mandatory personal presence in a specific time or space, these assemblies can take place transnationally and thus can include party members who are in other countries or elsewhere at the time of the meeting. Using titanpad, everything written on it can be read and modified in real time. There is no need to register and all participants can participate at the same level. Meetings also can be coordinated by a moderator who could establish a procedure regarding the topics covered, the time to be invested, and how to carry out voting or the collective decision making.

Podemos uses Loomio—an online tool for collaborative decision making (https://www.loomio.org)—to create clear, intuitive, and effective discussions. Loomio enables consensus to be reached faster and pluralistically.

Lastly, with Appgree (http://www.appgree.com/), another free user-friendly application, one can create polls quickly via cellphone or on the Web. Appgree enables thousands of users to participate. Thus, an administrator of a group could pose a question and an unlimited number of people could express a response. It is noteworthy that each user can propose different answers and, in just a few minutes, the application's smart algorithm can create a ranking of the proposals most widely accepted and yield results that tend to have a 90% consensus.

NEW "FORMS" TO PARTICIPATE IN POLITICS

One of the most unique features of this new political party is its community-organizing system of "circles," or spaces for citizen participation. These "circles" are groups of people and have no leaders, are free and spontaneously created, and the protagonists are the citizens. Under the banner of the party, these open public assemblies have a flexible organization; they are open to all who wish to participate and only require from participants to be considered "circles" of the party that they have a minimum of participants and not duplicate other existing circles.

Currently there are more than 600 "circles" or groups of persons networked with different themes and scopes, but all under the label *Podemos*. There are 42 sectorial "circles" on diverse topics such as disability, feminism, education, self-employment, culture, *Animalista* (animal protection), Unionists, sports, unemployed, information and communications technologies (ICTs), senior citizens, officials, health, transport, and journalism—to name a few—and territorial "circles" that cover not only the Spanish territory, but the entire globe. There are more than 500 circles nationwide, currently constituted and supported by *Podemos*, and more than 30 foreign circles throughout the world, in Europe, Africa, the United States, and Asia.

Each of these circles in turn relies on online social networking tools for their internal and external operation. In principle, each user has a Facebook page, and some also have a Twitter account, Web page, or blog linked to the circle. The purpose of these tools is also informative in nature. These tools utilize almost the same strategies used by *Podemos* Central, which publishes links to news from other media or other *Podemos* circles, disseminates political ideas, conducts activities, and shares information on face-to-face or virtual meetings, including their dates and times. It is worth noting, however, that these are not fan pages created by the party or its leaders, as one would find in other traditional political parties or even in *Podemos* Central. On the contrary, the pages are not created by *Podemos* professionals or communication officers, but by anonymous citizens who voluntarily decided to create them, so that they come into existence as autonomous initiatives, like many other virtual communities.[46]

Therefore, one of the most promising aspects of these "circles" is that they are citizen initiatives; that is, they are created by groups of individuals altruistically and voluntarily. These multitudes of anonymous citizens connected through circles carry out all sorts of outreach strategies typical of political marketing, but without expecting anything in return. In the process, they become virtual communities of content creators,[47] which, as is evident, is very beneficial for the party. Namely, this virtual troop, independent and free, distributes content and political messages without motivation other than being a part of the collective and expressing their discontent with the current political system in the country.

In this regard, it is curious to note how *Podemos* has motivated—perhaps unintentionally, perhaps strategically—the creation of a huge international network of people that is constantly debating ideas. This network aims to change the social and political foundations of discontent by taking the charge of political action, that is to say, by participating online and offline. In this respect, it could be said that the logic of cyberactivism—which above all has tried to spread a message making use of new technologies and leveraging its advantages in speed, virality, horizontality—has been reversed. Now the citizens form an organized multitude and are managing online participation.

The "circles" also use tools such as e-mail or Whatsapp for instant communication among members. For internal functioning, circles use tools such as Google groups and Google drive, and others apps such as Loomio or Doodle. Depending on need, different tools are used to facilitate decision making, to distribute public announcements, and to publish relevant information for members. It should be reiterated that the party does not control any of these internal and external procedures; they are performed by anonymous citizens, and carried out independently within each "circle." In this respect, it is a totally new phenomenon for thousands of people to be involved in political activism, motivated to a great extent by other citizens' messages posted on social networks.

Although it might seem that the majority of political activism or the participation promoted by *Podemos* and by the citizens at the service of the political party takes place in the virtual world, it must be remembered that the "circles" have a commitment to meet regularly in face-to-face meetings. Thus, all members can learn and discuss their topics of interest in person. In this sense it seems that the Internet reinforces the political practices of previously mobilized citizens[48] who—once gathered—select which online tools might be more useful and interesting for their communicative purposes.

FACEBOOK AT THE SERVICE OF THE "CIRCLES" AND *PODEMOS*

Given the interest in exploring how the Internet and social networks enrich democratic participation, a qualitative methodology was designed to tackle this question. Moreover, because the political party under study is based on citizen participation through "circles" or assemblies, the activity of three of these circles on Facebook was analyzed. In this sense, the study explored how average citizens connect and participate in democracy through networks and via the Internet. For this study, three circles of different scope were selected for examination: a sectorial "circle," *Podemos* Education, having 13,102 fans; a national "circle," *Podemos* Lavapies in Madrid, having 1,443 fans; and a foreign "circle," *Podemos* Chile, in Santiago de Chile, having 793 fans. The study examined all of the publications of *Podemos* Education (135) dating from February 12, 2014, to December 17, 2014; all *Podemos* Lavapies' posts (103), from its inception on January 26, 2014, through December 12, 2014; and all of *Podemos* Chile's entries (100), since its inception on January 25, 2014, to December 17, 2014.

To understand political action online through Facebook a number of categories were established to measure the activities and political functions of the three circles' fan pages. The categories of political action defined for content analysis were:

- Appeal to direct participation;
- Link to the news media, Web sites, or other *Podemos* circles;
- Publication of records and internal documents of the organization;

- Information about constituent process, organizational chart, and topics of interest to participants;
- Dissemination of political ideas by creating slogans, messages, and images;
- Requests of information for users; and
- Other functions.

POLITICAL FUNCTIONS OF FAN PAGES

In light of the analysis, significant differences were found in the political use of the three Facebook pages under analysis. These differences show that Facebook is a tool for the citizens who use them, and not a mere extension of the political party or its leaders. Moreover, as stated, these are citizen initiatives, pages that have been created by individuals or groups of individuals, pages that have increasingly added more participants. As for the differences found in the use of the three circles, it is clear that the main objective of the *Podemos* Lavapies fan page is to make a clear and direct appeal to participation (61), and *Podemos* Education's fan page focuses on the dissemination of political ideas through posts and messages created by its administrators or users (44). In a different vein, Chile's fan page simply posts links to news from different media, Web sites, and other sources (59), functioning more as an informational medium.

The *Podemos* Lavapies' Facebook wall calls for clear and straightforward participation using simple slogans such as "Join" or "Participating is in our hands," accompanied by activities. These calls are made by posting news, creating events, or by merely changing the profile picture, which gives the site users the opportunity to include the event information next to the picture. Most announcements and notices were for on-site activities, such as calls for regular assemblies, calls for special assemblies, meetings with the circle coordinator, committee meetings, campaign events, meetings of the city council, calls to collect food, group events, seminars, workshops, neighborhood meetings, gatherings, lectures, putting up posters, demonstrations, bicycle rides, searches for representatives for elections, and searches for volunteers for different causes. Although the vast majority of posts called for face-to-face meetings, two calls for virtual meetings were also made—a conference with party leaders (via reddit) and an invitation to download the application @appgree to be able to vote virtually.

The direct call for participation is much less evident in *Podemos* Education, which focuses on the dissemination of political ideas and encourages debate and reflection more than face-to-face political action. Posts include ideas and quotes from thinkers, politicians, and ordinary citizens, synthesized in clear, direct statements. It is also common to find pictures or drawings illustrating a slogan or a policy proposal. In these posts, *Podemos* is very explicit about what it wants as a political party and what the citizens—as members of the party—demand. It is a circle dedicated mainly to education and entries reflect upon and concentrate on this topic. Unlike the other two circles—*Podemos* Lavapies and *Podemos* Chile—which do not include ideological debate, *Podemos* Education posts slogans, quotes, or messages to encourage discussion, debate, and reflection.

Unlike the other two "circles" discussed here, *Podemos* Chile focuses on the dissemination of information. In fact, few messages (11) that encouraged participation were calls to convene on regular assemblies. There were very few posts (2) in which a proposed slogan or a clear political message from the public was promoted. Undoubtedly, this is due to geography. Although the Lavapies circle has a strategic connection with the country geographically and with other local community associations in Madrid, the Chilean "circle" is like an island of involvement abroad, with little chance of connecting with national activism. At the same time, being a "circle" with lower levels of participation, and many less active members, it is difficult for it to embark on more creative ventures, such as designing posters with slogans or finding discussion threads developed by the users.

The main function of the *Podemos* Chile fan page has been to publish and disseminate party or political information. It mainly includes links to videos, songs, and interviews with leaders, or links to other traditional news media or organizations (59). This information has also been supplemented with the publication of information about the grassroots constituent process, the organizational structure of the party, or information from other circles (16).

To a lesser extent, the *Podemos* Education fan page shares links on initiatives, monitoring of court cases, and news on current topics (42); or the *Podemos* Lavapies fan page might publish links to news from digital media, disseminate information through documentaries, or explore the social networking apps and tools used by the circle or party (10). In all three cases, the information published by the fan page site administrators focused on the party and its representatives, retrospectives on face-to-face activities, or real-time reports on activities that were taking place. Very few of the publications dealt with different issues, and the only connection to international conflicts were expressions of solidarity, such as regarding the Gaza conflict. Messages of support for the circle and the party also were found in the Assemblies' minutes for both Podemos Lavapies and Podemos Chile.

IS IT POSSIBLE TO HAVE A WORLDWIDE DEBATE?

At first glance, it is overwhelming to think that in the virtual space there are hundreds of political debates taking place in the more than 600 pages of

Facebook's "circles," in assemblies on reddit, through Appgree *#RuedaDeMasas*, or via Loomio. It also is hard to imagine how it is possible that all these virtual conversations converge on a common integrated space, or can be transformed into a true shared dialogue. This unstoppable network, which continues to expand, and these Facebook walls, which continue growing every second, precisely show some of the limitations or challenges that cyberactivism must solve. Thus, the question is how can all these micro-debates—taking place simultaneously in different parts of the world from and towards *Podemos*—be combined?

In fact, it is nearly—if not completely—impossible for all the people involved to actually be connected. As the Hong Kong "circle" proposes alternatives to vote from abroad on its Facebook page, the Beirut "circle" recaps how the Lebanese press deals with the recent cases of political corruption in Spain. In Isla Cristina, Huelva (Spain), the "circle" calls for the affiliation of new supporters. The Lugo, Galicia, circle shares news about a group of workers in a hotel who are going through a difficult employment situation. Thus, the "circle" in Lavapies is concerned about the neighbors, and *Podemos* Education proposes a deep reflection on public education, and *Podemos* in Chile debates the rights of expatriates returning to Spain or the difficulties of voting from abroad.

In this sense, it is difficult to find unity among the citizens or create a global dialogue, as it is evident that "circles" end up becoming islands of participation at the local level, which work only for members' immediate interests. Yet, participation dynamics of exclusion play out between those involved and those who cannot get involved (because they lack economic resources, Internet access, or cognitive resources to use the proposed social networking tools); and, for example, among those who are members of some circles but not of others. These islands of participation occur because the citizens who are grouped together or who engage in dialogue end up focusing on their individual or local issues and lose sight of the universal problems. Thus, it seems that participation is more individualistic than it appears at first glance.

Despite the limitations posed, the changes in democratic participation—in the context of this case study—show that activism and citizen mobilization generated through the Internet bring more information, greater transparency, and an increase in new forms of citizen participation. Ultimately, at a qualitative and a quantitative level, it should be evaluated as a positive trend, which contrasts with the alleged disaffection, disinterest, and apathy of citizens in relation to politics, often denounced or criticized by academics and governments, or evidenced in surveys or in percentages of electoral abstention.

CONCLUSIONS AND FUTURE CHALLENGES

First, in light of the analysis presented here one could conclude that, indeed, as it was suggested initially, the Internet and new online tools can enhance citizen participation. The use of Facebook is key to being part of a "circle," because all the information regarding online and offline participation is published there. Answers are still needed about how such participation increases, however, and whether there is a direct connection with the use of Facebook and face-to-face participation. In this regard, there is a need for further case studies and further empirical research that might offer suggestions about broader trends.

Secondly, after analyzing how the "circles" operate, and how participants use social networks, it is clear that using Facebook encourages communication among its members. This is especially evident in the case of the "circles" analyzed, which function as alternative media in which political ideas can be formed, and information on activities or details about the programmatic basic structure or the party's operation are provided.

Lastly, rather than being a debate on fan pages of the circles analyzed, the proposed publications allow citizens to form a critical opinion; reflect on political ideas; share with others relevant information to actively participate in various activities, such as assemblies, workshops, and demonstrations; and to question much of the content appearing in traditional media. In this sense the participation, which occurs in the virtual space, has a direct connection to the participation that takes place offline. The Facebook pages of the three circles under study, more than creating discussions that were irrelevant to reality, consolidated as "circles," as they organized assemblies and face-to-face meetings.

It is too early to know the real extent of social networks and online tools on *Podemos*'s electoral results in the general elections of 2015 but, certainly, new forms of participation motivated by the party and developed precipitously by anonymous citizens invite a reassessment of traditional concepts of participation and citizenship. Taking into account that these dynamics show that citizens are increasingly aware and interested in politics, one can no longer speak of political passivity.

NOTES

1. Félix Requena, *Redes sociales y sociedad civil* (Madrid: CIS, 2008); David de Ugarte, *El poder de las redes* (Barcelona: Ediciones El Cobre, 2007); David de Ugarte, Pere Quintana, and Arnau Fuentes, *De las naciones a las redes* (Barcelona: Ediciones El Cobre, 2009).

2. Ramón Cotarelo, "La dialéctica de lo público, lo privado y lo secreto en la ciberpolítica," in *La comunicación política y las nuevas tecnologías*, edited by Ramón Cotarelo and Ismael Crespo (Madrid: Catarata, 2012), 15–28.

3. Víctor Manuel Pérez Martínez, *El ciberespacio: la nueva agora* (Santa Cruz de Tenerife: Idea, 2009).

4. In this chapter *"Podemos"* (in Spanish: [po'ðemos], translated in English as "We can") is not translated.

5. Joan Subirats, "Los dilemas de una relación inevitable. Innovación democrática y Tecnologías de la información y de la comunicación," in *Democracia Digital. Límites y oportunidades*, edited by Heriberto Cairo (Madrid: Trotta, 2002).

6. Cotarelo and Crespo, *Comunicación política*.

7. Stefano Rodotá, *La democracia y las nuevas tecnologías de la comunicación* (Buenos Aires: Losada, 2000).

8. Cotarelo and Crespo, *Comunicación política*, 8; Andrew Chadwick, *The Hybrid Media System: Politics and Power* (New York: Oxford University Press, 2013).

9. Martin Hagen, "Digital Democracy and Political Systems," in *Digital Democracy: Issues of Theory and Practice*, edited by Ken L. Hacker and Jan Van Dijk (London: Sage, 2000); Subirats, "Dilemas"; Jan Van Dijk, "Models of Democracy and Concepts of Communication," in *Digital Democracy: Issues of Theory and Practice*, edited by Ken L. Hacker and Jan Van Dijk (London: Sage, 2000); Benjamin R. Barber, *A Place for Us: How to Make Society Civil and Democracy Strong* (New York: Hill and Wang, 1998); Franco Bifo Berardi, ed., *La rete come paradigma e la reivenzione della democrazia* (Roma: Castelvecchi, 1996).

10. Rabia Karakaya, "The Internet and the Political Participation," *European Journal of Communication* 20 (4) (2005): 435–59; Víctor Sampedro and José María Sánchez-Duarte, "Epilogue," in Víctor Sampedro, *Cibercampaña. Cauces y diques para la participación. Las elecciones generales de 2008 y su proyección tecnopolítica* (Madrid: Editorial Complutense, 2011), http://www.ciberdemocracia.es, accessed December 2014; José Manuel Robles Morales, Óscar Molina Molina, and Stefano de Marco, "Participación política digital y brecha digital política en España. Un estudio de las desigualdades digitales," *Arbor. Ciencia, Pensamiento y Cultura* 188 (756) (2012): 795–810.

11. Stuart Allan, *Citizen Witnessing* (Cambridge: Polity, 2013); Castells, *Redes*; Henry Jenkins, *Convergence Culture: Where Old and New Media Collide* (New York: NYU Press, 2006); José Manuel Robles, *Ciudadanía digital. Una introducción a un nuevo concepto de ciudadano* (Barcelona: UOC, 2008); Josep Lluís Micó and Andreu Casero-Ripollés, "Political Activism Online: Organization and Media Relations in the Case of 15M in Spain," *Information, Communication & Society* 17 (4) (2013): 858–71, doi: 10.1080/1369118X.2013.830634, accessed November 18, 2014; Rodotá, *Democracia*.

12. De Ugarte, *Poder*; Shadrin citado en Víctor N. Rudenko, "La ciberrepública y el futuro de la democracia directa," *Contribuciones desde Coatepec* 16, January–June (2009): 165–76; Cotarelo and Crespo, *Comunicación política*.

13. Clelia Colombo Villarrasa, *E-participación. Las TIC al servicio de la innovación democrática* (Barcelona: UOC, 2007), 54.

14. Cotarelo, *Dialéctica*, 18–19.

15. Howard Rheingold, *Multitudes inteligentes. La próxima revolución social* (Barcelona: Gedisa, 2002); Ramón Cotarelo, "La expansión de la ciberpolítica," in *España en crisis. Segunda legislatura de Rodríguez Zapatero*, edited by César Colino and Ramón Cotarelo (Valencia: Tirant lo Blanc, 2012), 331–57; Víctor Sampedro and José López

Rey, "Nunca Máis y la cara oculta de la esfera pública: la visibilidad mediática de un movimiento social en un contexto de control de la información," in *Medios de Comunicación y solidaridad: reflexiones en torno a la des/articulación social*, edited by Eloísa Nos Aldás (Castelló de la Plana: Publicacions de la Universitat Jaume 1, 2006).

16. Serge Moscovici, *Un tratado histórico de psicología de masas* (México: Fondo de Cultura Económica, 1985).

17. Nicholas A. Christakis and James H. Fowler, *Conectados* (Madrid: Taurus, 2010), 27.

18. Cotarelo and Crespo, *Comunicación política.*

19. Philippe J. Maarek, "Politics 2.0: New Forms of Digital Political Marketing and Political Communication," *Trípodos* 34 (2014): 13–22; Mayo Fuster and Joan Subirats, "Crisis de representación y de participación. ¿Son las comunidades virtuales nuevas formas de agregación y participación ciudadana?," *Arbor. Ciencia, Pensamiento y Cultura* 188 (756) (2012): 641–56.

20. Cotarelo, *Dialéctica*, 19.

21. Mario Tascón and Yolanda Quintana, *Ciberactivismo. Las nuevas revoluciones de las multitudes conectadas* (Madrid: Catarata, 2012), 64.

22. Linda Harasim, "Global Networks, an Introduction," in *Global Networks, Computers and International Communication*, edited by Linda Harasim (Cambridge, MA: MIT Press, 1993), 3–14; María Elena Martínez Torres, "The Internet: Post-Modern Struggle by the Dispossessed of Modernity" (paper presented at Congreso LASA, Guadalajara, México, 1997).

23. Rosa Borge and Ana Cardenal, "Surfing the Net: A Pathway to Participation for the Politically Uninterested?" *Policy & Internet* 3(1) (2011): 1–29.

24. Geoffrey Scott Aikens, "Deweyan Systems in the Information Age," in *Digital Democracy. Discourse and Decision Making in the Information Age*, edited by Barry N. Hague and Brian D. Loader (London: Routledge, 1999), 179–94; Malcolm Gladwell, "Small Changes: Why the Revolution Will Not Be Tweeted," *New Yorker*, December 8, 2014.

25. Josep Maria Llauradó, *Democràcia digital: información, participación, transparencia* (Illes Balears: Edicions UIB, 2006), 78; Evgeny Moroz, *El desengaño de Internet* (Barcelona: Destino, 2010).

26. Manuel Martínez Nicolás, "De la brecha digital a la brecha cívica. Acceso a las tecnologías de la comunicación y participación ciudadana en la vida pública," *Telos. Cuadernos de Comunicación e Innovación* 86 (2011): 24–36.

27. Eszter Hargittai and Amanda Hinnant, "Digital Inequality Differences in Young Adults' Use of the Internet," *Communication Research* 35(5) 2008, doi: 10.1177/0093650208321782, accessed December 2, 2014.

28. Russell Neuman, Bruce Bimber, and Matthew Hindman, "Internet and Four Dimensions of Citizenship," in *The Oxford Handbook of American Public Opinion and the Media*, edited by Robert Y. Shapiro and Lawrence R. Jacobs (Oxford: Oxford Handbooks, 2010), 22–42.

29. Jennifer Oser, Marc Hooghe, and Sofie Marien, "Is Online Participation Distinct from Offline Participation? A Latent Class Analysis of Participation Types and Their Stratification," *Political Research Quarterly* 66(1) (2013): 91–101.

30. Jennifer Earl and Alan Schussman, "The New Site of Activism: Online Organizations, Movement Entrepreneurs, and the Changing Location of Social Movement Decision-Making," in *Consensus Decision Making, Northern Ireland and Indigenous Movements*, edited by Patrick Coy (London: JAI Press, 2003), 155–88; Jeroen Van Laer and Peter Van Aelst, "Internet and Social Movement Action Repertorios," *Information, Communication & Society* 13 (8) (2010): 1146–71.

31. Luis Arroyo, "The Cyber Utopia Was This? The Side Effects of the Internet. Sofactivism, Tribalism, Trivialization and New Censorship," Seminar, Public Communication in the Evolving Media Landscape: Adapt or Resist? Brussels, March 22, 2013, http://consilium.europa.eu/publicom, accessed November 2, 2014.

32. Mario Diani, "Social Movement Networks Virtual and Real," *Information, Communication & Society* 3(3) (2000): 386–401.

33. Morozov, *Desengaño*.

34. Eva Anduiza, "The Internet, Election Campaigns and Citizens: State of Affairs," *Quaderns del Cac* 33 (2009): 5–12; Carlos Cunha, Irene Martin, James Newell, and Luis Ramiro, "Southern European Parties and Party Systems, and the New ICTs," in *Political Parties and the Internet: Net Gain?*, edited by Rachel Gibson, Paul Nixon, and Stephen Ward (London and New York: Routledge, 2003); Wainer Lusoli, "A Second-Order Medium? The Internet as a Source of Electoral Information in 25 European Countries," *Information Polity* 10 (2005): 247–65; Víctor Sampedro, Celia Muñoz, José Luis Dader, and Eva Campos Domínguez, "Spain's Cyber-Campaign: Only for a Dissatisfied, Yet Very Active, Minority, Following the Trend of Other Southern European Countries," *Catalan Journal of Communication & Cultural Studies* 3 (1) (2011): 3–20.

35. Richard Kimberlee, "Why Don't British Young People Vote at General Elections?," *Journal of Youth Studies* 5 (1) (2002): 85–98; Tracey Skelton and Gill Valentine, "Political Participation, Political Action and Political Identities: Young D/Deaf People's Perspectives," *Space and Polity* 7(2) (2003): 117–34; Ryan E. Carlin, "The Decline of Citizen Participation in Electoral Politics in Post-Authoritarian Chile," *Democratization* 13(4) (2006): 632–51; Therese O'Toole et al., "Tuning Out or Left Out? Participation and Non-Participation among Young People," *Contemporary Politics* 9(1) (2003): 45–61; Edward Phelps, "Young Citizens and Declining Electoral Turnout: Trends and Explanations" (paper presented at the annual conference The Elections, Public Opinion and Parties, EPOP, Nottingham University, 2006); Nelson Wiselman, "Get Out the Vote—Not: Increasing Effort, Declining Turnout," *Option Politiques*, February (2006): 18–23.

36. Lance Bennet, "New Media Power: The Internet and Global Activism," in *Contesting Media Power*, edited by Nick Couldry and James Curran (Lanham, MD: Rowman & Littlefield, 2003), 17–38; Andrew Chadwick, *Internet Politics* (New York: Oxford University Press, 2006); Peter Dahlgren, "The Internet, Public Spheres, and Political Communication: Dispersion and Deliberation," *Political Communication* 22 (2005): 147–62; Pippa Norris, "The Impact of the Internet on Political Activism: Evidence from Europe," *International Journal of Electronic Government Research* 1(1) (2005): 20–39; Pippa Norris, Stefaan Walgrave, and Peter Van Aelst, "Who Demonstrates? Anti-State Rebels, Conventional Participants, or Everyone?," *Comparative*

Politics 37(2) (2005): 189–205; Clay Shirky, "The Political Power of Social Media: Technology, the Public Sphere, and Political Change," *Political Affairs* 28 (2011), http://www.foreignaffairs.com/articles/67038/clay-shirky/the-political-power-of-social-media, accessed June 3, 2015.

37. Jorge Resina de la Fuente, "Ciberpolítica, redes sociales y nuevas movilizaciones en España: el impacto digital en los procesos de deliberación y participación ciudadana," *Mediaciones Sociales. Revista de Ciencias Sociales y de la Comunicación* 7 (2010): 143–64.

38. Stefano Rodotá, *La democracia y las nuevas tecnologías de la comunicación* (Buenos Aires: Losada, 2000), 9.

39. Martí, José Luis. "Alguna precisión sobre las nuevas tecnologías y la democracia deliberativa y participativa," *IDP. Revista de Internet, Derecho y Política* 6 (2008), 6, http://www.uoc.edu/idp/6/dt/esp/marti.pdf, accessed February 2014.

40. Center for Sociological Research.

41. Llauradó, *Democracia*, 86.

42. José Luis Martí, "Alguna precisión sobre las nuevas tecnologías y la democracia deliberativa y participativa," *IDP. Revista de Internet, Derecho y Política* 6 (2008), 5, http://www.uoc.edu/idp/6/dt/esp/marti.pdf, accessed February 2014.

43. Martínez Nicolás, *Brecha digital*, 5.

44. Rudenko, "Ciberrepública," 174.

45. For example, Pablo_Iglesias_ has 718K compared to 644K of marianorajoy. The Twitter account of Ahorapodemos Party has 466K versus 185K of the Popular Party (data retrieved December 17, 2014).

46. Ismael Peña-López, "Quin impacte tenen les xarxes socials en la participació?," *Coop. Revista de la Federació de Cooperatives de Treball* 3 (2014): 56–57.

47. Fuster and Subirats, *Crisis*.

48. Borge and Cardenal, "Surfing the Net"; Zizi Papacharissi, "The Virtual Sphere 2.0: The Internet, the Public Sphere and Beyond," in *Handbook of Internet Politics*, edited by Andrew Chadwick and Philip N. Howard (London: Routledge International Handbooks, 2009); Dahlgren, "Internet"; Shelley Boulianne, "Does Internet Use Affect Engagement? A Meta-Analysis of Research," *Political Communication* 26 (2009): 193–211; Bruce Bimber, "Information and Political Engagement in America: The Search for Effects of Information Technology at the Individual Level," *Political Research Quarterly* 54(1) (2001): 53–67.

BIBLIOGRAPHY

Aikens, Geoffrey Scott. "Deweyan Systems in the Information Age." In *Digital Democracy. Discourse and Decision Making in the Information Age*, edited by Barry N. Hague and Brian D. Loader, 179–94. London: Routledge, 1999.

Allan, Stuart. *Citizen Witnessing*. Cambridge: Polity, 2013.

Anduiza, Eva. "The Internet, Election Campaigns and Citizens: State of Affairs." *Quaderns del Cac* 33 (2009): 5–12.

Arroyo, Luis. "The Cyber Utopia Was This? The Side Effects of the Internet. Sofactivism, Tribalism, Trivialization and New Censorship." Seminar. Public Communication in the Evolving Media Landscape: Adapt or Resist? Brussels, March 22, 2013. http://consilium.europa.eu/publicom.

Barber, Benjamin R. *A Place for Us: How to Make Society Civil and Democracy Strong.* New York: Hill and Wang, 1998.

Barber, Benjamin R. *Strong Democracy: Participatory Politics for a New Age.* Berkeley: University of California Press, 1984.

Bellamy, Christine. "Modelling Electronic Democracy: Towards Democratic Discourses for an Information Age." In *Democratic Governance and New Technology*, edited by Jens Hoff, Ivan Horrocks, and Pieter Tops, 33–53. London: Routledge, 2000.

Bennet, Lance. "New Media Power: The Internet and Global Activism." In *Contesting Media Power*, edited by Nick Couldry and James Curran, 17–38. Lanham, MD: Rowman & Littlefield, 2003.

Berardi, Franco Bifo, ed. *La rete come paradigma e la reivenzione della democrazia.* Roma: Castelvecchi, 1996.

Bimber, Bruce. "Information and Political Engagement in America: The Search for Effects of Information Technology at the Individual Level." *Political Research Quarterly* 54 (1) (2001): 53–67.

Borge, Rosa, and Ana Cardenal. "Surfing the Net: A Pathway to Participation for the Politically Uninterested?" *Policy & Internet* 3 (1) (2011): 1–29.

Boulianne, Shelley. "Does Internet Use Affect Engagement? A Meta-Analysis of Research." *Political Communication* 26 (2009): 193–211.

Buchstein, Hubertus. "Online Democracy, Is it Viable? Is it Desirable? Internet Voting and Normative Democratic Theory." In *Electronic Voting and Democracy: A Comparative Analysis*, edited by Norbert Kersting and Harald Baldersheim, 39–58. Basingstoke: Palgrave Macmillan, 2005.

Carlin, Ryan E. "The Decline of Citizen Participation in Electoral Politics in Post-Authoritarian Chile." *Democratization* 13 (4) (2006): 632, 651.

Castells, Manuel. *Redes de indignación y esperanza. Los movimientos sociales en la era de Internet.* Madrid: Alianza Editorial, 2012.

Castells, Manuel. *Communication Power.* Oxford: Oxford University Press, 2009.

Castells, Manuel. *La sociedad red: una visión global.* Madrid: Alianza Editorial, 2006.

Chadwick, Andrew. *The Hybrid Media System: Politics and Power.* New York: Oxford University Press, 2013.

Chadwick, Andrew. *Internet Politics.* New York: Oxford University Press, 2006.

Christakis, Nicholas A., and James H. Fowler. *Conectados.* Madrid: Taurus, 2010.

Colombo Villarrasa, Clelia. *E-participación. Las TIC al servicio de la innovación democrática.* Barcelona: UOC, 2007.

Cotarelo, Ramón. "La dialéctica de lo público, lo privado y lo secreto en la ciberpolítica." In *La comunicación política y las nuevas tecnologías*, edited by Ramón Cotarelo and Isamel Crespo, 15–28. Madrid: Catarata, 2012.

Cotarelo, Ramón. "La expansión de la ciberpolítica." In *España en crisis. Segunda legislatura de Rodríguez Zapatero*, edited by César Colino and Ramón Cotarelo, 331, 357. Valencia: Tirant lo Blanc, 2012.

Cotarelo, Ramón, and Ismael Crespo, eds. *La comunicación política y las nuevas tecnologías.* Madrid: Catarata, 2012.

Cunha, Carlos, Irene Martin, James Newell, and Luis Ramiro. "Southern European Parties and Party Systems, and the New ICTs." In *Political Parties and the Internet: Net Gain?* edited by Rachel Gibson, Paul Nixon, and Stephen Ward. London and New York: Routledge, 2003.

Dahlgren, Peter. "The Internet, Public Spheres, and Political Communication: Dispersion and Deliberation." *Political Communication* 22 (2005): 147–62.

De Ugarte, David, Pere Quintana, and Arnau Fuentes. *De las naciones a las redes.* Barcelona: Ediciones El Cobre, 2009.

De Ugarte, David. *El poder de las redes.* Barcelona: Ediciones El Cobre, 2007.

Diani, Mario. "Social Movement Networks Virtual and Real." *Information, Communication & Society* (3) (2000): 386–401.

Earl, Jennifer, and Alan Schussman. "The New Site of Activism: Online Organizations, Movement Entrepreneurs, and the Changing Location of Social Movement Decision-Making." In *Consensus Decision Making, Northern Ireland and Indigenous Movements,* edited by Patrick Coy, 155–88. London: JAI Press, 2003.

Eyerman, Ron, and Andrew Yamison. *Social Movements. A Cognitive Approach.* Cambridge: Polity Press, 1991.

Fuster, Mayo, and Joan Subirats. "Crisis de representación y de participación. ¿Son las comunidades virtuales nuevas formas de agregación y participación ciudadana?" *Arbor. Ciencia, Pensamiento y Cultura* 188 (756) (2012): 641–56.

Gladwell, Malcolm. "Small Changes: Why the Revolution Will Not Be Tweeted." *New Yorker,* December 8, 2014.

Hagen, Martin. "Digital Democracy and Political Systems." In *Digital Democracy: Issues of Theory and Practice,* edited by Ken L. Hacker and Jan Van Dijk, 54–70. London: Sage, 2000.

Harasim, Linda. "Global Networks, an Introduction." In *Global Networks, Computers and International Communication,* edited by Linda Harasim, 3–14. Cambridge, MA: MIT Press.

Hargittai, Eszter, and Amanda Hinnant. "Digital Inequality Differences in Young Adults' Use of the Internet." *Communication Research* 35 (5) 2008. doi:10.1177 /0093650208321782. Accessed December 2, 2014.

Innerarity, Daniel. *El nuevo espacio político.* Madrid: España, 2006.

Jenkins, Henry. *Convergence Culture: Where Old and New Media Collide.* New York: NYU Press, 2006.

Karakaya, Rabia. "The Internet and the Political Participation." *European Journal of Communication* 2 (4) (2005): 435–59.

Kimberlee, Richard. "Why Don't British Young People Vote at General Elections?" *Journal of Youth Studies* (1) (2002): 85–98.

Llauradó, Josep Maria. *Democràcia digital: información, participación, transparencia.* Illes Balears: UIB, 2006.

Lusoli, Wainer. "A Second-Order Medium? The Internet as a Source of Electoral Information in 25 European Countries." *Information Polity* 10 (2005): 247–65.

Maarek, Philippe J. "Politics 2.0.: New Forms of Digital Political Marketing and Political Communication." *Trípodos* 34 (2014): 13–22.

Martí, José Luis. "Alguna precisión sobre las nuevas tecnologías y la democracia deliberativa y participativa." *IDP. Revista de Internet, Derecho y Política* 6 (2008). http://www.uoc.edu/idp/6/dt/esp/marti.pdf. Accessed February 2014.

Martínez Nicolás, Manuel. "De la brecha digital a la brecha cívica. Acceso a las tecnologías de la comunicación y participación ciudadana en la vida pública." *Telos. Cuadernos de Comunicación e Innovación* 86 (2011): 24–36.

Martínez Torres, María Elena. "The Internet: Post-Modern Struggle by the Dispossessed of Modernity." Paper presented at Congreso LASA, Guadalajara, México, 1997.

Micó, Josep Lluís, and Andreu Casero-Ripollés. "Political Activism Online: Organization and Media Relations in the Case of 15M in Spain." *Information, Communication & Society* 17 (4) (2013): 858–71. doi: 10.1080/1369118X.2013.830634. Accessed November 18, 2014.

Milner, Henry. *The Internet Generation: Engaged Citizens or Political Dropouts.* Medford, MA: Tufts University Press, 2010.

Morozov, Evgeny. *The Netdesilusion. The Dark Side of Internet Freedom.* New York: Public Affairs, 2011.

Morozov, Evgeny. *El desengaño de Internet.* Barcelona: Destino, 2010.

Moscovici, Serge. *Un tratado histórico de psicología de masas.* México: Fondo de Cultura Económica, 1985.

Neuman, Russell, Bruce Bimber, and Matthew Hindman. "Internet and Four Dimensions of Citizenship." In *The Oxford Handbook of American Public Opinion and the Media*, edited by Robert Y. Shapiro and Lawrence R. Jacobs, 22–42. Oxford: Oxford Handbooks, 2010.

Norris, Pippa. "The Impact of the Internet on Political Activism: Evidence from Europe." *International Journal of Electronic Government Research* 1 (1) (2005): 20–39.

Norris, Pippa. *Digital Divide. Civic Engagement, Information Poverty, and the Internet Worldwide.* Cambridge: Cambridge University Press, 2001.

Norris, Pippa, Stefaan Walgrave, and Peter Van Aelst. "Who Demonstrates? Anti-State Rebels, Conventional Participants, or Everyone?" *Comparative Politics* 37 (2) (2005): 189–205.

O'Toole, Therese, Michael Lister, Dave Marsh, Su Jones, and Alex McDonagh. "Tuning Out or Left Out? Participation and Non-Participation among Young People." *Contemporary Politics* 9(1) (2003): 45–61.

Oser, Jennifer, Marc Hooghe, and Sofie Marien. "Is Online Participation Distinct from Offline Participation? A Latent Class Analysis of Participation Types and Their Stratification." *Political Research Quarterly* 66 (1) (2013): 91–101.

Papacharissi, Zizi. "The Virtual Sphere 2.0: The Internet, the Public Sphere and Beyond." In *Handbook of Internet Politics*, edited by Andrew Chadwick and Philip N. Howard. London: Routledge International Handbooks, 2009.

Pasrurenzi, Pedro. *Tecnopolítica. La democracia y las nuevas tecnologías de la comunicación.* Buenos Aires: Losada, 2009.

Peña-López, Ismael. "Quin impacte tenen les xarxes socials en la participació?" *Coop. Revista de la Federació de Cooperatives de Treball* 3 (2014): 56–57.

Pérez Martínez, Víctor Manuel. *El ciberespacio: la nueva agora*. Santa Cruz de Tenerife: Idea, 2009.

Phelps, Edward. "Young Citizens and Declining Electoral Turnout: Trends and Explanations." Paper presented at the Annual Conference on The Elections, Public Opinion and Parties, EPOP, Nottingham University, 2006.

Requena, Félix. *Redes sociales y sociedad civil*. Madrid: CIS, 2008.

Resina de la Fuente, Jorge. "Ciberpolítica, redes sociales y nuevas movilizaciones en España: el impacto digital en los procesos de deliberación y participación ciudadana." *Mediaciones Sociales. Revista de Ciencias Sociales y de la Comunicación* 7 (2010): 143–64.

Rheingold, Howard. *Multitudes inteligentes. La próxima revolución social*. Barcelona: Gedisa, 2002.

Robles, José Manuel. *Ciudadanía digital. Una introducción a un nuevo concepto de ciudadano*. Barcelona: UOC, 2008.

Robles Morales, José Manuel, Óscar Molina Molina, and Stefano de Marco. "Participación política digital y brecha digital política en España. Un estudio de las desigualdades digitales." *Arbor. Ciencia, Pensamiento y Cultura* 188 (756) (2012): 795–810.

Rodotá, Stefano. *La democracia y las nuevas tecnologías de la comunicación*. Buenos Aires: Losada, 2000.

Rodríguez, Roberto, and Daniel Ureña (2011). "Diez razones para el uso de Twitter como herramienta en la comunicación política y electoral." *Comunicación y pluralismo* 10 (2011): 89–106.

Rudenko, Víctor N. "La ciberrepública y el futuro de la democracia directa." *Contribuciones desde Coatepec* 16, January–June (2009): 165–76.

Sampedro, Víctor. *Multitudes online*. Madrid: Los libros de la Catarata, 2005.

Sampedro, Víctor, and José López Rey. "Nunca Máis y la cara oculta de la esfera pública: la visibilidad mediática de un movimiento social en un contexto de control de la información." In *Medios de Comunicación y solidaridad: reflexiones en torno a la des/articulación social*, edited by Eloísa Nos Aldás. Castelló de la Plana: Publicacions de la Universitat Jaume I, 2006.

Sampedro, Víctor, Celia Muñoz, José Luis Dader, and Eva Campos Domínguez. "Spain's Cyber-Campaign: Only for a Dissatisfied, Yet Very Active, Minority, Following the Trend of Other Southern European Countries." *Catalan Journal of Communication & Cultural Studies* 3 (1) (2011): 3–20.

Sampedro, Víctor, and José María Sánchez-Duarte. Epilogue of *Cibercampaña. Cauces y diques para la participación. Las elecciones generales de 2008 y su proyección tecnopolítica*, by Víctor Sampedro. Madrid: Editorial Complutense, 2011. http://www.ciberdemocracia.es. Accessed December 2014.

Shirky, Clay. "The Political Power of Social Media: Technology, the Public Sphere, and Political Change." *Political Affairs* 28 (2011). http://www.foreignaffairs.com/articles/67038/clay-shirky/the-political-power-of-social-media.

Skelton, Tracey, and Gill Valentine. "Political Participation, Political Action and Political Identities: Young D/Deaf People's Perspectives." *Space and Polity* 7(2) (2003): 117–34.

Subirats, Joan. "Los dilemas de una relación inevitable. Innovación democrática y Tecnologías de la información y de la comunicación." In *Democracia Digital. Límites y oportunidades*, edited by Heriberto Cairo, 89–111. Madrid: Trotta, 2002.

Tascón, Mario, and Yolanda Quintana. *Ciberactivismo. Las nuevas revoluciones de las multitudes conectadas*. Madrid: Catarata, 2012.

Ureña, Daniel. "Diálogo para un candidato 2.0." *Comunicación política 2.0. Cuadernos de Comunicación* 4 (2011): 29–33.

Van Dijk, Jan. "Models of Democracy and Concepts of Communication." In *Digital Democracy: Issues of Theory and Practice*, edited by Ken L. Hacker and Jan Van Dijk, 30–53. London: Sage, 2000.

Van Laer, Jeroen, and Peter Van Aelst. "Internet and Social Movement Action Repertorios." *Information, Communication & Society* 13 (8) (2010): 1146–71.

Wiselman, Nelson. "Get Out the Vote—Not: Increasing Effort, Declining Turnout." *Option Politiques*, February (2006): 18–23.

9

Will the Revolution Be Tweeted? Activism, Politics, and the Internet

Lázaro M. Bacallao-Pino

INTRODUCTION

Newspapers announce the "Twitter Revolution": a singular social phenomenon characterized by a "spirit of liberty" and mass demonstrations—protests, rioting, and any other popular expression of discontent—that are "brought to the world in real time through social-media networks and online video."[1] Although some publications underscore the importance of a collective action that is coordinated using Twitter to plan the manifestations, mobilize the participants, and update the news about it to all around the world, other media outlets have speculated about the real impact of social media. These other outlets mentioned certain examples of what has happened in some of the countries where social media–based collective actions took place, such as the electoral victory of prime minister Recep Erdogan in Turkey, or the chaos and civil war in Libya.[2]

The debates on the interrelationships between the Internet and collective action began 20 years ago by the Zapatista movement, originating in the remote mountains of the Mexican Southeast. The Zapatista Army has been called the "first informational guerrilla movement"[3] and the "first social netwar."[4] Since 1994, there have been many examples of the uses of the information and communication technologies (ICTs) as part of collective action. A brief time line of these examples includes the Battle of Seattle, in November 1999 and the creation of the Indymedia network; the first edition of the World Social Forum (WSF), in 2001; the Arab Spring; the England riots, in 2011; the Spanish 15M Movement; Occupy Wall Street; the Mexican #YoSoy132 (English: #Iam132); and the protests in Brazil in 2013 and 2014.

Consequently, the increasing role of the ICTs in social movements has become a relevant topic of research. The analyses on the issue have underlined, on the one hand, the contribution of the ICTs for enabling participatory processes and improving democracy this way. Conversely, other approaches have criticized certain tendencies as leading to an ICTs-based technological neo-determinism and a new digital utopianism. Those discussions have increased in the case of the analyses of social media, because these technological platforms provide a greater openness and horizontality, and a richer user experience and participation.

A number of questions surround these debates. A transversal question is: Does the Internet provide a source for citizen participation or mobilization? Other questions refer to the interrelationships between online and offline dimensions, or ask how social movements deal with the tension among articulation and visualization—two of the possible uses of these technological resources as part of collective action. This chapter aims to discuss some ideas about the uses of the Internet by social mobilizations and the enabling conditions for the development of sustained processes of sociopolitical change as a result of this social media–based collective action.

INFORMATION AND COMMUNICATION TECHNOLOGIES AND COLLECTIVE ACTION: DEBATES AND CHALLENGES

Social movements and mobilizations, as part of their collective action, put into practice certain repertoires of contention[5] that cannot be seen as simply a public display of disruptive action or be limited to a set of various protest-related tools and actions available in a given period, but rather as certain performances and claims that make routines within a cultural frame of performance for collective action.[6] Over the last two decades, as part of the process of change of the repertoires of collective action, ICTs have become a particularly relevant resource of protests and public dissent.

At the early stages, ICTs were considered tools for the cyber-diffusion of contention,[7] from a perspective that regarded these technologies as a channel for making visible the actions that take place offline. Subsequent approaches moved toward a more complex perspective, underscoring the emergence of certain specific cyber-repertoires of action, understood as Internet-based actions that only exist in the cyberspace. This repertoire includes online sit-ins, signing online petitions, hacking, defacing Web pages, e-mail floods, releasing viruses and worms, and performing data theft or destruction—configuring what has been defined as "cyber-protests" or "cyberactivism."[8] In this new scenario for collective action—the digital scenario—the configuration of an electronic repertoire of contention becomes an intricate process, given the complexity of how its development, innovation, and diffusion take place. This all occurs in a context characterized by the rapid dynamics of cyber-diffusion, the impact of digital innovation even outside the realm of an online protest, and the use of the new tactics even by nonactivists.[9]

The rise of social media and of online social networks in particular, seems to give an added twist to this communicational context, increasing the enabling conditions for democratic dynamics through these Web 2.0 platforms. This is significant for debates on the articulation between ICTs and democracy—a topic that has been a relevant aspect in the analyses of the social impact of ICT from its very beginning. In that regard, discussions have ranged from analyses of the consequences of the Internet on traditional forms and structures of representative democracy—such as electoral campaigns, for example[10]—to the contribution of the ICTs to a new kind of participatory democracy through the revitalization of the public sphere,[11] and the enabling conditions for the configuration of an online scenario for democracy and, consequently, the emergence of what has been called "cyberdemocracy."[12]

The Internet has been considered as a resource of community building and as particularly important in the scenario of an increasing lack of political engagement.[13] Different authors have emphasized the articulations between radical democracy and Internet practice, arguing how the latter could contribute to advancing democratic dynamics beyond its conceptualizations and practices within the liberal-capitalist political context,[14] and considering the extension and pluralization of the public sphere as a consequence of the online practices.[15] According to these approaches there is a close interrelationship among some characteristics of the ICTs and certain dimensions of the democratic dynamics and the processes of configuration of the public sphere, such as their shared interactive, deliberative, and decentralized natures.[16]

In the view of other authors, however, there also is an obscure side of the Internet—a technological neo-determinism that mediates the understanding of ICTs by arguing that the development of certain technologies determines social progress.[17] Two tendencies resulting from this neo-determinism are "cyber-utopianism"—referring to the inability to see the negative consequences of the Internet for social relationships, assuming only the emancipatory nature of the uses of the ICTs; and "Internet-centrism"—the trend to identify ICTs as the main cause of all political and social changes in present societies.[18] In the specific case of the uses of the Internet for collective action, this perspective is present in some analyses that consider the role played by social media in social protests. It is cited as evidence of a new technological fetishism, through which the focus on the ICTs becomes a distraction from the most important conflicts of contemporary capitalist societies.[19]

Although some studies argue that social media has produced singular communicational dynamics—creating innovative forms of interaction and expression that stimulate civic participation, bringing new dynamism to public mobilization, and consequently bringing about processes of sociopolitical transformation.[20] Conversely, previous research on recent social mobilizations has identified a certain tendency toward a hierarchical structure in the appropriation of social media by its participants, as indicated by trends of information centralization and patterns of popularity growth.[21]

These debates lead to a complex issue: The interrelationships between the uses of the Internet—and, in particular, the new technological platforms associated to the Web 2.0—by the participants in social mobilizations and the processes of sociopolitical change that are configured as part of the collective action. In general, the interrelationships among social mobilizations and social change "can seem too obvious to need explanation. Rather than questioning whether there is a connection between protest and change, it is far more likely that it will be presumed that social change and social movements are simultaneously occurring."[22] Social movements have been defined in terms of social change, as agents that try to carry it out, nullify it, or avoid it;[23] and although it has been recognized that certain social movements focus on individual change, it has been emphasized the social dimension of this process of transformation.[24]

The definition of social movements as a "set of non-institutionalized collective actions consciously oriented towards social change (or resisting such changes) and possessing a minimum of organization"[25] highlights another particularly relevant issue: the interrelationships between social change and the organizational dimension of social movements. There has been an important trend to propose a relationship between social movements and social change that is based on their purpose of transforming the world without taking power, moving away from the traditional two-step revolutionary strategy that has dominated the anticapitalist projects during the 19th and 20th centuries (first, take the power—the state—and second, change the world).[26] From this perspective, social change takes place within social movements as part of their internal practices and far from the eyes of the state, configuring "new practices and social relations" through which "a new social organization it is collectively built."[27]

Given the importance of studying these collective actors as agents of social change, "discussion concerning what determines the outcomes they achieved has been central to the debate on social movements."[28] Their effects are part of an integral analysis of their actions, focused on dimensions such as the relationships between the level of radicalism of their proposals of social change and their successfulness, the comparison between radical social movements and moderate movements, the pros and cons of using violence to achieve social transformation, and the positive or negative consequences of a centralized and bureaucratic organization for social movements and their projects of social transformation.

In this regard, communication is considered one of the dimensions in which social change takes place within social movements. The impact of social movements is not limited to the enactment of politics through protests and cultural contestation, but also to the generation of a diverse knowledge—called "knowledge-practices,"[29]—an approach that leads to a point of view of social movements as processes through which knowledge is generated, modified, and mobilized. From this point of view, social movements are "more than collectively organized action: they also consist of collectively constructed and share meanings, interpretations, rituals, and identities," in a perspective that reinforces an idea of collective and individual transformation in social movements that is much closer to a "greater recognition of diffuse expressions of social activism" than to a "programmatic and conventionally political agendas for social change."[30]

SOCIAL NETWORKS AND COLLECTIVE ACTION: PARTICIPATORY ARTICULATION VERSUS MOBILIZING VISIBILITY

Recent social mobilizations throughout the world—such as the Arab Spring, the "BlackBerry riots" in the United Kingdom, the Chilean Winter, the 15-M Spanish Movement, the Occupy Wall Street movement, the Mexican #YoSoy132, and the occupation of the Taksim Square in Turkey—are relevant for discussing the articulation between social movements, social mobilizations, and the Internet (particularly social media), in the light of the interrelationships among social mobilizations and social change. All of these mobilizations have made a wide use of social media, to the extent of being able to be defined in terms of social media–based mobilizations. The notions of "Twitter and Facebook revolutions" have been coined as a result of such episodes of collective action, for naming and underscoring the potential of a popular mobilization emerged through—or based upon—online social networks for framing processes of deep sociopolitical change.

The particularities of the appropriations of these Web 2.0 platforms during these mobilizations, however, as well as their long-term development, provide a significant analytical scenario for understanding the complexities of this potential. There are two main tensions mediating the appropriations of ICTs by protesters: (1) the tension between an articulating or a visibility-centered appropriation and (2) the tension between online and offline collective action. This chapter examines how these tensions mediate the interrelationships between politics, activism, and the Internet, particularly in the case of social media–based mobilizations. It also provides a relevant perspective for overcoming two opposite extreme points of view on the communicative dimension of social movements: an instrumental approach that considers communication merely as a tool for other purposes, and a perspective that overestimates the role of the communication by placing it at the core of social movements' practices.

Particularly in the context of highly concentrated media systems, online social media provide new possibilities for the development of an autonomous communication. In that regard, previous studies have examined how the use and appropriation of ICTs—particularly online social networks—by certain groups can be considered the basis for a process of democratization of mainstream media, fostering pluralism and triggering important processes related to a given political culture.[31] Coherently, social media–based mobilizations have been considered influential actors in politics that have shown how collective action can emerge and evolve—through the use of these technological resources—to communicate certain concerns and organize protests. According to some authors, social media, particularly Twitter—have had an increasing importance for sociopolitical activism, to the extent of influencing governmental decision making and shaping the interrelationships between governments, politicians, citizens, and other stakeholders.[32]

The Web 2.0 is considered a significant technological resource that has produced relevant changes in political communication, its structures, and dynamics[33] by providing certain technological resources to citizens that enable them not only to consume but also to produce their own communicative products and spaces.[34] It empowers people, who then can find new spaces of autonomy on the Internet[35] by opening the information environment to multiple groups that can create or incorporate new issues or topics into the public debate.[36] This fosters transparency, facilitates many-to-many communication processes, and promotes citizen's interactivity.

Sometimes, however, this positive perspective does not take into account the complexities of the interrelationships between social media, social mobilizations, and processes of sociopolitical change. The tension among two possible main goals in the use of these technological resources as part of collective action—visualization of dissent or articulation of participatory processes—is a central tenant for how social media–based mobilizations can become—or fail to become—sustainable processes of sociopolitical change.

Participants in these episodes of collective action underscore their criticism of traditional dynamics and structures of representative democracy—questioning its legitimacy, making visible the opposition to it, and demanding a real democracy. This trend is summarized, for instance, by the slogans "Real Democracy NOW!" or "They call it 'democracy,' but it is not," by the Spanish 15M Movement. It is recognized that social media play a core role in visualizing the opposition and mobilizing people by encouraging them to participate in protests, occupations, and all collective actions. Some specific communicative contexts—for instance, the Mexican or the Turkish media systems, in the case of the #YoSoy132 and the protests in Taksim Square, respectively—reinforce certain uses of social media as resources for overcoming mainstream media's censorship of news about collective action—strengthening the tendency toward a use of social media that is focused on the visualization of protests and occupations.

At the same time, however, among the participants there is frequent and explicit concern about the limits of a collective action focused on this visualization through social media. Protesters are aware of the impossibility of a maintained high level of collective action over a prolonged period—"we must reach a point where we stop. We cannot be marching every day," said a participant in the #YoSoy132.[37] In their discourses are frequent calls to advance toward long-term strategies by creating spaces for the development of organizational dynamics, and the proposal of a future agenda on their specific demands regarding public policies, so "we will be capable of organizing ourselves very well and consolidate us as a movement."[38] These concerns point to the long-term tension between protest and proposal in social movements, a tension that has been at the center of some of the most important debates about these collective agents.

Prioritizing a visualization-centered use of social media instead of an articulation-focused use becomes a handicap in advancing toward the configuration of sustained spaces of social participation. Although visualization through social media is a core enabling condition for the spreading of collective action from below, at the same time this emphasis can produce an outside-oriented perspective on the uses of social media. Although Web 2.0 platforms enable users to create private spaces of interaction—for instance, sending private messages on Twitter or Facebook, and to create private groups in the latter—the inherently open and bidirectional nature of social media makes social media mobilizations face the visualization/articulation tension in a way unlike that in any other case. The open bidirectional articulation processes based on social media frequently are visible to general audiences, and not just individuals participating in the collective action.

Given this particular nature of social media, the challenge for the participants in collective action is to develop a communicative repertoire of action that joins articulating visualizations and visible articulations, instead of assuming a perspective that emphasizes participatory articulation or mobilizing visibility. Although both goals—visualization and articulation—are equally important, any trend favoring one over the other will have negative consequences for social media–based collective action. In the case of a visibility-type use of social media, social mobilizations might not succeed in becoming sustained dynamics of social participation (i.e., a social movement).

Conversely, if focused on articulation, then these episodes of social action can be criminalized more easily by mainstream media.

To avoid using social media that focuses on mobilizing visibility against censorship by mainstream media, and to achieve a dual and complex point of view that tries to join articulation and visualization, participants in collective action emphasize the necessity of overcoming the online dimension by creating offline spaces that are associated not only with mobilizations and protests, but also with organizational processes. Participants highlighted, for instance, that "there is a need to develop assemblies, meetings, conventions and other face-to-face collective spaces for discussing issues of internal organization and long-term demands."[39] This idea leads to another tension in the analysis of social media–based mobilizations as sources of social change: online/offline dualism.

ONLINE AND OFFLINE COLLECTIVE ACTION: MEDIATIONS FOR SOCIAL CHANGE

The online/offline tension traditionally has been a core point at issue in the analysis on the uses of the ICTs, but recent social mobilizations have added some complexities to this debate. Some of these episodes of collective action have emerged from social media and, in all cases, these technological resources are significant in the resistance of participants and the spreading of collective action. The importance of social media for collective action is explicitly recognized by participants in their public discourses. It is recognized that social media is "a fabulous way of fostering social mobilizations" and also a particularly emotional resource that "has led me to mobilization and solidarity."[40]

The importance of the digital dimension, however, is not limited to providing certain technological resources for visualizing collective action. As noted, the Internet also has become an important scenario for collective action in current societies. An increasing number of cyberactions—actions that specifically and solely take place in cyberspace—have been included in the repertoires of collective action. But ICTs are not only resources for collective action, the digital experiences also are considered by the participants in social mobilizations to be a preview of the alternative sociopolitical order they want to achieve, mainly by highlighting the horizontality and participatory nature of the interactions within social media: "In social networks no one is more important than anyone else; we all are 140 characters," stated a protester of the #YoSoy132.[41]

Online experiences are regarded as a prefiguration of this alternative sociopolitical order. "I see this park as the incarnation of Twitter. People retweet the information they receive, respond to it and save what they like most in their list of favorites," stated one of the participants in the occupation of Taksim Square.[42] In some cases, this sense is reinforced, because some protesters only participated in the online collective action, so their experience of the mobilizations are focused on the digital context: "I did not have time to participate in assemblies and demonstrations, so my activism was completely on the digital social networks," explained a participant in the #YoSoy132.[43]

Despite the importance given to the online repertoire of collective action as well as to the digital spaces generated thanks to these technological resources, however, there also is a critical position regarding the limits of the online action. Participants declared that they "know that hashtags are ephemeral and that it would be very good if people moved into the streets instead of protesting only on social networks."[44] This ephemerality is seen as being opposite to the need of a long-term action to bring about a sustainable process of sociopolitical transformation and to face "the challenge of taking the opportunity to promote profound changes in the relationship between citizens, the media, politicians and governments," according to a member of the #YoSoy132.[45]

The tension between online and offline collective action is another significant mediation in the transition from social media–based mobilizations toward social movements. Conversely, participants assume there is continuity of certain characteristics of social media and the sustainability of collective action; for instance by considering that "a decentralized movement is impossible to be cancelled or infiltrated."[46] On the other hand, participants also agree that it is not enough to put "pressure on corporations or governments with collective actions through social media, but it is also needs a mass of people in the street telling a unified story."[47] The importance given to the online collective action in the case of social media–based mobilizations, however, has consequences in this transition from digital to offline scenarios. It finds expression in the particular communicative and symbolical nature of the repertoires of offline collective actions (occupations, manifestations); for instance, participants in the #YoSoy132 held a symbolic funeral march to symbolically bury Mexico's democracy.

Besides this general communicative "contamination" of the practices of collective action as a result of its original communicational nature, there is another important risk for social media–based mobilizations. By seeing the online experiences as a foreshadowing of the sociopolitical change they want to achieve, participants can focus their action on the digital spaces, configuring what has been called "communicational happy islands"[48] and underestimating the offline dimension of the action. At the same time, this focus on the online dimension can lead to a perspective of the long-term actions that prioritize the communicative dimension in the strategies for becoming sustained sources of sociopolitical participation. Consequently, once the period of most intense collective action has passed, one of the most important dimensions through which action continues is through the creation of alternative communication spaces. Participants in the Taksim protests, for instance, created the Web site Taksim Dayanışması (Taksim Solidarity);[49] the members of #YoSoy132 created a digital space for communication (Colectivo Másde131);[50] and the participants in the protests of Occupy Wall Street also founded a Web site.[51]

How such online/offline tension is resolved by those participating in episodes of collective action mediates—in a central way—the enabling conditions for social mobilizations to become social movements and thus be agents of sustained processes of sociopolitical change. At the same time, it also mediates the sense of success within participants and regarding their emotional involvement with the collective action. In that regard, in some cases there exist clear feelings of disappointment about the long-term outcomes of the social media–based mobilizations.

> "I was [a member of the #YoSoy] 132. I witnessed how it emerged on a Thursday night from a trending topic on Twitter. I lived the collective feeling of a march for the first time, I embraced a political banner. I participated in internal meetings at my school, I even was a spokeswoman and I gave interviews. I organized, I tweeted, I voted and wrote . . . but the genesis of energy change was not constant."[52]

In some cases, the illusion remains even when individuals assume that what they demand "will not happen," but think "here we are, enjoying the moment; we know that this is a good thing, we are all together, but then what? We do not know." Opposite this positive view of the collective action as it takes place, however, after the mobilizations have ended, the participants consider contradictions that can arise and "everything becomes very ambiguous."[53] There are expulsions and fighting between different factions, thus participants quit the collective action as a response to "the attempt to impose a single point of view," arguing that they are disappointed. "I did not join the movement for this. . . . I am not here to fight with peers. It is profoundly absurd [after] this entire effort to fight each other, instead of dedicating ourselves to change the situation," declared an individual who has been expelled from the 15M Movement.[54]

CONCLUSIONS

The appropriations of ICTs for collective action do not stand in stark contrast, but rather are in a gray area; it is here that the tensions reside. On the one side is articulation and visualization, and on the other are the online and offline dimensions. The way that participants deal with these tensions will mediate the development of social media–based mobilizations and the configuration of certain enabling conditions for them to become social movements—this is, sustained spaces for processes of sociopolitical change. Although the traditional idea of revolution—the so-called two-step revolutionary theory—implies a moment of particular energy in a collective action, a moment that leads to the taking of power, the sense of what being revolutionary means for current social movements implies a permanent, step-by-step process through which individuals try to configure the transformation from the bottom up.

Some recent episodes of collective action that have been centrally associated to a widened use of social media show the complexities of the process. These mobilizations demonstrate that there is a clear sense among participants that, although social media is an important resource for visualizing their actions as well as for spreading the protests, these Web 2.0 platforms alone are not enough for achieving success and guaranteeing the continuity of the collective action after the demonstrations have ended. Online social networks are considered fabulous tools for both organization and diffusion, but these two uses seem to be somewhat contradictory or, at minimum, difficult to make complement each other within collective action.

Particularly in contexts that lack spaces for expressing social dissent, focusing the use of social media on visualization could have negative consequences. It could, for instance, lead to a collective action focused on the online dimension, creating digital happy islands even though the offline experiences remain unchanged. It could also produce a tendency to overestimate the communicative dimension in the long-term strategies of collective action, taking the form of the creation of spaces of alternative communication as the main action of continuity of social participation.

Although protesters usually are aware of the possible negative consequences and try to avoid them, many participants note the importance given to a visibility-centered use of social media as compared with an articulating use. Paradoxically, this use of social media as part of collective action seems to be closer to the traditional idea of what a revolution is. The highly emotional nature of a visibility-centered use of these Web 2.0 platforms during the most active moment of the social mobilization is in line with an approach to social change that is focused on a moment—the seizure of power—instead of on a process for building a new sociopolitical order, step by step, in a long-term configuration. The immediacy of these technological resources seems to contradict the long-term nature of a process of social change from here and now and from below. Perhaps, after all, such a deep transformation—a revolution—cannot be tweeted.

The author wishes to thank the CONICYT/FONDECYT, Postdoctoral Program, Project #3150063.

NOTES

1. "Editorial: Iran's Twitter Revolution," *Washington Times*, June 16, 2009, http://www.washingtontimes.com/news/2009/jun/16/irans-twitter-revolution/.

2. Rakesh Sharma, "Is the Twitter Revolution Dead?" *Forbes*, January 4, 2014, http://www.forbes.com/sites/rakeshsharma/2014/04/01/is-the-twitter-revolution-dead/.

3. Manuel Castells, *The Information Age: Economy, Society and Culture Volume II: The Power of Identity* (Malden, MA: Blackwell Publishers, 1997), 79.

4. David Ronfeldt, John Arquilla, Graham Fuller, and Melissa Fuller, *The Zapatista Social Netwar in Mexico* (Santa Monica, CA: RAND, 1998), 1.

5. Sydney Tarrow, *Power in Movement: Social Movements and Contentious Politics* (Cambridge: Cambridge University Press, 1998).

6. Donatella Della Porta, "Repertoires of Contention," in *The Wiley-Blackwell Encyclopedia of Social and Political Movements*, edited by D. A. Snow, D. Della Porta, B. Klandermans, and D. McAdam (Malden, MA: Blackwell, 2013), doi: http://dx.doi.org/10.1002/9780470674871.wbespm178.

7. Jeffrey M. Ayres, "From the Streets to the Internet: The Cyber-Diffusion of Contention," *The Annals of the American Academy of Political and Social Science* 566, no. 1 (1999): 132–43, doi: 10.1177/000271629956600111.

8. Jeroen Van Laer and Peter Van Aelst, "Cyber-Protest and Civil Society: The Internet and Action Repertoires in Social Movements," in *Handbook on Internet Crime*, edited by Y. Jewkes and M. Yar (Portland, OR: Willan Publishing, 2009), 230–54.

9. Brett Rolfe, "Building an Electronic Repertoire of Contention," *Social Movement Studies* 4, no. 1 (2005): 65–74, doi: 10.1080/14742830500051945.

10. Steve Davis, Larry Elin, and Grant Reeher, *Click on Democracy: The Internet's Power to Change Political Apathy into Civic Action* (Boulder, CO: Westview Press, 2002).

11. Alinta L. Thornton, "Does the Internet Create Democracy?" *Ecquid Novi: African Journalism Studies* 22, no. 2 (2001): 126–47, doi:10.1080/02560054.2001.9665885.

12. Mark Poster, "Cyberdemocracy: Internet and the Public Sphere," in *Internet Culture*, edited by D. Porter (New York: Routledge, 1997), 201–18.

13. Davis et al.

14. Lincoln Dahlberg and Eugenia Siapera (eds.), *Radical Democracy and the Internet: Interrogating Theory and Practice* (Basingstoke, UK: Palgrave Macmillan, 2007).

15. Peter Dahlgren, "The Internet, Public Spheres, and Political Communication: Dispersion and Deliberation," *Political Communication* 22, no. 2 (2005), 147–62, doi: 10.1080/10584600590933160.

16. Ibid.

17. Sally Wyatt, "Technological Determinism Is Dead; Long Live Technological Determinism," in *Philosophy of Technology: The Technological Condition: An Anthology*, edited by R. C. Scharff and V. Dusek (Sussex, UK: John Wiley & Sons, 2013), 456–66.

18. Evgeny Morozov, *The Net Delusion: The Dark Side of Internet Freedom* (New York: Public Affairs, 2011).

19. Christian Fuchs, "Social Media, Riots, and Revolutions," *Capital & Class* 36, no. 3 (2012): 383–91, doi: 10.1177/0309816812453613.

20. Minavere V. Bardici, "A Discourse Analysis of the Media Representation of Social Media for Social Change—The Case of Egyptian Revolution and Political Change." Master's Thesis (Malmö University, 2012), http://muep.mah.se/handle /2043/14121.

21. Javier Borge-Holthoefer, Alejandro Rivero, Iñigo García, Elisa Cauhé, Alfredo Ferrer, Darío Ferrer, David Francos, David Iñiguez, María Pilar Pérez, Gonzalo Ruiz, Francisco Sanz, Fermín Serrano, Cristina Viñas, Alfonso Tarancón, and Yamir Moreno, "Structural and Dynamical Patterns on Online Social Networks: The Spanish May 15th Movement as a Case Study," *PLoS ONE* 6, no. 8 (2011): e23883, doi:10.1371/journal.pone.0023883.

22. Tim Jordan, "Social Movements and Social Change," *CRESC Working Papers*, London, CRESC, 2005, http://www.archive.cresc.ac.uk/documents/papers /wp7.pdf.

23. Joachim Raschke, "Sobre el concepto de movimiento social [The Concept of Social Movement]," *Zona Abierta*, no. 69 (1994): 121–34.

24. Federico Javaloy, Álvaro Rodríguez, and Esteve Espelt, *Comportamiento colectivo y movimientos sociales [Collective Behavior and Social Movements]* (Madrid: Pearson Educación, 2001).

25. J. Craig Jenkins, "Resource Mobilization Theory and the Study of Social Movements," *Annual Review of Sociology* (1983): 533.

26. John Holloway, *Change the World Without Taking Power: The Meaning of Revolution Today* (London: Pluto Press, 2002).

27. Raúl Zibechi, "Los movimientos sociales latinoamericanos: tendencias y desafíos [Latin American Social Movements: Trends and Challenges]," *OSAL: Observatorio Social de América Latina*, no. 9 (2003): 187.

28. Donatella Della Porta, and Mario Diani, *Social Movements: An Introduction* (Malden, MA: Blackwell Publishing, 2006), 226.

29. María I. Casas-Cortés, Michal Osterweil, and Dana E. Powell, "Blurring Boundaries: Recognizing Knowledge-Practices in the Study of Social Movements," *Anthropological Quarterly* 81, no. 1 (2008): 17–58, doi: 10.1353/anq.2008.0006.

30. Joseph E. Davis, "Narrative and Social Movements: The Power of Stories," in *Stories of Change: Narrative and Social Movements*, edited by Joseph E. Davis (Albany: State University of New York Press, 2002), 8.

31. Rodrigo Gómez García, and Emiliano Treré, "The #YoSoy132 Movement and the Struggle for Media Democratization in Mexico," *Convergence: The International Journal of Research into New Media Technologies*, 1354856514541744 (2014), doi: 10.1177/1354856514541744.

32. Rodrigo Sandoval-Almazan and Juan R. Gil-Garcia, "Cyberactivism Through Social Media: Twitter, YouTube, and the Mexican Political Movement I'm Number 132," in *Proceedings of the 46th Hawaii International Conference on System Sciences (HICSS)* (Wailea, HI: IEEE, 2013), 1704–13, doi: 10.1109/HICSS.2013 .161.

33. Ramón Feenstra and Andreu Casero-Ripollés, "Democracy in the Digital Communication Environment: A Typology Proposal of Political Monitoring Processes," *International Journal of Communication* 8, no. 21 (2014), http://ijoc.org/index .php/ijoc/article/view/2815.

34. Ivan Dylko and Michael McCluskey, "Media Effects in an Era of Rapid Technological Transformation: A Case of User-Generated Content and Political Participation," *Communication Theory* no. 22 (2012), 250–78, doi:10.1111/j.1468 -2885.2012.01409.x.

35. Manuel Castells, *Communication Power* (Oxford: Oxford University Press, 2009).

36. Andrew Chadwick, "The Political Information Cycle in a Hybrid News System: The British Prime Minister and the 'Bullygate' Affair," *International Journal of Press/Politics* 16, no. 1 (2011), 3–29, doi: 10.1177/1940161210384730.

37. Israel Carreón, quoted in Susana Moraga, "#YoSoy132 busca ser la sombra del poder en Mexico [#YoSoy132 Aims to be the Shadow of the Power in Mexico]," *ADN Político*, July 24, 2012, http://www.adnpolitico.com/2012/2012/07/21/el-yosoy132 -busca-ser-la-sombra-del-poder-en-mexico.

38. Alina Rosa Duarte, quoted in Susana Moraga, "#YoSoy132 busca ser la sombra del poder en Mexico [#YoSoy132 Aims to be the Shadow of the Power in Mexico]," *ADN Político*, July 24, 2012, http://www.adnpolitico.com/2012/2012/07/21/el-yosoy132 -busca-ser-la-sombra-del-poder-en-mexico.

39. Antonio Attolini, "Por una democracia auténtica, #YoSoy132 [For a Real Democracy, #YoSoy132]," *ADN Político*, September 19, 2012, http://www.adnpolitico .com/opinion/2012/09/19/antonio-attolini-por-una-democracia-autentica -yosoy132.

40. Diego Dante, quoted in Flor Goche, "Yo Soy 132, movimiento del siglo XXI [I am 132, a 21st century movement]," *Contralínea*, September 11, 2012, http://contra linea.info/archivo-revista/index.php/2012/09/11/yo-soy-132-movimiento-del-siglo -xxi/.

41. Rodrigo Serrano, quoted in Flor Goche, "Yo Soy 132, movimiento del siglo XXI [I am 132, a 21st century movement]," *Contralínea*, September 11, 2012, http:// contralinea.info/archivo-revista/index.php/2012/09/11/yo-soy-132-movimiento-del -siglo-xxi/.

42. Engin Onder, quoted in José M. Calatayud, "Los jóvenes del Parque Gezi [Youth in Gezi Park]," *El País*, June 9, 2013, http://internacional.elpais.com/internacional /2013/06/09/actualidad/1370781813_691701.html.

43. Tania, quoted in Héctor de Mauleón, "De la red a las calles [From the Web to the Streets]," *Nexos Online*, September 1, 2012, http://elecciones2012mx.wordpress .com/2012/09/01/de-la-red-a-las-calles-hector-de-mauleon-blog-nexos-en-linea/.

44. Ibid.

45. Regina Santiago, quoted in De Mauleón, "De la red a las calles [From the Web to the Streets]," *Nexos Online*, September 1, 2012, http://elecciones2012mx .wordpress.com/2012/09/01/de-la-red-a-las-calles-hector-de-mauleon-blog-nexos-en -linea/.

46. Member of the Spanish 15M Movement, quoted in Eva Cavero, "Esto está lleno de infiltrados [This is full of infiltrators]," *El País*, June 5, 2011, http://elpais .com/diario/2011/06/05/ madrid/1307273054_850215.html.

47. "About Us," *Occupy Wall Street*, http://occupywallst.org/about/, accessed November 20, 2014.

48. Rosa M. Alfaro, "Culturas populares y comunicación participativa [Popular Cultures and Participatory Communication]," *Revista Caminos*, no. 20 (2000): 13–20.

49. *See* http://taksimdayanisma.org/.

50. *See* http://www.colectivo131.com.mx/.

51. *See* http://occupywallst.org/.

52. Gisela Pérez de Acha, "La democracia de #YoSoy132 [The Democracy of the #YoSoy132]," *Animal Político*, September 19, 2012, http://www.animalpolitico.com /blogueros-blog-invitado/2012/09/19/la-democracia-de-yosoy132/#ixzz33PD7zyBc.

53. Gokce Gunac, quoted in José M. Calatayud, "Los jóvenes del Parque Gezi [Youth in Gezi Park]," *El País*, June 9, 2013, http://internacional.elpais.com/internacional /2013/06/09/actualidad/1370781813_691701.html

54. Carlos Paredes, quoted in María J. Hernández, "Cuando asoma el desencanto [When Disenchantment Peeps out]," *El Mundo*, May 15, 2012, http://www.elmundo .es/elmundo/2012/05/10/espana/1336646985.html.

BIBLIOGRAPHY

Alfaro, Rosa M. "Culturas populares y comunicación participativa [Popular Cultures and participatory communication]." *Revista Caminos* no. 20 (2000), 13–20.

Attolini, Antonio. "Por una democracia auténtica, #YoSoy132 [For a real democracy, #YoSoy132]." *ADN Político*. http://www.adnpolitico.com/opinion/2012/09/19 /antonio-attolini-por-una-democracia-autentica-yosoy132. September 19, 2012.

Ayres, Jeffrey M. "From the Streets to the Internet: The Cyber-Diffusion of Contention." *The Annals of the American Academy of Political and Social Science* 566, no. 1 (1999): 132–43. doi: 10.1177/0002716299566600111.

Borge-Holthoefer, Javier, Alejandro Rivero, Iñigo García, Elisa Cauhé, Alfredo Ferrer, Darío Ferrer, David Francos, David Iñiguez, María Pilar Pérez, Gonzalo Ruiz, Francisco Sanz, Fermín Serrano, Cristina Viñas, Alfonso Tarancón, and Yamir Moreno. "Structural and Dynamical Patterns on Online Social Networks: The Spanish May 15th Movement As a Case Study." *PLoS ONE* 6, no. 8 (2011): e23883. doi:10.1371/journal.pone.0023883.

Carreón, Israel, quoted in Susana Moraga. "#YoSoy132 busca ser la sombra del poder en Mexico [#YoSoy132 Aims to Be the Shadow of the Power in Mexico]." *ADN Político* (2012, July 24). http://www.adnpolitico.com/2012/2012/07/21 /el-yosoy132-busca-ser-la-sombra-del-poder-en-mexico.

Casas-Cortés, María I., Michal Osterweil, and Dana E. Powell, "Blurring Boundaries: Recognizing Knowledge-Practices in the Study of Social Movements." *Anthropological Quarterly* 81, no. 1 (2008): 17–58. doi: 10.1353/anq.2008.0006.

Castells, Manuel. *Communication Power* (Oxford, Oxford University Press, 2009).

Castells, Manuel. *The Information Age: Economy, Society and Culture Volume II: The Power of Identity* (Malden, MA: Blackwell Publishers, 1997), 79.

Chadwick, Andrew. "The Political Information Cycle in a Hybrid News System: The British Prime Minister and the 'Bullygate' Affair." *International Journal of Press /Politics* 16, no. 1 (2011): 3–29. doi: 10.1177/1940161210384730.

Dahlberg, Lincoln, and Eugenia Siapera (eds.). *Radical Democracy and the Internet: Interrogating Theory and Practice* (Basingstoke, UK: Palgrave Macmillan, 2007).

Dahlgren, Peter. "The Internet, Public Spheres, and Political Communication: Dispersion and Deliberation." *Political Communication* 22, no. 2 (2005): 147–62. doi: 10.1080/10584600590933160.

Dante, Diego, quoted in Flor Goche. "Yo Soy 132, movimiento del siglo XXI [I am 132, a 21st century movement]." *Contralínea.* September 11, 2012. http://contralinea.info/archivo-revista/index.php/2012/09/11/yo-soy-132-movimiento-del-siglo-xxi/.

Davis, Joseph E. "Narrative and Social Movements: The Power of Stories." In *Stories of Change: Narrative and Social Movements*, edited by Joseph E. Davis (Albany: State University of New York Press, 2002), 8.

Davis, Steve, Larry Elin, and Grant Reeher. *Click on Democracy: The Internet's Power to Change Political Apathy into Civic Action* (Boulder, CO: Westview Press, 2002).

Della Porta, Donatella. "Repertoires of Contention." In *The Wiley-Blackwell Encyclopedia of Social and Political Movements*, edited by D. A. Snow, D. Della Porta, B. Klandermans, and D. McAdam (Malden, MA: Blackwell, 2013). doi: http://dx.doi.org/10.1002/9780470674871.wbespm178.

Della Porta, Donatella, and Mario Diani. *Social Movements: An Introduction* (Malden, MA: Blackwell Publishing, 2006), 226.

Dylko, Ivan, and Michael McCluskey, "Media Effects in an Era of Rapid Technological Transformation: A Case of User-Generated Content and Political Participation." *Communication Theory*, no. 22 (2012): 250–78. doi:10.1111/j.1468-2885.2012.01409.x.

Feenstra, Ramón, and Andreu Casero-Ripollés. "Democracy in the Digital Communication Environment: A Typology Proposal of Political Monitoring Processes." *International Journal of Communication* 8, no. 21 (2014). http://ijoc.org/index.php/ijoc/article/view/2815.

Fuchs, Christian. "Social Media, Riots, and Revolutions." *Capital & Class* 36, no. 3 (2012): 383–91. doi: 10.1177/0309816812453613.

García, Rodrigo Gómez, and Emiliano Treré. "The# YoSoy132 Movement and the Struggle for Media Democratization in Mexico." *Convergence: The International Journal of Research into New Media Technologies* (2014). doi: 10.1177/1354856514541744.

Holloway, John. *Change the World Without Taking Power: The Meaning of Revolution Today* (London: Pluto Press, 2002).

Javaloy, Federico, Álvaro Rodríguez, and Esteve Espelt. *Comportamiento colectivo y movimientos sociales* [*Collective Behavior and Social Movements*] (Madrid: Pearson Educación, 2001).

Jenkins, J. Craig. "Resource Mobilization Theory and the Study of Social Movements." *Annual Review of Sociology* (1983): 533.

Jordan, Tim. "Social Movements and Social Change." *CRESC Working Papers*, London, CRESC, 2005. http://www.archive.cresc.ac.uk/documents/papers/wp7.pdf.

Member of the Spanish 15M Movement. Quoted in Eva Cavero. "Esto está lleno de infiltrados [This is full of infiltrators]." *El País* (2011, June 5). http://elpais.com/diario/2011/06/05/madrid/1307273054_850215.html.

Minavere V. Bardici. "A Discourse Analysis of the Media Representation of Social Media for Social Change—The Case of Egyptian Revolution and Political Change." Master's Thesis (Malmö University, 2012). http://muep.mah.se/handle/2043/14121.

Morozov, Evgeny. *The Net Delusion: The Dark Side of Internet Freedom* (New York: Public Affairs, 2011).

Occupy Wall Street. "About Us." http://occupywallst.org/about/. Accessed November 20, 2014.

Onder, Engin. Quoted in José M. Calatayud, "Los jóvenes del Parque Gezi [Youth in Gezi Park]." *El País*. (2013, June 9). http://internacional.elpais.com/internacional/2013/06/09/actualidad/1370781813_691701.html.

Paredes, Carlos. Quoted in María J. Hernández. "Cuando asoma el desencanto [When disenchantment peeps out]." *El Mundo* (2012, May 15). http://www.elmundo.es/elmundo/2012/05/10/espana/1336646985.html.

Pérez de Acha, Gisela. "La democracia de #YoSoy132 [The democracy of the #YoSoy132]." *Animal Político* (2012, September 19). http://www.animalpolitico.com/blogueros-blog-invitado/2012/09/19/la-democracia-de-yosoy132/#ixzz33PD7zyBc.

Poster, Mark. "Cyberdemocracy: Internet and the Public Sphere." In *Internet Culture*, edited by D. Porter (New York: Routledge, 1997), 201–18.

Rakesh, Sharma. "Is The Twitter Revolution Dead?" *Forbes*. http://www.forbes.com/sites/rakeshsharma/2014/04/01/is-the-twitter-revolution-dead/. Modified January 4, 2014.

Raschke, Joachim. "Sobre el concepto de movimiento social [The concept of social movement]." *Zona Abierta* no. 69 (1994): 121–34.

Rolfe, Brett. "Building an Electronic Repertoire of Contention." *Social Movement Studies* 4, no. 1 (2005): 65–74. doi: 10.1080/14742830500051945.

Ronfeldt, David, John Arquilla, Graham Fuller, and Melissa Fuller. *The Zapatista Social Netwar in Mexico* (Santa Monica, CA: RAND, 1998), 1.

Sandoval-Almazan, Rodrigo, and Juan R. Gil-Garcia. "Cyberactivism through Social Media: Twitter, YouTube, and the Mexican Political Movement I'm Number 132." In *Proceedings of the 46th Hawaii International Conference on System Sciences (HICSS)* (Wailea, HI: IEEE, 2013), 1704–1713. doi: 10.1109/HICSS.2013.161.

Tania. Quoted in Héctor de Mauleón, "De la red a las calles [From the Web to the Streets]." *Nexos Online* (2012, September 1). http://elecciones2012mx.wordpress .com/2012/09/01/de-la-red-a-las-calles-hector-de-mauleon-blog-nexos-en-linea/.

Tarrow, Sydney. *Power in Movement: Social Movements and Contentious Politics* (Cambridge: Cambridge University Press, 1998).

Thornton, Alinta L. "Does the Internet Create Democracy?" *Ecquid Novi: African Journalism Studies* 22, no. 2 (2001): 126–47. doi:10.1080/02560054.2001.9665885.

Van Laer, Jeroen, and Peter Van Aelst. "Cyber-Protest and Civil Society: The Internet and Action Repertoires in Social Movements." In *Handbook on Internet Crime*, edited by Y. Jewkes and M. Yar (Portland, OR: Willan Publishing, 2009), 230–54.

Washington Times. "Iran's Twitter Revolution," Editorial. http://www.washington times.com/news/2009/jun/16/irans-twitter-revolution/. Modified June 16, 2009.

Wyatt, Sally. "Technological Determinism Is Dead; Long Live Technological Determinism." In *Philosophy of Technology: The Technological Condition: An Anthology*, edited by R. C. Scharff and V. Dusek (Sussex, UK: John Wiley & Sons, 2013), 456–466.

Zibechi, Raúl. "Los movimientos sociales latinoamericanos: tendencias y desafíos [Latin American social movements: trends and challenges]." *OSAL: Observatorio Social de América Latina*, no. 9 (2003): 187.

10

You Say You Want a Revolution?
The Internet's Impact on Political Discussion,
Activism, and Societal Transformation

James D. Ponder and Rekha Sharma

INTRODUCTION

There are two primary positions when it comes to perceptions of the Internet: utopian and dystopian. Indeed, at the dawn of the Internet age many thought that the instantaneous information opportunities afforded by this new mass media platform would shepherd in a new utopian age in politics. Journalists no longer would be limited by the traditional constraints of the information-gathering process (e.g., finding stories, seeking out information) and could seek out new areas for inquiry, as the Internet would provide both content and opportunities for identifying sources. Citizens no longer would be constrained by the limitations of their own region's newsgatherers. Citizens would freely seek out unbiased information from various sources, or use a variety of sources to ensure that the information they retrieved was credible. Additionally, the numerous opportunities for interaction provided by the Internet would allow people to thoroughly vet their ideas and expose themselves to differing views.

Conversely, the dystopian view suggests that the rise of the Internet would lead to the downfall of society. People would become more insular in terms of their interactions with other people and with the information they would seek. Journalists would rely on unreliable sources for information and could more easily fabricate information, resulting in numerous false but salacious stories. Further, arguments from the dystopian side state that advances in the Internet would allow for more monitoring of the populace by the government and additional governmental control over the types of information available to the public.

Although this polarization has been used to frame the present anthology, the chapters herein provide more nuanced perspectives on the influence of the Internet. This is because the truth lies somewhere in the middle of that spectrum. Rather than restricting arguments to one side of that binary (i.e., whether the Internet is "good" or "bad"), scholars or anyone interested in the role of the Internet in politics instead should focus on the strengths and weaknesses of the medium. What does the Internet do well? What does it not do well? Answers to these questions will help determine whether, how, or to what extent the Internet impacts political change. Further, for the sake of this particular chapter, the focus is on the effects of the Internet on politics that have led to positive social outcomes.

Scholars have long lamented the decline of American democracy, particularly citing low voter turnout, increased apathy, increased cynicism, and decreased knowledge.[1] Some scholars have indicated that the Internet could play an important role in revitalizing democracy and enhancing civic engagement.[2] Further, the Internet serves as a catalyst for political activism, including acts of insurgency, revolution, and regime change. This chapter traces the history of political candidates' use of the Internet, and covers current research on how political candidates and groups use the Web to inform, persuade, and mobilize the populace.

A BRIEF HISTORY OF POLITICAL CAMPAIGNS ONLINE

The Internet first crept its way into the U.S. political scene in 1992, when Bill Clinton's presidential campaign used the medium as a sort of online brochure. Those who went to the Web site could see the candidates' biographies, read an explanation of stance on particular policies, and read the full text of candidate speeches.[3] Not surprisingly, Clinton's Internet strategy largely was overlooked by the public, journalists, and even his opponent because many U.S. households were not online. In fact, less than 20 percent of the overall U.S. population had a computer in the household.[4]

Clinton, however, apparently realized the potential of the Internet, as it played a key role in his presidency. He even called his terms as president "the first Administration of the Internet Age," whereby Clinton and Gore worked together to unveil the first White House Internet page in 1994, something that was quickly replicated throughout different branches of the government. Clinton's stated goal was to expand citizen access to government information and services online.

In 1996, the Internet began to further make its way into presidential campaigning as it became more well known and was used more; the overall U.S. Internet user population grew to around 20 million subscribers.[5] Further, as a result of Clinton's focus on online activities, his campaign began to use its online portal as more than just a brochure, offering a few donors the opportunity to make donations that resulted in $10,000 being raised for the campaign.[6] Interestingly, Bob Dole was the candidate who first brought the Internet into the limelight when referencing his Web site ("dolekemp96org") in his closing statements of the presidential debate. As a sign of the inherent unfamiliarity of the system, Dole accidentally misdirected viewers to a different site when he forgot to include the period before the ".org" part of the URL. Most users were able to recognize this error; however, the Dole Web site soon crashed after the announcement because it was unprepared for the increased traffic that his closing statement generated.[7]

In the 2000 presidential election, candidate Web sites included position papers, rebuttals against opponents' statements, and appeals for donations.[8] Major news organizations (i.e., CNN, C-SPAN) streamed live audio and images from the 2000 party conventions to interested individuals.[9] John McCain was one of the pioneers in the use of the Internet as a campaign tool. In his 2000 presidential primary campaign, McCain raised more than $5.6 million and recruited more than 18,000 volunteers from the two battleground states of Michigan and California.[10]

HOWARD DEAN'S INTERNET SUCCESS

In 2004, Democratic hopeful Howard Dean used a diverse network of bloggers and donors to rise from a relative unknown to a frontrunner for the nomination in only a few months.[11] In fact, the Dean campaign was hailed by political and media scholars as the first digital campaign.[12, 13, 14] It was Dean's willingness to relinquish control over his campaign that empowered Internet opinion leaders to support and strengthen it. Still, the Dean campaign presented an enigma to political scholars because his vast success in the early stages of the Democratic primary failed to result in the Democratic nomination.

Dean's story can really be understood by closely examining the changing demographics of Internet users; more specifically, those who use the Internet for political purposes. From the onset of the Internet revolution, Web access and usage closely adhered to social demographics. The rich and educated used the Internet more than those with less money and education; women lagged behind men; and Hispanics and African Americans trailed in overall use and knowledge about the Internet.[15] Though most of these usage gaps had narrowed, large inequalities remained.

Liberals dominated the audience for politics online across a wide range of politically relevant activities; from gathering news online to visiting government Web sites, liberals outpaced conservatives by a wide margin.[16] Among self-identified Democrats, frequent visitors to political Web sites were dramatically more liberal than the party as a whole, were more educated, and were disproportionately young. Although some of these characteristics included ideal demographics, the Dean campaign's reliance on young voters—who did not turn out to vote—ended up being one of the campaign's biggest oversights.

Dean additionally attributed most of his success in the presidential primary to his ability to empower bloggers to serve as virtual opinion leaders. Dean believed that the best way to empower the disenchanted voters was to do so via the Internet. He acknowledged listening and paying attention to bloggers, reactions to his speeches and stances on certain issues, and changed those aspects that bloggers did not like. Dean also used the Internet to raise money and organize his "ground game." Through the Internet, Dean organized thousands of volunteers who would go door to door, write personal letters to likely voters, host meetings, and distribute flyers to the voting public. As successful as Dean was in organizing voters and supporters alike through the Internet, he could not win his party's nomination for president.

BARACK OBAMA'S 2008 PRESIDENTIAL CAMPAIGN

Barack Obama's presidential run can be attributed to a multitude of different factors: a stagnant economy, a vastly unpopular president, an unorganized campaign by his general election challenger, and many other factors. Obama's road to the White House was helped by a diverse constituency that included many young voters, people of different ethnic backgrounds, and disgruntled voters—most of whom were reached through a computer-mediated network. Obama used the many different features of the Internet, along with its vast social networking capabilities, to organize and grow his campaign. He was registered on a multitude of social networking sites (i.e., Facebook, MySpace, Flickr, Digg, Twitter, Eventful, LinkedIn, BlackPlanet, FaithBase, Eons, Glee, MiGente, MyBatanga, AsianAve, DNC, PartyBuilder) and used those sites as virtual meeting rooms where he could provide strategies to his many "friends." Obama's campaign staff constantly updated his profile as well and kept adding "friends"; at the end of his campaign, Obama had more than 120,000 followers on Twitter alone, a site where his profile was updated daily.[17]

Obama also used the Internet for fundraising purposes. Obama raised more than $640 million for his campaign using the Internet. In comparison, his opponent, John McCain, raised slightly more than $200 million with the help of federal funding. Obama achieved this large number by taking smaller contributions from his constituency and giving his supporters an easy online process for donating money. To his credit, more than 3 million people donated to his campaign online.[18]

Obama also was unafraid to use the Internet Web site YouTube to post all of his speeches and messages for anyone to view. He also encouraged his supporters to post videos as well, including individuals such as the "Obama Girl" and Will.i.am (of the musical group The Black Eyed Peas). Obama's focus on using the Internet to reach out to young voters was evident, as more than 70 percent of young and first-time voters reported voting for Barack Obama.[19]

THE YOUTUBE DEBATES

One feature of the 2008 election campaign that was unique was the CNN-YouTube debates. Originally the brainchild of CNN Washington Bureau chief David Bohrman, the CNN YouTube debates originated from a desire to reach more people.[20] Building from the fact that YouTube videos had played a significant role in airing videos of candidates behaving badly, YouTube also sought to position itself as a major player in the political arena by teaming with CNN to host a debate with questions asked by the subscribers. In July (for Democrats) and November (for Republicans), candidates participated in two-hour-long debates, moderated by CNN's Anderson Cooper, and which featured citizens asking the candidates questions via 30-second videos posted online. In all, more than 8,000 people participated, with 60 percent of the videos coming from people less than 30 years old.[21]

Questions in the debate seemed to be selected as much for the entertainment value as for their potential to educate voters. For example, some questions featured a talking snowman asking the candidates' views on global warming, and others came from a "lifetime member of the NRA" asking candidates their position on gun control after firing a round from his AK-47. McKinney and Rill,[22] however, found that viewers of both the Democratic and Republican CNN-YouTube debates experienced a significant reduction in overall political cynicism; but it was not a significant difference as compared with traditional forms of viewing debates. McKinney and Rill concluded by saying that this particular type of strategy—aimed at younger citizens—did seem to engage the younger citizens in the political process.

THE 2012 U.S. PRESIDENTIAL ELECTION

Each political race gave candidates, strategists, journalists, and voters a better understanding of the increasing role of the Internet, as well as the enhanced possibilities of social media and mobile technologies for informing, persuading, and mobilizing people. During the 2012 presidential election, Obama's campaign raised $690 million online, often through from donors contributing small amounts of money repeatedly.[23] Many of those donations were the direct result of personalized fundraising e-mails with intriguing taglines.[24] A team of approximately 20 writers tested 18 versions of donation solicitation messages to predict which messages would be most likely to get people to donate to the campaign.[25] The subject line deemed to be most effective, "I will be outspent," exceeded strategists' expectations, as it raked in $2,673,278 for the Obama campaign.[26] Due to the rapid pace of electronic communication, however, every message has a "shelf life," and political campaigns must adapt once a message loses its persuasive power. To that end, strategists recognized the need to keep messages fresh and to communicate to potential donors and voters via the myriad social media channels online.

Compared to the $52 million that the Obama-Biden campaign spent on Web ads, the Romney-Ryan campaign spent only $26 million.[27] According to Moffatt,[28] the Romney-Ryan campaign raised $182 million online between May and November of 2012, and 96 percent of donations were for less than $250. The campaign also reported that 38 million people visited MittRomney.com, and more than 900,000 campaign-related items were purchased via the online store.[29] Romney gained 10,433,367 Facebook "likes," 1,302,785 Twitter followers, and 25,363,122 YouTube views; Ryan's Facebook page totaled 5,183,712 "likes," and he gained 546,738 Twitter followers.[30]

THE INTERNET AND ITS IMPACT ON POLITICAL KNOWLEDGE

This section discusses the role of the Internet and its impact on political knowledge in four specific areas: Overall information available, navigating the available information, determining where people turn to for political information, and examining how people use information available through the Internet in conversations (either online or offline).

In terms of the overall amount of information available to voters, there is no debate that voters today have access to the most information about politicians, political movements, legislation, and even issues than was available ever before. Additionally, the contemporary Internet enables users to choose from far more political news sources than anyone ever had imagined. Recently, scholars have sought to identify how people navigate this vast terrain through the use of online filter systems. [31, 32, 33] In this vein of research, scholars have worried that the vastly expanded information environment would lead to a more polarized public sphere. Even in the face of communication theories such as selective exposure theory or even cognitive dissonance, however, researchers have been unable to confirm that people routinely seek out information that supports their own opinions and avoid information that is contradictory to their views. Instead, scholars have mixed opinions as to whether people today routinely expose themselves to more information supportive or contradictory to their own political views[34] or seek out information that is free from some sort of obvious political bias.[35]

Garrett,[36] however, found that when people do select news stories that challenge their views, they are more likely to spend more time reading such news articles. This lends support to the argument that many posited at the onset of the Internet age: People might spend more time exposing themselves to differing views. Thus, one of the outcomes of having more information available is that people are willing to expose themselves to a variety of differing opinions. Researchers have yet to determine whether this exposure leads to changes in knowledge or if people are merely exposing themselves to counterarguments in an effort to identify potential weaknesses in the arguments of those holding opposing views.

Another key component in the learning process is how users evaluate particular sources' credibility and how that evaluation influences later recall of information. Although television still is the dominant information source for much of the electorate, the Internet has risen to a level that challenges television's place as main source for political information for people in the United States.[37] A recent survey conducted by the Pew Research Center for the People and the Press[38] found that 67 percent of those surveyed reported the television as a main source of political information, whereas 47 percent reported the Internet as a main source of political information. Additionally, although television is the most widely preferred source across age groups, the Internet is the second most popular choice in all age groups except for people aged 65 years or older.

In assessing potential voters' learning from TV news, newspapers, and online news sources during the 2000 election, Eveland, Seo, and Marton[39] found that recall for news stories from television and print sources was greater than that for online newspapers, but that election knowledge (mentally grouping election stories) was greater in the online newspaper condition than it was for traditional media. Increased attention was more significant in the television news condition than in the online or print newspaper conditions. The researchers concluded that print and television news was associated with greater levels for overall news recall as compared to Web exposure, but online exposure helped participants structure election-related knowledge better than the other two types of media in the study did. As compared to television, Internet use also was associated with more accurate recall of election-based information.

In addition to mere exposure to information, scholars also have determined that the reasons people use the Internet have a significant impact on the knowledge gained. For instance, Althaus and Tewksbury[40] found that although people did use the Internet regularly for surveillance—and passing time was another popular reason—political knowledge was positively correlated with using the Internet for surveillance purposes. Further, political knowledge and desire for control were more strongly related to Web use than to traditional news media use. Finally, people who were heavier users of the Internet found the medium to be a superior news source than traditional media.

Additionally, Internet users have more mediated avenues for seeking political information to use to discuss politics with others and to mobilize toward other efforts. Social media sites, along with other media, provide countless channels for people to access information and discuss politics. Much of contemporary political communication scholarship has focused on this fact. Communication mediation models[41] position communication with other people as an important mediator in the acquisition of political knowledge, suggesting that people seek others to help them make sense of the complex political information presented through the media. This social dimension of mediated communication warrants investigation as Delli Carpini and Keeter[42] argued.

> Political learning is affected not only by individual factors, such as one's interest in politics, but also—and often profoundly—by forces external to the individual: the information environment and, more generally, the political context in which learning occurs. The distinctive patterns observed with knowledge of local, state, and national politics suggest that differences in the availability of political information at the three levels of politics interact with variations in individual interest and motivation.

Thus, one of the benefits of the Internet is that it has allowed interested users to obtain political information from a variety of sources in a timely and convenient fashion. Users also can navigate the wealth of information more easily and convert it into support for online and offline argumentation, as well as actionable knowledge for making decisions and engaging in the political process. The Internet provides tools and opportunities for people to locate, amalgamate, and verify a wealth of data, but it also affords people the freedom to incorporate that information into their preexisting knowledge structures selectively or expand their minds to entertain a diversity of viewpoints.

THE INTERNET AND PERSUASION

In its brief history in the political realm, the Internet has made a powerful and lasting impact.[43] This section highlights the research and popular press information that have positioned the Internet as a key component in the political persuasion process. Specifically, it focuses on the role of the Internet in lobbying and the role the Internet has on crafting legislation.

The Internet and social media have also enabled special interest groups to hire public relations firms to mine data on the behaviors, characteristics, preferences, and social networks of people who communicate online. More specific than demographic research typically employed by advertisers and public relations practitioners, the construction of more detailed "psychographic" profiles grew out of the "hypermedia campaign" style that emerged in the 1990s, alongside the Internet and mobile technologies. This new style of campaigning allowed advertisers to target ads to niche audiences on the Internet at far less expense than a television or print ad published for a larger market.[44]

The possibility of communicating anonymously online also has meant that journalists must be especially careful to verify information and source attribution, because the Internet complicates ethical, social, and legal issues in a news environment. The potential obfuscation of the sources of some political messages requires online news outlets to take precautions in their information-gathering practices, their editorial practices, and their reader-feedback policies. On the Internet, individuals can post comments under pseudonyms, so it can be difficult to identify particular users or to hold them accountable for inaccurate or disparaging comments. News outlets must also articulate policies about the transparency of journalists' identities when they choose to share their opinions online. In violation of their publications' ethics policies, a senior editor at *New Republic* magazine as well as a blogger from the *Los Angeles Times* invented "sock puppets"—fabricated identities to confront critics and attack other media outlets.[45] Although journalists have the right to express individual opinions, they also must remember that their actions represent their media organization, so it is imperative that they understand the ethical limits of anonymity in online contexts.

The Internet enables users to cloak their identities using a variety of technological features. For instance, anonymous remailer programs remove information about the origins of an electronic message before forwarding the message to different users. Chains of remailers could protect political activists or dissidents by helping them to expose injustices perpetrated by corporations, government, or media without the threat of retribution. That same technology, however, could enable criminals to operate in cyberspace with impunity. The anonymity the Internet often provides could let people exert profound influence with no fear of backlash or recrimination.[46]

CRAFTING LEGISLATION

People have recently also sought to give more power to average citizens by developing programs that allow Internet users to work together along with interested parties, lawmakers' staff, and others to draft legislation and to write a report explaining how the legislation came to be. The goal of this approach is to reconnect citizens with their government. This phenomenon has started to gain traction across the globe.[47,48] Although these are the initial steps in this direction, it does offer citizens an entirely new opportunity to become even more invested in local, regional, and even national politics.

THE INTERNET AND CIVIC MOBILIZATION

Recent history repeatedly has demonstrated the power of the Internet to unite protestors and activists toward a common purpose, enable grassroots movements to amplify their voices in the marketplace of ideas, and advocate for political change. Kahn and Kellner[49] noted some prominent moments in online oppositional politics, including the EZLN Zapatista movement in Mexico in the early 1990s as well as protests against neoliberalism and multinational corporations (e.g., the "Carnival Against Capital!" demonstrations and the "Battle for Seattle" protests against the World Trade Organization). In the early 2000s, several groups established an online presence to oppose the war in Iraq and related policies and actions implemented by U.S. president George W. Bush.[50]

Tech-savvy activists have armed themselves with mobile devices including pagers, personal digital assistants, notebook and tablet computers, and cell phones equipped with access to global positioning systems and the Internet, and have used these devices to do everything from sharing information to organizing flash mobs.[51] Today, activists also use blogs and wikis to spread awareness of political ideas, manipulate search engines to display links related to their agendas, and even influence mainstream media coverage of political events and viewpoints.[52]

West[53] stated that just as telephones, postal letters, Citizens Band radio, and other forms of communication did in previous decades, the Internet could facilitate social protest. But he worried that commercial forces had dominated most of the important societal conversations online, rendering the Internet a mere façade for democratic discourse and action rather than a true incubator for the goals of the masses. "Hacktivist" groups, however, have exposed government's attempts at electronic surveillance, developed open-source software, and consolidated lists of free wireless Internet access points to allow the public to skirt government and corporate attempts to control cyberspace.[54] If tracing the history of computer-mediated communication provides insight into the evolution of online politics, then it is equally vital to consider emerging channels that undoubtedly will shape politics offline as well. Protestors have connected with one another and with mainstream media outlets when the regimes they opposed limited communication via other channels to silence any dissent and to disrupt coordinated movements of people. As people and candidates in the United States learn to communicate with these new forms of technology to achieve specific political ends, researchers should examine the potential for emerging information networks to provide insight into the way people use the Internet to achieve a variety of political outcomes.

The Internet also has been an instrumental tool in mobilizing people abroad. Its continued growth in use—coupled with globalization—has highlighted the potential for the Internet to serve as a means for social change. At the same time, some governments around the world have sought to limit citizen access to the Internet in efforts to curb such broader political participation and inhibit change. Myanmar, for example, in 2007 blocked access to the Internet to curb demonstrations. China has blocked access to YouTube, Twitter, and Facebook, and in 2009 even shut down access to the Internet in one region during fighting between Uighurs and Han.[55] To operate in China, Google succumbed to the Chinese government's demands and have censored content. China also has blocked content from mobile phone text messaging.

More recently, many of the movements that made up the "Arab Spring" relied on the Internet as a way to connect with protestors, organize demonstrations, and disseminate the word to outside press organizations. Although a number of factors led to the protests and events that toppled authoritarian regimes and forced others to institute significant political reforms, social media such as Facebook, YouTube, and Twitter provided a vehicle for large numbers of people to communicate before, during, and after the protests and uprisings.[56] Even today, a variety of ICTs and social networking sites (such as Facebook and Twitter) are used to spread political ideas, foster and advance social and political movements, and institute change. It therefore is difficult to argue that the revolutions during the Arab Spring would have been successful without the Internet.

In the United States, political "third parties"—such as the Tea Party movement—underscored the importance of the Internet and social media to empower the individual and spread ideas of counterrevolution. O'Hara[57] offered practical advice for Americans who wished to organize their own Tea Party chapters, saying that

> the power of social networking cannot be underestimated. Only with tools like Twitter, Facebook, and e-mail were the tea parties as wildly successful as they were. . . . There simply is no better way to get information to more people more efficiently than through these tools on the Internet. . . . I had 40- and 50-year-olds telling me that they barely navigated e-mail but made themselves open Facebook accounts to be plugged into the tea parties. Counterradicals of every age group must embrace this technology and make it their own.

Despite such declarations, the mobilization potential of the Internet sometimes is met with skepticism. Although West[58] acknowledged the Internet's ability to mobilize groups of individuals efficiently, citing the 2011 "Occupy Wall Street" movements as an example, he argued that "the ability of the new electronic media to transform those movements into lasting social change, or to use the new media as a public sphere whose discourse must be reckoned with, is not yet evident."

Social media enables the acquisition of information such as the location and timing of events, the number of people expected to participate, and the potential for danger (e.g., fires, tear gas, violence).[59] Social media, however, also can trigger emotions such as anger, or feelings of injustice, or a group identity that could motivate people to participate in protests, as was the case in the demonstrations centering on allegations of racial bias of law enforcement in Ferguson, Missouri, and in several other American cities.[60]

For instance, following a grand jury's decision not to indict a police officer for the death of a black man in New York City in 2014, demonstrators used Facebook, Twitter, and Tumblr to organize protests in Times Square, to stage a "die-in" in Grand Central Terminal, and to coordinate similar activities nationwide.[61] The Internet allowed for another form of protest when the group Anonymous claimed responsibility for hacking the Web site of the city of Cleveland, purportedly because the city did not train its police officers sufficiently.[62] The hacking came after Cleveland police responded with lethal force to a call that someone was waving a gun at a playground, though the individual in question turned out to be a 12-year-old African American boy who had been playing with a toy gun.[63]

Tucker[64] noted that people could receive the same types of information and emotional motivation from traditional media sources, but pointed out five features of social media that are unique in political protest contexts:

- Social media can cover multiple developments in a political situation simultaneously and rapidly.
- Social media let individuals plan events, which they cannot do via traditional media.
- Social media are easier to filter, letting people search for information tailored to their own concerns.
- The information selected has been vetted by social networks with which the individual has affiliated, so the information is shared among "friends."
- Social media provide a forum for political issues and protests that have been ignored by mainstream media.

THE FUTURE OF POLITICS IN THE INTERNET ERA

The Internet is not positive or negative per se; it is an amalgamation of tools. Thus, the focus of this chapter is to investigate what is uniquely effective about the Internet in political contexts. The various tools and channels for communication on the Internet—especially when combining social media with mobile technology—make it useful for disseminating information, connecting people with each other, and enabling people to avoid government surveillance. Technology evolves, however, making it important to revisit the issues and questions raised in this chapter as new methods of informing, persuading, and mobilizing emerge.

For instance, protestors typically must stay ahead of the technology curve and utilize the most innovative forms of communication to avoid attempts by government to control the flow of information via mainstream channels. If government entities try to silence voices of dissent or block access to certain media outlets, then protestors must be able to come up with new ways to retrieve and share information. If those in power use technology to track, identify, and punish opponents, then it is important to investigate the promise of anonymity via technology. Users sometimes operate under the illusion of invisibility, believing that communicative acts on social media remain private. Failure to understand the limits of privacy online, coupled with the development of increasingly sophisticated data-mining techniques, might render even the most advanced encryption protocols moot, given enough time. The tug-of-war between those in power and those trying to assert their opposition will play out in zero-sum fashion as technology changes and the people and groups using it adapt to its strengths and weaknesses.

This is an important point to underscore because—like the Internet itself—regime change might not be positive or negative. As a political endgame, revolution might be considered victorious only if the uprising does not replace one form of tyranny with another. Therefore, it is imperative to recognize that the Internet is a neutral conduit for human communication. Among many other uses, people can employ the tools of the Internet to share information, advocate for beliefs, and mobilize toward political action. The effects of those communicative acts, however, only can be characterized as positive or negative as people continue to deliberate, discuss, and debate the ideas being disseminated.

NOTES

1. Michael Delli Carpini and Scott Keeter, *What Americans Know about Politics and Why It Matters* (New Haven, CT: Yale University Press, 1996).

2. Bruce Bimber, "Information and Political Engagement in America: The Search for Effects of Information Technology at the Individual Level," *Political Research Quarterly* 54 (2001).

3. John A. Hendricks and Lynda L. Kaid, *Techno Politics in Presidential Campaigning: New Voices, New Technologies, and New Voters* (New York: Routledge, 2011).

4. National Telecommunications and Information Administration, "A Nation Online: How Americans Are Expanding Their Use of the Internet," http://www.ntia .doc.gov/legacy/ntiahome/dn/anationonline2.pdf, accessed January 4, 2015.

5. Farhad Manjoo, "Jurassic Web: The Internet of 1996 Is Almost Unrecognizable Compard with What We Have Today," *Slate* (February 24, 2009), http://www.slate .com/articles/technology/technology/ 2009/02/jurassic_web.html, accessed January 2, 2015.

6. Hendricks and Kaid, *Techno Politics*.

7. Robert J. Klotz, *The Politics of Internet Communication* (Lanham, MD: Rowman and Littlefield, 2003).

8. Caroline J. Tolbert and Ramona S. McNeal, "Unraveling the Effects of the Internet on Political Participation," *Political Research Quarterly* 56 (2003).

9. Ibid.

10. Stephen. J. Farnsworth and Diana Owen, "Internet Use and the 2000 Presidential Election," *Electoral Studies* 23 (2004).

11. Matthew Hindman, "The Real Lessons of Howard Dean: Reflections of the First Digital Campaign," *Perspectives* 3 (2005).

12. Ibid.

13. Matthew R. Kerbel and Joel David Bloom, "Blog for America and Civic Involvement," *Press/Politics* 10 (2005).

14. Andrew Paul Williams, Kaye D. Trammell, Monica Postelnicu, Kristen D. Landreville, and Justin D. Martin, "Blogging and Hyperlinking: Use of the Web to Enhance Viability during the 2004 US Campaign," *Journalism Studies* 6 (2005).

15. Hindman, "Howard Dean."

16. Ibid.

17. "Barack Obama Wins Online Campaign," www.kabissa.org, accessed November 6, 2008.

18. Colgan, *One Way Obama Has Changed the World*, http://blogs.news .com.au/starsnsnipes/index.php/ news/comments/the_one_change_obama_has_already _achieved/44108, accessed November 6, 2008.

19. "Barack Obama Wins."

20. Scott Leith, "YouTube, CNN to Quiz Hopefuls for President; Network to Host Debates, Use Questions Submitted on Web," *The Atlanta Journal Constitution*, June 15, 2007: 1A.

21. Mitchell S. McKinney and Leslie A Rill, "Not Your Parents' Presidential Debates: Examining the Effects of the CNN/YouTube Debates on Young Citizens' Civic Engagement," *Communication Studies* 60 (2009).

22. Ibid.

23. "Political Strategists Praise Power of Online Fundraising: Obama, Rubio, Paul Cited As Politicians Already Seeing Benefits," *The Center for Public Integrity*, http:// www.publicintegrity.org/2013/04/10/ 12478/political-strategists-praise-power-online -fundraising, accessed March 10, 2014.

24. Joshua Green, "The Science behind Those Obama Campaign E-Mails." *Bloomberg Businessweek* (November 29, 2012), http://www.businessweek.com/articles/2012-11-29 /the-science-behind-those-obama-campaign-e-mails, accessed March 10, 2014.

25. Green, "Obama E-mails."

26. Ibid.

27. B. J. Lutz, "Obama Nearly Doubles Romney in Online Ad Spending." NBC Chicago (November 5, 2012), http://www.nbcchicago.com/blogs/ward-room/obama -romney-2012-election-online-advertising-money-spent-177306601.html, accessed January 2, 2014.

28. Zac Moffatt, "Successes of the Romney and Republican Digital Efforts in 2012," Targeted Victory (December 11, 2012), http://www.targetedvictory.com/2012 /12/11/success-of-the-romney-republican-digital-efforts-2012/, accessed March 10, 2014.

29. Ibid.

30. Ibid.

31. Michael Beam and Gerald M. Kosicki, "Personalized News Portals: Filtering Systems and Increased News Exposure," *Journalism and Mass Communication Quarterly* 91 (2014).

32. Ivan Dylko, Michael A. Beam, Kristen D. Landreville, and Nicholas G. Geidner, "Filtering 2008 Presidential Election News on YouTube by Elites and Nonelites: An Examination of the Democratizing Potential of the Internet," *New Media and Society* 14 (2012).

33. R. Kelly Garrett, "Politically Motivated Reinforcement Seeking: Reframing the Selective Exposure Debate," *Journal of Communication* 59 (2009).

34. Ibid.

35. Beam and Kosicki, "Personalized News Portals."

36. R. Kelly Garrett, "Echo Chambers Online? Politically Motivated Selective Exposure Among Internet News Users," *Journal of Computer-Mediated Communication* 14 (2009).

37. "Low Marks for the 2012 Election," http://www.people-press.org/2012/11/15 /section-4-news-sources-election-night-and-views-of-press-coverage/, accessed December 31, 2014.

38. Ibid.

39. William P. Eveland, Mihye Seo, and Krisztina Martin, "Learning from the News in Campaign 2000: An Experimental Comparison of TV News, Newspapers, and Online News," *Media Psychology* 4 (2002).

40. Scott L. Althaus and David Tewksbury, "Patterns of Internet AND Traditional News Media Use in a Networked Community," *Political Communication* 17 (2000).

41. William P. Eveland Jr., Andrew F. Hayes, Dhavan Shah, and Nojin Kwak, "Understanding the Relationship between Communication and Political Knowledge: A Model Comparison Approach Using Panel Data," *Political Communication* 22 (2005).

42. Delli Carpini and Keeter, *What Americans Know*, 176.

43. Andrew Paul Williams, Kaye D. Trammell, Monica Postelnicu, Kristen D. Landreville, and Justin D. Martin, "Blogging and hyperlinking: Use of the Web to Enhance Viability during the 2004 US Campaign," *Journalism Studies* 6 (2005).

44. Phillip N. Howard, "Digitizing the Social Contract: Producing American Political Culture in the Age of New Media," *The Communication Review* 6 (2003).

45. Andrew Keen, *The Cult of the Amateur: How Today's Internet Is Killing Our Culture* (New York: Doubleday, 2007).

46. James Cowan, "The Orwellian Power of Anonymity; the Web Has Changed the Rules of Politics and Journalism," *National Post*, December 31, 2008, A6.

47. Jennifer McKenzie, "Crowdsourced Internet Freedom Bill a First for Filipino Lawmakers," July 31, 2013, http://techpresident.com/news/wegov/24226/crowd-sourced-internet-freedom-bill-first-philippine-lawmakers, accessed January 1, 2015.

48. Andy, "Finland Wants to Kill Crowdsourced Copyright Law," https://torrent freak.com/finland-wants-to-kill-crowdsourced-copyright-law-141009/, last modified October 9, 2014.

49. Richard Kahn and Douglas Kellner, "New Media and Internet Activism: From the 'Battle of Seattle' to Blogging," *New Media & Society* 6 (2004).

50. Ibid.

51. Ibid.

52. Ibid.

53. Mark D. West, "Is the Internet an Emergent Public Sphere?" *Journal of Mass Media Ethics* 28 (2013).

54. Kahn and Kellner, "Internet Activism."

55. Garth Jowett and Victoria O'Donnell, *Propaganda and Persuasion* (5th ed.) (Thousand Oaks, CA: Sage, 2012).

56. James D. Ponder, Paul Haridakis, and Gary Hanson, "Social Networking in Political Campaigns," in *Encyclopedia of Social Network Analysis and Mining*, edited by Reda Alhajj and Jon Rokne (New York: Springer, 2014).

57. John M. O'Hara, *A New American Tea Party: The Counterrevolution Against Bailouts, Handouts, Reckless Spending, and More Taxes* (Hoboken, NJ: John Wiley & Sons, 2010), 249.

58. West, "Emergent Public Sphere," 158.

59. Joshua Tucker, "Tweeting Ferguson: How Social Media Can (and Cannot) Facilitate Protest," *Washington Post* (November 25, 2014), http://www.washing tonpost.com/blogs/monkey-cage/wp/2014/11/25/tweeting-ferguson-how-social -media-can-and-can-not-facilitate-protest/, accessed January 2, 2015.

60. Ibid.

61. Jennifer Smith and Andrew Tange, "Social Media Help Fuel Protests after New York Officer Not Indicted over Death of Eric Garner," *Wall Street Journal* (December 4, 2014), http://www.wsj.com/articles/social-media-help-fuel-protests-after-new-york-officer-not-indicted-over-death-of-eric-garner-1417662999, accessed January 3, 2015.

62. Nikki Ferrell and Homa Bash. "Anonymous Claims Responsibility for Taking Down Cleveland's Website after Tamir Rice Shooting," NewsNet5 Cleveland (November 24, 2014), http://www.newsnet5.com/news/local-news/cleveland-metro /anonymous-claims-responsibility-for-taking-down-clevelands-website-after-tamir -rice-shooting, accessed January 3, 2015.

63. Ibid.

64. Tucker, "Tweeting Ferguson."

BIBLIOGRAPHY

Althaus, Scott L., and David Tewksbury. 2000. "Patterns of Internet and Traditional News Media Use in a Networked Community." *Political Communication* 17: 21–45.

Andy. Finland Wants to Kill Crowdsourced Copyright Law. Last modified October 9, 2014. https://torrentfreak.com/finland-wants-to-kill-crowdsourced-copyright-law -141009/. Accessed July 24, 2015.

Beam, Michael, and Gerald M. Kosicki. 2014. "Personalized News Portals: Filtering Systems and Increased News Exposure." *Journalism and Mass Communication Quarterly* 91: 59–77.

Beckel, Michael. 2014, May 12. Political Strategists Praise Power of Online Fundraising: Obama, Rubio, Paul Cited as Politicians Already Seeing Benefits. The Center for Public Integrity. http://www.publicintegrity.org/2013/04/10/12478 /political-strategists-praise-power-online-fundraising. Accessed March 10, 2014.

Bimber, Bruce. 2001. "Information and Political Engagement in America: The Search for Effects of Information Technology at the Individual Level." *Political Research Quarterly* 54 (1): 53–67.

Breslow, Harris. 1997. "Civil Society, Political Economy, and the Internet." In *Virtual Culture: Identity & Communication in Cybersociety,* edited by Steve Jones, 236–57. Thousand Oaks, CA: Sage.

Colgan, P. 2008, November 6. One Way Obama Has Changed the World. http://blogs .news.com.au/starsnsnipes/index.php/news/comments/the_one_change_obama _has_already_achieved/44108. Accessed November 6, 2014.

Cowan, James. The Orwellian Power of Anonymity; the Web Has Changed the Rules of Politics and Journalism. *National Post*, December 31, 2008, A6.

Delli Carpini Michael, and Scott Keeter. 1996. *What Americans Know about Politics and Why It Matters.* New Haven, CT: Yale University Press.

Dylko, Ivan B., Michael A. Beam, Kristen D. Landreville, and Nicholas G. Geidner. 2012. "Filtering 2008 Presidential Election News on YouTube by Elites and Nonelites: An Examination of the Democratizing Potential of the Internet." *New Media and Society* 14: 832–49.

Eveland, William P., Andrew F. Hayes, Dhavan Shah, and Nojin Kwak. 2005. "Understanding the Relationship between Communication and Political Knowl- edge: A Model Comparison Approach Using Panel Data." *Political Communication* 22: 423–46. doi: 10.1080/10584600500311345.

Eveland, William P., Mihye Seo, and Krisztina Martin. 2002. "Learning from the News in Campaign 2000: An Experimental Comparison of TV News, Newspa- pers, and Online News." *Media Psychology* 4: 355–80.

Farnsworth, Stephen. J., and Diana Owen, D. 2004. "Internet Use and the 2000 Pres- idential Election." *Electoral Studies* 23: 415–29.

Ferrell, Nikki, and Homa Bash. "Anonymous Claims Responsibility for Taking Down Cleveland's Website after Tamir Rice Shooting." *NewsNet5 Cleveland* (November 24, 2014). http://www.newsnet5.com/news/local-news/cleveland -metro/anonymous-claims-responsibility-for-taking-down-clevelands-website -after-tamir-rice-shooting. Accessed January 3, 2015.

Garrett, R. Kelly. 2009a. "Echo Chambers Online? Politically Motivated Selective Exposure among Internet News Users." *Journal of Computer-Mediated Communication* 14: 265–85.

Garrett, R. Kelly. 2009b. "Politically Motivated Reinforcement Seeking: Reframing the Selective Exposure Debate." *Journal of Communication* 59 (4): 676–99.

Green, Joshua. The Science behind Those Obama Campaign E-Mails. *Bloomberg Businessweek.* November 29, 2012. http://www.businessweek.com/articles/2012-11-29/the-science-behind-those-obama-campaign-e-mails. Accessed March 10, 2014.

Hendricks John A., and Lynda L. Kaid. 2011. *Techno Politics in Presidential Campaigning: New Voices, New Technologies, and New Voters.* New York: Routledge.

Hindman, Matthew. 2005. "The Real Lessons of Howard Dean: Reflections of the First Digital Campaign." *Perspectives* 3: 121–28.

Howard, Phillip N. 2003. "Digitizing the Social Contract: Producing American Political Culture in the Age of New Media." *The Communication Review* 6: 213–45.

Jowett, Garth, and Victoria O'Donnell. 2012. *Propaganda and Persuasion* (5th ed.). Thousand Oaks, CA: Sage.

Kahn, Richard, and Douglas Kellner. 2004. "New Media and Internet: From the 'Battle of Seattle' to Blogging." *New Media & Society* 6: 87–95.

Keen, Andrew. 2007. *The Cult of the Amateur: How Today's Internet Is Killing Our Culture.* New York: Doubleday.

Kerbel, Matthew R., and Joel David Bloom. 2005. "Blog for America and Civic Involvement." *Press/Politics* 10: 3–27.

Leith, Scott. 2007, June 15. "YouTube, CNN to Quiz Hopefuls for President; Network to Host Debates, Use Questions Submitted on Web." *The Atlanta Journal Constitution.* 1A.

Lutz, B. J. "Obama Nearly Doubles Romney in Online Ad Spending." *NBC Chicago* (November 5, 2012). http://www.nbcchicago.com/blogs/ward-room/obama-romney-2012-election-online-advertising-money-spent-177306601.html. Accessed January 2, 2014.

Manjoo, Farhad. 2009. "Jurassic Web: The Internet of 1996 Is Almost Unrecognizable Compard with What We Have Today." http://www.slate.com/articles/technology/technology/ 2009/02/jurassic_web.html. Accessed January 2, 2015.

McKenzie, Jennifer. 2013, July 31. "Crowdsourced Internet Freedom Bill a First for Filipino Lawmakers." http://techpresident.com/news/wegov/24226/crowdsourced-internet-freedom-bill-first-philippine-lawmakers. Accessed January 1, 2015.

McKinney, Mitchell S., and Leslie A Rill. 2009. "Not Your Parents' Presidential Debates: Examining the Effects of the CNN/YouTube Debates on Young Citizens' Civic Engagement. *Communication Studies* 60: 392–406.

Moffatt, Zac. 2012, December 11. "Successes of the Romney and Republican Digital Efforts in 2012." *Targeted Victory.* http://www.targetedvictory.com/2012/12/11/success-of-the-romney-republican-digital-efforts-2012/. Accessed March 10, 2014.

National Annenberg Election Survey. 2008. "Internet As Political Tool Popular, But Television Still Dominates, Annenberg Survey Finds." www.annenbergpublicpolicycenter.org/ NewsDetails.aspx?myId=272.

National Telecommunications and Information Administration (2002). "A Nation Online: How Americans Are Expanding Their Use of the Internet." http://www .ntia.doc.gov/legacy/ntiahome/dn/anationonline2.pdf. Accessed January 4, 2015.

O'Hara, John M. 2010. *A New American Tea Party: The Counterrevolution Against Bailouts, Handouts, Reckless Spending, and More Taxes.* Hoboken, NJ: John Wiley & Sons.

Pew Research Center for the People and the Press. 2012. "Low Marks for the 2012 Election." http://www.people-press.org/2012/11/15/section-4-news-sources-election -night-and-views-of-press-coverage/. Accessed December 31, 2014.

Ponder, James D., Paul Haridakis, and Gary Hanson. 2014. "Social Networking in Political Campaigns." In *Encyclopedia of Social Network Analysis and Mining*, edited by Reda Alhajj and Jon Rokne. New York: Springer.

Smith, Jennifer, and Andrew Tangel. 2014, December 4. "Social Media Help Fuel Protests after New York Officer Not Indicted over Death of Eric Garner." *Wall Street Journal.* Accessed January 3, 2015. http://www.wsj.com/articles/social-media -help-fuel-protests-after-new-york-officer-not-indicted-over-death -of-eric-garner-1417662999.

Smith, S. D. "How Many Volunteers Did Obama Have?" https://my.barackobama. com/page/community/ post/trishaifw/gGxZYv/commentary. Accessed November 5, 2008.

Tolbert, Caroline J., and Ramona S. McNeal. 2003. "Unraveling the Effects of the Internet on Political Participation." *Political Research Quarterly* 56: 175–85.

Tucker, Joshua. 2014, November 25. "Tweeting Ferguson: How Social Media Can (and Cannot) Facilitate Protest." *Washington Post.* http://www.washingtonpost. com/blogs/monkey-cage/wp/2014/11/25/tweeting-ferguson-how-social-media -can-and-can-not-facilitate-protest/. Accessed January 2, 2015.

Waldman, Steven. 2011. *The Information Needs of Communities: The Changing Media Landscape in a Broadband Age.* Washington, DC: Federal Communications Commission.

West, Mark D. 2013. "Is the Internet an Emergent Public Sphere?" *Journal of Mass Media Ethics* 28: 155–59. doi: 10.1080/08900523.2013.792702.

White House. (2009). "The Clinton-Gore Administration: A Record of Progress." http://clinton5.nara.gov/WH/Accomplishments/eightyears-09.html. Accessed January 4, 2015.

Williams, Andrew Paul, Kaye D. Trammell, Monica Postelnicu, Kristen D. Landreville, and Justin D. Martin. 2005. "Blogging and hyperlinking: Use of the Web to Enhance Viability during the 2004 US Campaign." *Journalism Studies* 6: 177–86.

Ground-Up Expert: Everyday People and Blogs

Richard J. Batyko

In August 2002, Julie Powell, a frustrated office worker who also was an aspiring gourmet cook, decided to prepare all 524 of Julia Childs' recipes found in Childs' book *Mastering the Art of French Cooking*[1] in 365 days. Powell also chose to document her journey in a blog. Over the course of that year, the blog attracted increasing amounts of attention, culminating in a 2005 Little, Brown and Company book, *Julie and Julia: 365 Days, 524 Recipes, 1 Tiny Apartment Kitchen*[2] and in the 2009 Sony Pictures movie adaptation of the book, directed by Nora Ephron, titled *Julie & Julia*.[3] In 2009, Powell was awarded an honorary diploma from Le Cordon Bleu, the same cooking school from which Child graduated in 1951.[4]

Although the trajectory of this individual—from an unknown person, to book author, to the subject of a major motion picture—might be an extreme case of blogging one's way to fame, it is an example of how individuals have used the tool to promote their passions and to find an audience. Such an accession into the limelight was difficult prior to the advent of the Internet, blogging, and social media. The barrier to finding an audience no longer is an affordable communications channel accessible to nearly all.

This chapter considers how some bloggers rose to fame from their keyboards. It also examines the ongoing debate about whether blogging is still a valid method of finding one's audience. Is the term "blogger" becoming passé? Additionally, it evaluates the impact that ever-changing technology has on the art of blogging.

BLOGGING TO FAME

It takes time, ability, and sometimes a little luck to go from an unknown to a must-read blogger on an international stage. One thing bloggers have going for them is that Internet users always are looking for content on the topics that interest them. According to *Social Media Today*, in 2013, 23 percent of Internet users' time was spent reading blogs and 77 percent of Internet users read blogs. Of course, there is a lot of "noise" in the blogosphere. In 2013, 6.7 million people blogged on blogging sites and 12 million used a social media channels to blog.[5] To put those numbers in context, the worldwide population as of January 2014 was 7.09 billion. Of that, 2.43 million use the Internet (35 percent penetration), 1.85 million are active on social networks (26 percent penetration), and 6.57 million are mobile subscribers (93 percent penetration).[6]

Some people not only have blogged their way to fame; they also have created their own income streams. Fourteen percent of bloggers earn their salaries by blogging, with an average salary of $24,086.[7] A few have parlayed their passion into multimillion-dollar operations. The following sections provide a sampling of the people who have found their audiences and then blogged their way to success.

Lisa Leake

Lisa Leake thought she was a conscientious mom providing healthy food choices for her family. But after she and her husband Jason saw an Oprah show in 2010 featuring author Michael Pollan—who wrote the book *In Defense of Food*[8]—Leake had an awakening. After reading Pollan's book, Leake decided to make changes in her family's diet. She, her husband, and their two small girls pledged to go 100 days without eating highly processed or refined foods—a challenge she opened to readers on her blog.

> So soon after we started making changes, I launched a blog called The Food Illusion . . . and began to build an audience. After a few months of blogging I decided it was time to do something big, something bold, and something that would get as many other people as we can to not only read about eating real foods, but to also make a commitment to this important change.[9]

The blog's audience grew to 4 million readers. Next came Leake's Web site, www.100daysofrealfood.com, followed by a book by the same name published by William Morrow Cookbooks.[10]

Michelle Mismas

Michelle Mismas has a thing for nail polish. She took her passion to the Web in 2007 as a nail-focused beauty blogger. Her blog caught the attention of the popular makeup blog, Blogdorf Goodman, and Andrea Lustig, a beauty editor at *Glamour*. It now gets 250,000 unique visitors each month. In 2009, Mismas was making enough money from ads and sponsorships to leave her job as an accountant and blog full-time.[11] In 2013, among fierce competition, Mismas' AllLacqueredUp.com site beat out 248 other nominees to take top honors in a contest conducted by *Marie Claire*.[12]

Brianne Manz

Brianne Manz is a "mommy blogger" focused on all things related to children. Her blog, *Stroller in the City*, is about city living, kids' fashion, and all things related to raising kids—especially in an urban environment. As Manz told *What a Brand*,

> Little did I know that this little blog I decided to start almost 5 years ago would turn into this space of influence, creativity, and tons of opportunity. I originally started my blog to reference the fun places and cool brands I was discovering being a mom. Back when my son was born, boys fashion was very hard to come by. Now, it has been a wonder[ful] outlet for me to share what it's like raising three kids in NYC.[13]

Manz frequently is cited in blogs and the press on mommy topics—including in a 2014 feature on back-to-school trends for *Extra TV*.[14]

Mark Sission

Mark Sission is a former elite endurance athlete who set a lofty goal in 2006: to change the lives of 10 million people. When he first published MarksDailyApple.com, Sission's followers numbered in the hundreds. His next goal was to build his Web site into one of the leading health resources on the Internet. Today, 150,000 unique visitors view his site every day and it ranked as the No. 1 blog for men by *The Modest Man*.[15]

Stephanie Le

"Stephanie Le used to claim she could make most dishes she tried at restaurants. Her husband challenged her to do it. And thus, the beginning of *I Am a Food* blog," wrote *Food & Wine* magazine about this blogger.[16] Stephanie calls Vancouver home, but at this writing is living in Tokyo. Her blog, which is just two years old, was named best food blog of 2014 by *Saveur*.[17] How did this happen so fast? Here is how Le answered that question in a reply comment on her site, "I'm not too sure how I got a following—in my eyes, I'm still a baby blog next to the giants of the food blog world. The followers did not come right away at all."[18] Only in the Internet world can someone go from being an unknown to one of the best known in a two-year period—in a "noisy" environment such as food—be considered slow! Stephanie's draw allowed her to publish her first book, *Easy Gourmet*, from Page Street Publishing.[19]

Alborz Fallah

Alborz Fallah's *Car Advice* blog (CarAdvice.com.au) is this entrepreneur's eighth blog and fourth business. Fallah's foray into commerce began with selling advertising space on Geocities at age 14. The Australian-based *Car Advice* site, which began as a blog in 2006 as a result of Fallah's passion for cars, has been ranked ninth in the world among Google Adsense earners.[20] Today it has grown into much more than a blog. In 2014, *Car Advice* had a staff of 20, a board of directors, revenues estimated at U.S. $5 million a year, and has been valued at approximately U.S. $20 million.[21]

THE FUTURE OF BLOGGING

When discussing individuals who have turned their blogs into a business from the ground up, it is worth considering how businesses are using blogs. The top-down view can be instructive when considering the future of blogging.

If current research is any indication, companies seem to be losing interest in hosting blogs on their sites, despite research that shows blogs can be helpful in driving traffic. According to *Social Media Today*, small businesses with blogs generate 126 percent more leads. Google's Hummingbird finds Web sites with blogs have 434 percent more indexed pages. In the United States, 61 percent of consumers have made a purchase based on a blog post and 81 percent of consumers trust advice and information from blogs.[22] With such compelling data, why would companies reconsider blogging's value?

The fact is that after years of adding blogs to their external Web sites, companies are slowing down. A 2014 study conducted by Nora Ganim Barnes, PhD, at The Center for Marketing Research, found 157 companies on the Fortune 500 list (31 percent) have public-facing corporate blogs, a decrease of 3 percent in the number of blogging companies from 2013 numbers.[23]

As the study explains, it is premature to try to determine whether this is the beginning of a Fortune 500 movement away from this iconic tool, toward replacing it with newer communications tools. At present, there is no indication that blogging in other business sectors is waning.[24]

One company that is in the business of content, *The New York Times*, is pulling back significantly on its blogs. In July 2014, *The New York Times* assistant managing editor Ian Fisher told The Poynter Institute, "We're going to continue to provide bloggy content with a more conversational tone. . . . We're just not going to do them as much in standard reverse-chronological blogs." The *Times* has been "moving away from blogs over the past year and a half," stated spokesperson Eileen Murphy. Of the paper's current blogs, "[a]lmost half of them will be gone as a blog or will have merged into something else."[25]

Some *New York Times* blogs are quite popular, but others "got very, very little traffic, and they required an enormous amount of resources, because a blog is an animal that is always famished." Fisher said he thinks that the "quality of our items will go up now, now that readers don't expect us to be filling the artificial container of a blog." Another issue is that "[v]ery few people went to the blog landing pages," said Fisher. Most enter "sideways," through a shared link or a link on the *Times'* homepage.[26]

In the same interview, Fisher credited blogging with training a generation of reporters to write for the Web, but the Web they trained for is not the same any longer. Freelance journalist and cultural critic Mel Campbell agrees. In a post on *The Guardian* Web site, she wrote, "[B]logging has indelibly influenced mainstream news reporting, which is now much more immediate, informal, link-rich and inclusive of reader comments. When I taught online journalism at Monash University from 2009–11, students published their assignments on WordPress blogs."[27]

It is clear that *The New York Times* has determined blogging is so similar to their regular feature and news reporting that they do not need as many blogs. Such a move makes sense for a content company, but what about companies in other industries?

NEW TECHNOLOGIES AND BLOGGING

The Fortune 500 consistently has shown a preference for Twitter over Facebook, and the Inc. 500 prefers Facebook. This continued in 2014 with 80 percent of the Fortune 500 on Facebook and 83 percent using Twitter, although the gap is narrowing. Last year, 70 percent of the Fortune 500 used Facebook and 77 percent had Twitter accounts.[28]

Of the Fortune 500, 413 companies (83 percent) have corporate Twitter accounts and have "tweeted" in the past 30 days. This represents a 6 percent increase from last year. Seven of the top 10 companies (Wal-Mart Stores, Exxon, Chevron, Phillips 66, General Motors, General Electric, and Ford Motors) consistently post on their Twitter accounts. Berkshire Hathaway, Apple, and Valero Energy do not tweet.[29]

Of the Fortune 500, 401 (80 percent) now are on Facebook. This represents a 10 percent increase from last year. All of the top 10 companies (Wal-Mart Stores, Exxon Mobil, Chevron, Berkshire Hathaway, Apple, Phillips 66, General Motors, Ford Motors, General Electric, and Valero Energy) have corporate Facebook pages.[30]

From the ground up to the top down, social media is making an impact on blogging. Harvard's Nieman Lab noted that

> [i]nstead of blogging, people are posting to Tumblr, tweeting, pinning things to their board, posting to Reddit, Snapchatting, updating Facebook statuses, Instagramming, and publishing on Medium. In 1997, wired teens created online diaries, and in 2004, the blog was king. Today, teens are about as likely to start a blog (over Instagramming or Snapchatting) as they are to buy a music CD. Blogs are for 40-somethings with kids.[31]

Another telltale sign of a change of behaviors in the blogosphere is that many bloggers are reporting that fewer comments are left by readers within the blogs. Comments are an important component of many blogs because it is the way readers engage with the author. The readers' comments also add to the visibility of the blog, which helps to build an audience. Why are readers leaving fewer well-thought-out comments? Benjie Moss, an editor at Webdesigner Depot, explains,

> There was a time when email was fun; back when you set up your first Hotmail account, and received your first mail. Then, about two minutes later, our inboxes were flooded with spam. What happened when sorting through the dross in our inboxes became too time-consuming? We turned to social networks for our messaging, and the same process is now being applied to blog comments; where once we posted a thoughtful response, now we tweet a short quip.[32]

Not everyone is excited about the decline in readers' comments at the expense of social media access. As Moss states:

> [O]ne thought keeps returning to me: That is, that an article with no comments is one individual's ideas; when comments are introduced the ideas become truly public domain. . . . When comments are attached to an article, it creates a repository for the ideas to be recorded and to grow. If social media commoditizes individuals, then comments commoditize ideas. If we value the community over individuals, we won't make the leap to social media just yet.[33]

THE IMPACT OF STREAMING AND SOCIAL MEDIA ON BLOGGING

Clearly, the blogosphere is changing. What seismic shift happened in the Internet world to cause the next generation to use new approaches to convey their thoughts and opinions? Two factors account for much of the shift—streaming and social media.

First, consider streaming. Many are familiar with streaming as it applies to music and movies. For those not familiar with the technology, streaming is real-time access to content. Consider music as an example. A person can purchase digital music from services such as iTunes and download a song. That is not streaming. A person also can connect to a streaming service such as Spotify and receive a steady "stream" of music that is not downloaded to a computer or device. That is streaming.

So what does streaming have to do with the way we share content through blogging? Betaworks' John Borthwick provides a good explanation.

> In the initial design of the web, reading and writing (editing) were given equal consideration—yet for 15 years the primary metaphor of the web has been pages and reading. The metaphors we used to circumscribe this possibility set were mostly drawn from books and architecture (pages, browser, sites etc.). Most of these metaphors were static and one-way. The stream metaphor is fundamentally different. It's dynamic, it doesn't live very well within a page and still [is] very much evolving. . . . A stream. A real-time, flowing, dynamic stream of information—that we as users and participants can dip in and out of and whether we participate in them or simply observe we are a part of this flow.[34]

Now imagine what this real-time, always moving environment does to blogs that are traditionally static, existing permanently in one place. In this world, a blog posted on a webpage is swept away in the stream in a microsecond. Add to this scenario the referenced reduction in the number of comments left by readers who viewed the blog—which is a key way the existence of a blog is shared through user networks. How does a blog find a larger audience? After all, many of the bloggers who went from unknown to famous made the transition by garnering a bigger audience.

Aspiring bloggers need not be concerned. Coming into play in the blogosphere is social media, as well as other technologies designed to embrace the stream. Erick Schonfeld of *The Atlantic* explains the interaction of the stream and the Web.

> Information is increasingly being distributed and presented in real-time streams instead of dedicated Web pages. The shift is palpable, even if it is only in its early stages. Web companies large and small are embracing this stream. It is not just Twitter. It is Facebook and Friendfeed and AOL and Digg and Tweetdeck and Seesmic Desktop and Techmeme and Tweetmeme and Ustream and Qik and Kyte and blogs and Google

Reader. The stream is winding its way throughout the Web and organizing it by nowness.[35]

Today, a blog that might see a hundred hits a day on average suddenly could see a swarm of readers—perhaps thousands—arrive one day and then be gone the next. How? A blog covering a particularly hot topic could become visible to a well-connected reader who passes it on through a social media site and then, almost instantaneously, the link is spread virally and draws a substantial number of readers. Although blogs of old built readership in a slow, linear fashion, today a blog could find an enormous audience in seconds.

Bloggers must consider this new environment when planning and executing blog content. In the past, readers who left comments under a blog post helped build the credibility of the blogger. Today, readers are less interested in the blogger's credibility and more focused on their own. Instead of leaving a comment under someone's blog, they will post a brief comment about it on their own social media networks and add a link to the blog, thereby building the reader's value to his or her own network and to a lesser extent also showcasing the blog.

Bloggers also must address a more fundamental question: What is a blog? Is it a 1,500-word essay posted on the author's or a third-party's Web site or is it a short social media post? If the intent is to draw readers to one's content, then perhaps a short social media push is enough of a "blog" to get readers to one's source. As Borthwick explains,

> Today context is provided mostly via social interactions and gestures. People send out a message—with some context in the message itself and then the network picks up from there. The message is often retweeted, favorited, liked or re-blogged, it's appropriated usually with attribution to creator or the source message—sometimes it's categorized with a tag of some form and then curation occurs around that tag—and all this time, around it spins picking up velocity and more context as it swirls.[36]

RUMORS OF BLOGGING'S DEATH COULD BE GREATLY EXAGGERATED

What does all this mean? Is blogging dead, dying, or simply evolving? The answer seems to be the latter. In 1996—an eon ago in Internet years—William Henry "Bill" Gates III, of Microsoft, popularized the phrase "content is king" in an essay.[37] However it is defined, packaged, or distributed, readers seek meaningful, thoughtful, and original content. Streaming and social media are paradigm-changing technologies that cut across many communication and distribution channels, but they are not a replacement for content. Regardless of whether the content is labeled "blog" seems irrelevant, and some even might say it is passé to use the term "blogging."

Harvard's Kottke says it well.

> So, R.I.P. The Blog, 1997–2013. But this isn't cause for lament. . . . All media on the web and in mobile apps has blog DNA in it and will continue to for a long while. Over the past 16 years, the blog format has evolved, had social grafted onto it, and mutated into Facebook, Twitter, and Pinterest and those new species have now taken over. No biggie, that's how technology and culture work.[38]

Campbell has a slightly different point of view, and states:

> Blogging persists, of course. But it's mostly for adults—professionalized to the point where the old "bloggers vs journalists" debates now seem hopelessly quaint. Maintaining a personal blog has become entrepreneurial: a job that earns an income through display advertising, network marketing, e-books and blog-to-book deals.[39]

The fact is that the stream and social media have opened up vast new channels for individuals to find an audience and build an income stream, if not a business. For that reason blogging is alive and well, even its name has changed. No matter the terminology or technology, finding an audience remains a key function of marketing, so there is an insight into marketers' plans for blogging. A 2014 study Michael Stelzner published in the *Social Media Examiner* found, "a significant 68 percent of marketers plan on increasing their use of blogging, making it the top area marketers will invest in for 2014."[40]

The lesson this chapter provides is that bloggers must be more savvy today than they've been in years past if they want to leverage the stream and social media. In doing so, bloggers are entering a world that is rather demanding. As Madrigal states,

> The great irony is that we got what we wanted from the stream: a way to read and watch outside the editorial control of editors, old Yahoo-style cataloging, and Google bots. But when the order of the media cosmos was annihilated, freedom did not rush into the vacuum, but an emergent order with its own logic. We discovered that the stream introduced its own kinds of compulsions and controls. Faster! More! Faster! More! Faster! More! And now, who can keep up? There is a melancholy to the infinite scroll. Wouldn't it be better if we just said . . . "Let's do something else? Let's have the web be a museum or a curio box or an important information filter or an organizing platform. Or maybe let's just let the web be the web again, a network of many times, not just now.[41]

Technology is a treadmill that never stops, so we must adapt to its current state and not wait for it to slow or stop so we can more comfortably jump on. With that in mind, here are some tips from a social media professional on how everyday people can take advantage of streaming and social media to bring visibility to their blogs.

HOW EVERYDAY PEOPLE CAN LEVERAGE NEW TECHNOLOGIES TO FIND AN AUDIENCE

Jessa Hochman, a social media, content marketing, and inbound marketing consultant offers her suggestions for getting your blog noticed.

When beginning your blogging/content marketing journey, it may be beneficial to master one or two social networks rather than diving into them all at once. No two social networks function the same way, so try not to copy and paste from one site to another. I'd suggest starting small depending on your industry. Once you've mastered a few networks, move onto another. I've prepared a list below with tips about how to use each social network successfully.

- Turn your blog into multiple tweets.

The thing about Twitter is that less than half of its users visit the site daily. So, when you tweet about a blog, don't just tweet about it once. Two of the world's fastest-growing Web sites, Upworthy and Buzzfeed, do something on Twitter that's proven to work. Check out these numbers: Buzzfeed receives over 150 million monthly unique visitors; Upworthy (the fastest-growing media company in history) entertains 28 million.[42] How do they do it? They "Repurpose." The average blog is 500 words and [T]witter limits users to only 140 characters (minus 23 for images and an additional 23 for outside links). Instead of trying to sum up your entire blog in 94 characters—separate it, and write several tweets about all the different topics your blog discusses, turn the topics into hashtags and write multiple, timeless tweets.
    Things to try:

    - Mentioned someone in your blog? Tweet it to them. They might retweet it.
    - Don't bombard followers with tweets about one blog all at once. Schedule tweets into the future using scheduling tools. I use Sprout Social.
    - Don't give too much away—leave readers wanting more so they click on that link.

- Boost posts on Facebook to reach your target audience.

Can you simply describe your ideal customer? This is something I ask all of my clients and normally they quickly describe a person based on interests, age, location, etc. Similarly, Facebook provides a tool for delivering your content to that ideal reader. Boosting your posts on Facebook is not "buying likes," it's targeting content. Imagine you write a blog about home decorating; using Facebook, you can target people based on everything from age and gender to which magazines they read and the HGTV TV shows they most enjoy. It's a powerful tool and it's relatively inexpensive (today) to use—so take advantage!

Things to try:

- Include an image in your boosted post but remember the 20-percent rule: Images in your ad on Facebook "may not include more than 20 percent text in the image to ensure people only see high-quality content."[43]
- Have some fun: Facebook is for entertainment. Don't just ask the audience to weigh-in, make it easy and amusing for them to do so. Post two different images and ask which the audience prefers. Is your blog structured as a list? Ask the audience what they'd like to add to the list—and then add it and give them credit. Content marketing is a community effort.
- Remember that Facebook is algorithm-driven. Facebook qualifies posts based on how many users interact with them—so be interactive!
- Allow users to find you on Pinterest.

If your content is focused on creative topics or project-driven concepts, I recommend using Pinterest. Pinterest functions in two ways: It allows users to visually "bookmark" content on the Web by pulling images from a site, and it allows content creators and bloggers to pin their own content to reach a wider audience. Fortunately, Pinterest is highly searchable. It accepts both hashtags like social networks and search terms like Google to allow users to find the content they're looking for.

Things to try:

- Use specific keywords and hashtags when describing your pins so that users can find you.
- Create infographics—it's easy! Adding some design elements and text to your pins can encourage users to click and "repin." Create infographics easily and inexpensively with programs like Canva and PicMonkey.

- Post teasers on Instagram.

Instagram was derived from a mobile app and then later (in 2011) became a true "social network." It still doesn't function like other networks; it has never allowed the user to add clickable links to posts. Unlike Pinterest, it's impossible to post an image from your blog and get "clicks" via Instagram. This doesn't mean you should rule it out entirely. Be creative.

Things to try:

- Post a "sneak peek" to an upcoming blog.
- Post a #TBT (throw-back Thursday), to a timeless blog that may be relevant to what people are posting about today. Seeing lots of #snowday posts? Why not #TBT to an image from the blog you wrote about winter styling trends?
- Always use lots of topical hashtags. This is a great way to accumulate like-minded followers.[44]

One more tip, this one not related to streaming or social media: manage your time wisely. As Steven wrote in *The New York Times*, "A tricky thing to avoid as a full-time blogger, considering that the Internet never sleeps, readers want fresh content daily and new social media platforms must be mastered and added to the already demanding workload the economic challenges of blogging full time."[45]

Emily Schuman, who writes the Cupcakes and Cashmere blog (and wrote a book by the same name), addressed this issue in a post on her blog:

Ultimately, it's normal to feel worn down from time to time. For me, even admitting that I'm going through a slump has proven to be effective, since I'm then able to take the necessary steps to fix it. And as my blog has evolved, it [has] forced me to realize that I can't do everything on my own, which is why I brought on some key people to help the business continue to grow.[46]

CONCLUSION

Everyday people have and will continue to find an audience when they produce content that others seek. Today there is an increasing number of channels to use to find that audience. To capitalize on these channels bloggers must understand the available technology. As J. D. Rucker writes on Soshable, "Great content. Great promotion. That's it. If your content is strong and you're able to promote your site properly and gain credibility on social media and search, your blog will grow and be successful."[47]

This chapter touched the tip of the iceberg of blogging, social media, and streaming, and as for any point-in-time writing on technology, things have changed since this book's publication. One thing will not change: Content really is king. Technology simply enables bloggers to travel from the ground up to international visibility.

NOTES

1. Julie Child and Simone Beck, *Mastering the Art of French Cooking*, new rev. ed. (New York: Knopf, 1983).

2. Julie Powell, *Julie and Julia: 365 Days, 524 Recipes, 1 Tiny Apartment Kitchen* (New York: Little, Brown and Co.) 2005.

3. *Julie & Julia* (film), 2009, directed by Nora Ephron (New York: Columbia Pictures).

4. "Le Cordon Bleu Welcomed Julie Powell," *Le Cordon Bleu*, http://www.cordon bleu.edu/ index.cfm?fa=NewsEventFrontMod.DisplayNewsPage&ElementID=282& SetLangID=1, accessed December 3, 2014.

5. Mike McGrail, "The Blogocomony: Blogging Stats," *Social Media Today*, http://www.socialmediatoday.com/content/blogconomy-blogging-stats-infographic, last modified August 28, 2013.

6. Arden Hepburn, "Global Digital Statistics 2014," *Digital Buzz*, http://www .digitalbuzzblog.com/slideshare-global-digital-statistics-2014-stats-facts-study -presentation/, last modified January 8, 2014.

7. Mike McGrail, "The Blogocomony: Blogging Stats."

8. Michael Pollan, *In Defense of Food* (New York: Penguin Books, 2009).

9. Lisa Leak, "About," *100 Days of Real Food*, http://www.100daysofrealfood.com /about/, accessed December 12, 2014.

10. Lisa Leak, *100 Days of Real Food* (New York: Harper Collins, 2014).

11. Alanna Okun, "Meet the 'Lacqueristas' Who Rule Nail Art on the Internet," *BuzzFeed*, http://www.buzzfeed.com/alannaokun/meet-the-lacqueristas-who-rule-nail -art-on-the-i#.gy7k18kly, last modified November 28, 2012.

12. Jennifer Goldstein, "Our First Annual Most Wanted Beauty Awards," *Marie Claire*, http://www.marieclaire.com/beauty/g1917/most-wanted-beauty-awards/?slide=2, last modified 2013.

13. "Blogging from New York—Interview with Stroller in the City," *What a Brand*, http://whatabrand.blogspot.com/2014/01/blogging-from-new-york-interview-with .html, last modified January 25, 2014.

14. Extra TV, "Go Back to School with H&M Fashion!," *Extra TV*, video, 1:00. August 6, 2014, http://extratv.com/2014/08/06/go-back-to-school-with-h-and-m-fashion/.

15. Brock McGoff, "Top 50 Blogs Every Man Should Know About," *Modest Man*, http://www.themodestman.com/top-50-blogs-every-man-should-know-about/, last modified June 25, 2012.

16. Kristin Donnelly, "How to Make Your Food Cute: Tips from I Am a Food Blog," *Food & Wine*, http://www.foodandwine.com/blogs/2014/2/25/i-am-a-food -blog-on-cute-overload-in-japan, last modified February 25, 2014.

17. "Fifth Annual Saveur Best Food Blog Awards," *Saveur*, http://www.saveur.com /content/best-food-blog-awards-2014-winners, last modified 2014.

18. Stephanie Le, May 22, 2014 (10:44 a.m.), reply comment to reader Hannah, "*I Am a Food Blog*" (blog), http://iamafoodblog.com/faq/.

19. Stephanie Le. *Easy Gourmet: Awesome Recipes Anyone Can Cook* (Salem, MA Page Street Publishing, 2014).

20. "Top Google Adsense Earners in the World," *Daily Dock*, http://www.dailydock .com/top-google-adsense-earners-world/, last modified December 23, 2014.

21. Yaro Starak, "Alborz Fallah: The Million Dollar Car Blogger," *Entrepreneurs-Journey.com*, http://www.entrepreneurs-journey.com/external-videos/alborz-fallah -the-million-dollar-car-blogger/, last modified May 24, 2014.

22. Laura Donovan, "Why Your Blog May Be a Waste of Time," *Business 2 Community*, http://www.business2community.com/blogging/blog-may-waste-time-0822891 #Fsy7hrbPCHlrYu5R.99, last modified March 25, 2014.

23. Nora Ganim Barnes and Ava M. Lescault, "The 2014 Fortune 500 and Social Media: LinkedIn Dominates as Use of Newer Tools Explodes," http://www.umassd.edu /cmr/socialmediaresearch/2014fortune500andsocialmedia/, last modified August 24, 2014.

24. Ibid.

25. Andrew Beaujon, " 'Almost Half' of the NYT's Blogs Will Close or Merge," *Poynter*, http://www.poynter.org/news/mediawire/256936/almost-half-of-the-nyts-blogs -will-close-or-merge/#.U6sq-Ul43bc.twitter, last modified June 25, 2014.

26. Ibid.

27. Mel Campbell, "Should We Mourn the End of Blogs?," *The Guardian*, http:// www.theguardian.com/commentisfree/2014/jul/17/should-we-mourn-the-end-of -blogs, last modified July 16, 2014.

28. Nora Ganim Barnes, "The 2014 Fortune 500 and Social Media: LinkedIn Dominates As Use of Newer Tools Explodes."

29. Ibid.

30. Ibid.

31. Jason Kotetke, "The Blog Is Dead, Long Live the Blog," *Nieman Lab*, http:// www.niemanlab.org/2013/12/the-blog-is-dead/, last modified December 19, 2013.

32. Benjie Moss, "Nothing Left to Say? The Decline and Fall of Blog Comments," Web Designer Depot, http://www.webdesignerdepot.com/2014/02/nothing-left-to -say-the-decline-and-fall-of-blog-comments/, last modified February 21, 2014.

33. Ibid.

34. John Borthwick, "Distribution . . . Now," *Think/Musing*, http://www.borthwick .com/weblog/2009/05/13/699/, last modified May 13, 2009.

35. Alexis Madrigal, "2013: The Year 'the Stream' Crested," *The Atlantic*, http:// www.theatlantic.com/technology/archive/2013/12/2013-the-year-the-stream-crested /282202/, last modified December 12, 2013.

36. John Borthwick, "Distribution . . . Now."

37. Craig Bailey, "Content Is King," *Craig Bailey* (blog), http://www.craigbailey .net/content-is-king-by-bill-gates/, last modified May 31, 2010.

38. Jason Kotetke, "The Blog Is Dead, Long Live the Blog."

39. Mel Campbell, "Should We Mourn the End of Blogs?"

40. Michael Stelzner, "Social Media Marketing Industry Report," *Social Media Examiner* (2014): 5.

41. Alexis Madrigal, "2013: The Year 'the Stream' Crested."

42. Jeff Bullas, "10 Content Marketing Lessons from the World's Fastest Growing Websites," *Jeffbullass* (blog RSS), September 29, 2014, accessed January 6, 2015.

43. "How Much Text Can I Include in My Ad? *Facebook* (Facebook Help Center)," https://www.facebook.com/help/468870969814641, accessed January 6, 2015.

44. Jessa Hochman, e-mail message to author, January 7, 2014.

45. Steven Kurutz, "When Blogging Becomes a Slog," *New York Times*, September 9, 2014, http://www.nytimes.com/2014/09/25/garden/when-blogging -becomes-a-slog.html?_r=1. Last modified September 24, 2014.

46. Emily Schuman. "Blogging Burnout," *Cupcakes and Cashmere* (blog), 2014, http://cupcakesandcashmere.com/series-stories/thoughts-on-blogging-burnout, last modified October 28.

47. J. D. Rucker, "Building Up a Blog with Social Promotion," *Soshable*, http:// soshable.com/building-up-a-blog-with-social-promotion/, last modified January 2, 2015.

BIBLIOGRAPHY

Bailey, Craig. "Content Is King." *Craig Bailey* (blog). http://www.craigbailey.net/con-tent-is-king-by-bill-gates/. Last modified May 31, 2010.

Barnes, Nora Ganim, and Ava M. Lescault. "The 2014 Fortune 500 and Social Me-dia: LinkedIn Dominates as Use of Newer Tools Explodes." Last modified August 24, 2014, http://www.umassd.edu/cmr/socialmediaresearch/2014fortune500andso cialmedia/.

Beaujon, Andrew. " 'Almost Half' of the NYT's Blogs Will Close or Merge," *Poynter*. http://www.poynter.org/news/mediawire/256936/almost-half-of-the-nyts-blogs -will-close-or-merge/#.U6sq-Ul43bc.twitter. Last modified June 25, 2014.

Blogging from New York—Interview with Stroller in the City. *What a Brand*. http:// whatabrand.blogspot.com/2014/01/blogging-from-new-york-interview-with.html. Last modified January 25, 2014.

Borthwick, John. Distribution . . . Now. *Think/Musing*. http://www.borthwick.com /weblog/ 2009/05/13/699/. Last modified May 13, 2009.

Bullass, Jeff. 2014, September 29. "10 Content Marketing Lessons from the World's Fastest Growing Websites." Jeffbullass Blog RSS. Accessed January 6, 2015.

Campbell, Mel. "Should We Mourn the End of Blogs?" *The Guardian*. http://www .theguardian.com/commentisfree/2014/jul/17/should-we-mourn-the-end-of-blogs. Last modified July 16, 2014.

Child, Julia, and Simone Beck. *Mastering the Art of French Cooking*. New rev. ed. New York: Knopf, 1983.

Donnelly, Kristin. "How to Make Your Food Cute: Tips from I Am a Food Blog." *Food & Wine*. http://www.foodandwine.com/blogs/2014/2/25/i-am-a-food-blog-on-cute-overload-in-japan. Last modified February 25, 2014.

Donovan, Laura. "Why Your Blog May Be a Waste of Time." *Business 2 Community*. http://www.business2community.com/blogging/blog-may-waste-time-0822891 #Fsy7hrbPCHlrYu5R.99. Last modified March 25, 2014.

Extra TV. "Go Back to School with H&M Fashion!" *Extra TV*, video, 1:00. August 6, 2014. http://extratv.com/2014/08/06/go-back-to-school-with-h-and-m-fashion/.

"Fifth Annual Saveur Best Food Blog Awards." *Saveur*. http://www.saveur.com/content /best-food-blog-awards-2014-winners. Last modified April 14, 2014.

Goldstein, Jennifer. "Our First Annual Most Wanted Beauty Awards." *Marie Claire*. http://www.marieclaire.com/beauty/g1917/most-wanted-beauty-awards/?slide=2. Last modified 2013.

Hepburn, Arden. "Global Digital Statistics 2014." *Digital Buzz*. http://www .digitalbuzzblog.com/slideshare-global-digital-statistics-2014-stats-facts-study -presentation/. Last modified January 8, 2014.

Hochman, Jessa. 2014, January 7. E-mail correspondence with author.

Julie & Julia. Directed by Nora Ephron. 2009. New York: Columbia Pictures. Film.

Kotetke, Jason. "The Blog Is Dead, Long Live the Blog," *Nieman Lab*. http://www .niemanlab.org/2013/12/the-blog-is-dead/. Last modified December 19, 2013.

Kurutz, Steven. "When Blogging Becomes a Slog." *New York Times*. http://www .nytimes.com/2014/09/25/garden/when-blogging-becomes-a-slog.html?_r=1. Last modified September 24, 2014.

"Le Cordon Bleu Welcomed Julie Powell." *Le Cordon Bleu*. http://www.cordonbleu .edu/index.cfm?fa=NewsEventFrontMod.DisplayNewsPage&ElementID=282&S etLangID=1. Accessed December 3, 2014.

Le, Stephanie. 2014. *Easy Gourmet: Awesome Recipes Anyone Can Cook*. Salem, MA: Page Street Publishing.

Le, Stephanie. 2014, May 22 (10:44 a.m.). Reply comment to reader Hannah, "I Am a Food Blog" (blog). http://iamafoodblog.com/faq/.

Leake, Lisa. 2014. *100 Days of Real Food*. New York: Harper Collins.

Leake, Lisa. 2014. *100 Days of Real Food: How We Did It, What We Learned, and 100 Easy, Wholesome Recipes Your Family Will Love*. New York: William Morrow Cookbooks.

Leake, Lisa. "About." *100 Days of Real Food*. http://www.100daysofrealfood.com /about/. Retrieved December 12, 2014.

Madrigal, Alexis. "2013: The Year 'the Stream' Crested." *The Atlantic*. http://www .theatlantic.com/technology/archive/2013/12/2013-the-year-the-stream-crested /282202/. Last modified December 12, 2013.

McGoff, Brock. "Top 50 Blogs Every Man Should Know About." *Modest Man*. http://www.themodestman.com/top-50-blogs-every-man-should-know-about/. Last modified June 25, 2012.

McGrail, Mike. "The Blogocomony: Blogging Stats." *Social Media Today*. http://www .socialmediatoday.com/content/blogconomy-blogging-stats-infographic. Last modified August 28, 2013.

Moss, Benjie. "Nothing Left to Say? The Decline and Fall of Blog Comments." Web Designer Depot. http://www.webdesignerdepot.com/2014/02/nothing-left-to-say -the-decline-and-fall-of-blog-comments/. Last modified February 21, 2014.

Okun, Alanna. "Meet the 'Lacqueristas' Who Rule Nail Art on the Internet." *Buzz-Feed.* http://www.buzzfeed.com/alannaokun/meet-the-lacqueristas-who-rule-nail -art-on-the-i#.gy7k18kly. Last modified November 28, 2012.

Pollan, Michael. *In Defense of Food: An Eater's Manifesto.* New York: Penguin Press, 2008.

Powell, Julie. 2005. *Julie and Julia: 365 Days, 524 Recipes, 1 Tiny Apartment Kitchen.* New York: Little, Brown and Co.

Rucker, J. D. "Building Up a Blog with Social Promotion." *Soshable.* http://soshable .com/building-up-a-blog-with-social-promotion/. Last modified January 2, 2015.

Schuman, Emily. "Blogging Burnout." *Cupcakes and Cashmere* (blog). http://cupcake sandcashmere.com/series-stories/thoughts-on-blogging-burnout. Last modified October 28, 2014.

Starak, Yaro. "Alborz Fallah: The Million Dollar Car Blogger," *Entrepreneurs-Journey.com.* http://www.entrepreneurs-journey.com/external-videos/alborz-fallah -the-million-dollar-car-blogger/. Last modified May 24, 2014.

Stelzner, Michael. Social Media Marketing Industry Report. *Social Media Examiner* (2014): 5.

"Top Google Adsense Earners in the World." *Daily Dock.* http://www.dailydock.com /top-google-adsense-earners-world/. Last modified December 23, 2014.

12

Self-Promotion for All! Content Creation and Personal Branding in the Digital Age

Justin Lagore

Decades before MySpace became the "place for friends," Facebook expanded beyond Harvard's campus, and Twitter enabled effortless stream-of-consciousness sharing, a cornerstone of the Internet's development was its potential for social interactivity. The "global village" phenomenon stemming from the evolution of social media shrank the world. Every voice became amplified, and geographic barriers melted away, making it easier than ever to connect with like-minded individuals and bond over mutual interests.

As communication evolved, personal connections online grew stronger, and social media assumed an intimate role in our lives. Facebook is the modern journal, allowing us to share our adventures, feelings, successes, and struggles. Twitter's rapid-fire nature encourages us to share our thoughts at a rate limited only by how quickly we can type. Instagram and its filters breathe new life into visual storytelling for the connected generation.

Perhaps one of the social Web's most profound effects on society, however, has been its role in the evolution of the "personal brand." Living in such a connected age presents countless opportunities for virtually anyone to become known for their passions, and for many, creating content around what they love has become a lucrative career opportunity.

This chapter examines applications of social media for self-expression and self-promotion, the benefits of establishing an authentic personal brand, and how an evolving online social ecosystem has impacted entertainment and changed the role of third-party media institutions in the digital age.

THE "SELFIE": FROM CLASSIC PORTRAITURE TO CAMERA PHONES

Centuries before the Internet existed, self-promotion was alive and well in media and the arts. Storytelling is arguably as old as humanity, and self-expression has always been a motive behind paintings, music, and literature. Life experiences, human emotion, and triumphs or failures of society were the inspiration for the poems, songs, and legends passed down both orally and in writing from generation to generation, and from continent to continent. Artists painted the likenesses of their muses to preserve the beauty and wonder of their human forms. Writers crafted novels and plays professing the realities of love and loss, and actors performed them for eager crowds entranced by the stories of others. These works were shared for a vast number of reasons—gaining adoration from others, reveling in successes of the past, coping with grief, and finding an ability to relate with one another despite vastly differing life stories.

One of the best-known works of late Italian Renaissance artist Parmigianino is his *Self Portrait in a Convex Mirror* (c. 1524 BCE). His recreation of his own reflection was rendered as viewed in a convex mirror on a custom-prepared panel, emphasizing the distortion of the image reflected in the unique shape of the mirror. It's said Parmigianino produced this piece to showcase his talents to potential customers and ideally gain an audience with the Pope. Parmigianino hoped demonstrating his impressive technical skills and creativity would earn him commissions from the church.[1]

Parmigianino's flashy display of technical finesse and his intentions behind it aren't far off from modern motivations of self-promotion. It's easy to label Parmigianino and other early artists as pioneers of the "selfie" and some of the first to work at establishing a personal brand through their expression in media.

The earliest use of the term "selfie" as we've come to know it reportedly dates back to 2002. After tripping down a set of stairs at a friend's party, a commenter from Australia posted to the ABC Online Forum a picture showing his bloody lip and other damages done to his face. He blamed the photo being a "selfie" for its poor quality and lack of focus.[2]

Adoption of the word was slow, gaining traction over several years. It wasn't until 2012 when the word's constant use in mainstream media popularized it and ignited a massive spike in its appearance as hashtags on Flickr, Instagram, and Twitter. Oxford Dictionaries named "selfie" 2013's Word of the Year after data scraped from the Web revealed a booming 17,000 percent increase in use over just 12 months,[3] quite recently after the word's acceptance into the dictionary:[4] "selfie, sel·fie, 'selfē (noun): a photograph that one has taken of oneself, typically one taken with a smartphone or webcam and shared via social media."[5]

Katie Warfield of Kwantlen Polytechnic University reported on a study she led in which young women were surveyed regarding the circumstances surrounding and process of taking selfies.

> Most young women said that apart from seeking an image that looked "good," they also sought "good lighting," "good background," and images that were not blurry or out of focus. Apart from lighting some young women used technical terminology like looking for the correct "ISO" and "exposure" of the picture and using specific filters to "highlight" and "increase contrast" to correct qualities that weren't—according to them—"right."[6]

Warfield's report continues with one respondent commenting, "with friends, no [I don't pose]. By myself absolutely, I imitate models and try out 'artsy' poses." The respondents also stressed the importance of making sure that in addition to looking attractive that their selfies also conveyed authenticity.[7]

The focus on aesthetics during the process of taking a good selfie manifests not simply as a symptom of vanity, but as a consorted effort to enhance what that selfie contributes to one's personal brand and the impression made online. The editing process—adjusting a photo's white balance, boosting the contrast, applying filters or even deciding which imperfections to leave untouched—represent the artistic liberties new technology affords as a channel of self-expression. Parmigianino exercised his artistic freedom when he painted his *Self Portrait*, opting for the convex mirror and its distortion effect to create visual interest. Just as the mirror guided Parmigianino's work, the front-facing cameras of smart phones aid with the composition of selfies. Parmigianino's distortion effect foreshadowed the numerous filters and photo effects available today. The choices we make as we pose for, edit, and post selfies all contribute in part to what we communicate to our audiences, and they play into our own motivations just as Parmigianino's *Self Portrait* did for him. Today, selfies are just one example of content used to build a personal brand and bolster visibility and individual presence online.

CONTENT CREATION: A HOME FOR ALL THINGS

In her book, *Seeing Ourselves through Technology: How We Use Selfies, Blogs and Wearable Devices to See and Shape Ourselves*, Jill Walker Rettberg of the University of Bergen, Norway, writes:

> Digital self-presentation and self-reflection is cumulative rather than presented as a definitive whole. A weblog or social media feed consists

of a continuously expanded collection of posts, each of which may express a micro-narrative, a comment that expresses an aspect of the writer or an image showing a version of themselves. This cumulative logic is built into the software and into our habits of reading and sharing online, and it acts as a technological filter that lets certain kinds of content seep through while others are held back, either never being expressed or finding other outlets.[8]

The diversity of social media available today means that it's easier than ever for creative content of all forms to find a home online. Instagram allows seasoned photographers and amateurs alike to tell stories visually. Sound Cloud has grown as an online hub for budding musicians and podcasters to share their work. Wordpress powers more than 74 million sites across the Internet and dominates as the world's largest blogging platform, making it easy for even the least Internet-savvy users to write about and share their passions with a community of millions.

Despite the array of channels available and their differing functionalities, however, the social Web remains highly connected across platforms. Blogs feature buttons enabling readers to quickly share content on Facebook, Twitter, Google+, and more. With just a few taps, Instagram photos can be effortlessly posted to other networks; Tweets can be shared automatically as updates on Facebook, and virtually anything—even rich media such as audio hosted on Sound Cloud—can be embedded easily.

RISE OF THE YOUTUBERS: EXPRESSING PASSIONS THROUGH VIDEO

Of all creative content on the Web, however, video has risen to the top as a powerhouse format for users to showcase their creativity. Since its launch in 2005, YouTube has enabled individuals with vastly varying interests to express themselves and share their passions.[9] The platform boasts more than 1 billion unique visitors watching more than 6 million hours of video every month. On average, 100 hours of video are uploaded to the site every minute. Millions of YouTube users subscribe to new channels every day,[10] and social media's interconnectivity gives users limitless options to build their brands online, cross-promote their content, and spark discussions around their work.

Some of the earliest YouTubers got their start by producing sketch-like comedy clips. Anthony Padilla began posting content in 2002 to Smosh.com, a site he created to share content with his friends in school.[11] His first projects included flash animations and lip-sync videos created in collaboration with Ian Hecox, the second half of the Smosh duo. Padilla and Hecox eventually started posting their content to YouTube in 2005 during the site's infancy.[12]

From 2006 on, Smosh began to expand its horizons, launching new channels with dubs of their videos in other languages, behind-the-scenes clips, question-and-answer videos, and more. In 2006, Smosh was recognized by *Time* magazine when the publication bestowed its accolade for Person of the Year on the millions of Internet content creators expressing themselves through digital media.[13]

Smosh has branched out beyond YouTube, leveraging other digital media to promote its original content. Smosh's mobile and Xbox One applications optimize access to everything available on Smosh.com and in its community forums. Smosh also boasts an original game for iOS and Android devices and, in September 2014, Lionsgate Media announced a Smosh movie in the works.[14] With more than 19 million subscribers and 3.9 billion video views, Smosh has found massive success on YouTube, and its enormous following has enabled the brand to grow into what is known as one of the most popular channels on YouTube.

Alternatively, several other users have gone on to achieve YouTube stardom without expanding into as many genres as Smosh. To date, Felix Arvid Ulf Kjellberg, better known online as "PewDiePie," is a YouTube celebrity known for his "let's play" video-game commentary channel. With more than 33 million subscribers, PewDiePie is the most subscribed channel on YouTube (as of January 2015). Despite the size of his following, however, Kjellberg's commitment since the beginning has been to fan engagement. YouTube's biggest personality has a reputation for interacting with his fans as much as possible through every social network manageable and also recognizing what channels aren't conducive to such commitment. In August of 2014, Kjellberg released a post-vacation video discussing his frustration with YouTube's comments feature and the high volume of spam making it difficult for him to read relevant posts from his viewers. Kjellberg announced he would be disabling the comments feature on his videos, opting to communicate with fans via hashtags on Twitter and threads on Reddit as he had done in the past.[15]

In 2013, his channel won a Social Star Award for best social media show thanks to the support of the channel's fiercely loyal fan base, which Kjellberg refers to as his "bro army."[16] Kjellberg has received mainstream media attention as well, appearing on South Park[17] and making headlines for leveraging his online following to benefit charities such as Save the Children,[18] for which Kjellberg and his subscribers raised more than $630,000.[19]

For some YouTube stars, achieving popularity online has proven a valuable marketing asset as they make the jump to success across other media. Comedian Grace Helbig's YouTube fame started with "Grace N Michelle," a channel she created with her roommate. Helbig's work landed her a job making videos for *DailyGrace*, a comedy Web series that she created in conjunction with the multichannel network MyDamnChannel. Helbig now posts to her own channel, "It's Grace," and has made her mark as one of the Internet's most versatile and well-known new celebrities.

Helbig's platform has given her the opportunity to plug her numerous projects, including the travel series *HeyUSA* she and fellow YouTuber Mamrie Hart produced with Astronauts Wanted. Her Internet success also has afforded Helbig the opportunity to leap into more traditional media endeavors, including the feature-length film *Camp Takota,* which she co-produced; and her *New York Times* bestselling book, *Grace's Guide: The Art of Pretending to Be a Grown-Up*. Helbig also partnered with E! Entertainment Television on *The Grace Helbig Project,* a hybrid comedy and talk show that premiered in April 2015.[20]

Helbig has used other social media—including Tumblr and Facebook—to source fan questions for Q&A-style videos and to interact with her audience. Twitter also has proven useful for Helbig as a means of reacting alongside and engaging with her excited fans when announcements about a new project surface on days that she isn't scheduled to post a video.

In a manner similar to the way Helbig segued into acting and comedy projects, South African YouTube personality Troye Sivan has been able to leverage his YouTube following to advance his music career. Sivan actually started singing before he began making videos in 2012, but on July 25, 2014, Sivan released his single, "Happy Little Pill," on his channel.[21] The single was from Sivan's five-song EP, which was released August 15, 2014. The EP debuted at No. 1 on iTunes in 55 countries[22] and made No. 5 on the Billboard 200 a week later.[23] The album also went gold as certified by the Australian Recording Industry Association.[24]

VLOGGING: THE INTERNET'S REALITY TELEVISION

Even video blogging, or "vlogging," has found success on YouTube. The term is a hybrid of the words "video" and "blogging," and this diary-style genre has become one of the most popular types of content on the site.

In his clip, "Me at the Zoo," YouTube cofounder Jawed Karim stands in front of elephants at the San Diego Zoo, and describes the animals for 18 seconds.[25] His clip was the first such video posted to YouTube, and is perhaps one of the earliest examples of what could be considered a vlog on the site. The clip has no intro or end card, no background music, and isn't edited with any jump cuts or transitions; it's raw. Vlogging has become more sophisticated, with content creators typically employing some degree of editing to keep content concise and interesting, but the genre in general remains bare bones in nature.

Outside Karim's video, it's difficult to determine exactly who the first YouTube vloggers were, but Philip DeFranco definitely was one of the first.

DeFranco created his main channel, "sxephil" during his junior-year finals at East Carolina University[26] and has been uploading content since 2006. DeFranco's channel focused on news commentary and satire. In 2007, he created a second channel, "PhilipDeFranco," for vlog-style content. It was well received by fans of his main-channel videos.

Shay Carl Butler, known online as "ShayCarl" has found massive success with this genre. Butler got his start after discovering YouTube and watching videos by early creators, including DeFranco. Butler began making videos of his own and entered a contest that earned him an endorsement from DeFranco promoting Butler's content, which also earned him new subscribers. Butler said in an interview with *Forbes*:

> It was amazing to me that some dude across the country was taking the time to send me an e-mail asking for more videos. I remember waking up and feeling like I got to make a video I don't want to let this guy down. I loved the feedback and the gratification was instant. I would film some random thought I had about hand sanitizer or gas prices or me dancing in my wife's old unitard and I would upload it and people were instantly there to tell me if they thought it was funny or not. I loved the communication and the community of it all.[27]

Butler's audience grew as he continued producing videos, and in 2008 he launched his second channel to document his weight-loss journey.[28] On March 5, 2009, Butler decided to start posting daily vlogs to the channel. The videos have since documented the daily lives of Butler, his wife, and their four kids, who are known as "YouTube's first family," and have been hugely successful.

Butler manages five YouTube channels and is producing and starring in *I'm Vlogging Here*, a 90-minute documentary about vloggers and the dramatic effect that posting their lives online has had on them.[29] Butler tapped into his YouTube following, securing funding through an online campaign to cover production costs of the film.

The community aspect of YouTube that Butler describes in his *Forbes* interview is not only one of the platform's greatest appeals for creators, but also is one of the most valuable assets it provides. The enormous pool of talent on YouTube means the community has limitless potential for collaboration, which has become a cornerstone of creativity and a best practice for content promotion on the site.

YouTube's Creator Academy—which functions as a resource center for content creators—features an entire section on collaboration and highlights the importance finding complementary and creative partners with whom to work. Collaboration comes in all shapes and sizes and could be something as intricate as concepting, writing, shooting, and editing a video together, or as simple as creators giving each other shout-outs in their video, similar to the publicity Butler received from DeFranco in the infancy of his YouTube career.

In 2009, Butler and DeFranco took collaboration to the next level when they joined forces with fellow YouTube personalities Lisa Donovan, Kassem Gharaibeh, Danny Zappin, Scott Katz, and others to create "The Station," a channel built entirely as a collaborative work.[30] Featuring sketches created by the high-profile cast of viral stars, the team was able to leverage the popularity of their already well-established personal channels to drive viewers to their collaborative projects.

The Station later evolved in 2009 into Maker Studios, Inc., one of the biggest multichannel networks that now helps aspiring content creators build their brands online. Maker Studios and other similar networks offer a wide range of tools such as analytics reporting, video management and distribution systems, and community forums that help connect creators who have similar interests and who might want to work on projects together.

CREATIVE CAREERS IN THE DIGITAL AGE

For many of these creators, making videos on YouTube has become so much more than a hobby. The niche audiences and communities they build around their content are assets they've been able to leverage and monetize to make a living as a full-fledged YouTube career.

The Google AdSense program enables creators to get paid for ads displayed with their videos by advertisers who have recognized and acted on laser-targeted marketing opportunities to engage with potential customers of very specific demographics and interests. The platform also lends itself to brand partnerships between advertisers and YouTube talent who will actually promote products or services in their videos. Sometimes it's as simple as YouTube giving a shout-out to sponsors and encouraging their audience to check out the brand. Other times, endorsements on YouTube come from product reviews. Tanya Burr, for example, is a U.K. vlogger known for her beauty channel and makeup tutorials. Burr has received free products from luxury brands like Chanel, Dior, and TSL who give out the freebies in hopes of being featured in one of Burr's videos or on her Instagram feed.[31] It is not guaranteed coverage, but if Burr uses and enjoys a product her endorsement has the potential to increase sales and convert her viewers into brand-loyal customers.

One of the largest brand partnerships with YouTube talent was a campaign by Ford called the Fiesta Movement. The auto manufacturer recruited some of YouTube's most visible stars as "Fiesta Agents" and gave them a car to drive. The auto manufacturer also covered the insurance and gas for the duration of the project. The Fiesta Agents in exchange created videos in response to challenges Ford issued that featured the car and showcased its versatility. The agents were asked to cross-promote their videos on Twitter with #fiestamovement, engaging their social networks even further and giving Ford's campaign an additional boost in visibility.

Merchandise also is a popular revenue opportunity for YouTubers. Channels across all genres—and especially musicians, comedians, and vloggers who have well-known catchphrases—have made money designing and selling branded apparel through sites such as District Lines and Spreadshirt.

Multichannel networks such as Maker Studios and Fullscreen work with YouTube talent not only to grow their channels and help the YouTubers produce great content, but also to take advantage of all the potential of all revenue streams available to creators. The networks frequently manage channels' advertising inventory on behalf of their partners, and often the networks are responsible for connecting brand representatives to the talent that might be interested in partnering with them. Some networks offer merchandise solutions such as in-house production or partnerships with production companies to create and sell apparel at a discounted rate. Networks also offer creators access to music and other assets to use in their videos free of royalties thanks to licensing agreements with major record labels. This gives creators clearance to monetize covers of songs or to use their favorite tracks to audibly enhance their projects.

Because the success of these partnerships or even the size of Google Ad-Sense checks depends on the size of each channel's audience, however, self-promotion and ensuring that creators' audiences are as engaged with their channels as possible is essential. Fortunately, social media provides Internet personalities with countless touchpoints that can be leveraged to interact with fans 24/7. Even YouTube makes it easy for creators to cross-promote their own content right on their channels with annotations that can be configured to link to other videos or to other social media. This makes it quick and effortless for YouTubers to promote their holistic presence on the Web and invite their fans to engage even when there might not be any new content posted.

EMOTIONALLY DRIVEN ENTERTAINMENT AND THE NEW ROLE OF THIRD-PARTY MEDIA

As entertainment moves online into the world of user-generated content, things seem to be classifiable into distinct categories. It also still is easy to see that the genres of content on YouTube and even Tumblr, Reddit, and other social networks differ vastly from those of traditional entertainment media. But what content does well on the social Web in general?

The answer can be found through an examination of "memes"—one of the earliest forms of viral online content. The word "memes" comes is an alteration of the word, "mimesis," meaning imitation or mimicry. Memes are static images, usually featuring a text overlay conveying humor or adding commentary, and are a prime example of the social Web's ability to spread content across the Internet rapidly. It seems since the dawn of social media, the digital world has been infatuated with cats—which were the subject of some of the earliest memes and continue to be popular today. A 2010 article from Mashable titled, "The Million Dollar Question: Why Does the Web Love Cats?," featured commentary from creators, viral marketing experts, and other industry professionals aiming to shed light on what made cats the Internet sensation they've become.[32] "Cats get themselves into all kinds of amusing predicaments," the article reads, "and when there's a human on hand to capture the moment . . . that's raw material just begging to be shared. And because lots of people have cats, that's lots of content."[33]

Memes haven't stopped with animals, however. In January 2012, the image that became the "Bad Luck Brian" meme hit Reddit—another news and social networking site—with the text overlay "Takes drivers test. . . . Gets first DUI." Kyle Craven, the subject of the photo explained Bad Luck Brian was born as the result of a high school yearbook photo botched intentionally. Craven had a reputation as a class clown, and bad yearbook photos were something he looked forward to every year. When the school principal made Craven retake the photo he didn't think much else of it, and the photo faded into obscurity until Ian Davies, a friend of Craven's since elementary school posted the image to Reddit. Users up-voted the photo, increasing its popularity and visibility until Bad Luck Brian became an Internet sensation. Users ran with the image, creating their own captions that kept the theme of a terrible streak of misfortune and sharing their creations all across the Web.[34]

Even YouTuber Laina Morris got her start as a meme. On June 6, 2012, Morris uploaded her video titled, "JB Fanvideo" in response to Justin Bieber's announcement of a sing-off promoting his new fragrance called "Girlfriend." The video depicted her performing her own rendition of Bieber's song, called "Boyfriend," with a clingy twist. Morris sings that, if she was Bieber's girlfriend, she would not let him out of her sight unless he wore a recording device on his shirtsleeve. She goes on to say that she would always be checking in on his where he is and who he is taking to, finally ending with, "Spend a day with your girl, I'll be calling you my husband."[35]

Morris' video, like the Bad Luck Brian image, sparked its own meme when posted to Reddit. The thread created around her video was called, "Overly Attached Girlfriend," and it spread like wildfire through social media. The video and users reactions to it were featured on several prominent news and

Internet culture blogs, including "The Daily What," "Jezebel," "Buzzfeed," and "The Daily Dot."

What makes memes so successful as online content? When it comes to cats, "they're the perfect distraction from our hectic lives," Know Your Meme cofounder Elspeth Rountree explains to "Mashable": "You don't need any explanation or prior knowledge to understand the slapstick humor that animals provide. Cat videos and images are a quick hit of pure, unfettered 'cute.' They're also entertainment in easily digestible doses."[36]

The same can be said for the popularity of Craven and Morris. They provide easily shareable, small doses of entertainment and humor easily understood by the lay user of the Internet. In the grander scheme of things, anything can go viral so long as it strikes a chord with its audience. In an interview with *AdWeek*, publisher Dao Nguyen explained what does well online based upon her experience working at "Buzzfeed."

> Basically content that stirs an emotion does well. It can range from a funny list about kids to an insightful commentary on current events to investigative stories about an injustice or a heartbreaking situation. It can be a list, it can be a long-form story, it can be an essay, but if a story stirs an emotion then it drives sharing.[37]

By Nguyen's explanation, any high-arousal emotion can encourage sharing, because if it makes the audience feel something, then it's relatable. New viral content is born every day—dancing across newsfeeds as links to videos, photos, and articles. For instance the mom who strung together witty one-liners from her four-year-old daughter[38] to promote her book on parenting, or the parents who perfectly perform to the soundtrack of *Frozen*, entertaining their child who listens from the car's backseat.[39] "Buzzfeed" headlines such as, "23 Times Rachel from 'Friends' Perfectly Summarized What It's Like to Be in Your Twenties"[40] are built on this entire concept of content relatable to the audience.

NEW MEDIA FOR NEWS MEDIA: TRADITIONAL PRESS AND THE MIGRATION TO SOCIAL MEDIA

If there's one industry that picked up on this trend after learning the hard way, it was traditional news media. On journalism and social media's implications on the profession, Nieman Fellow Michael Skoler wrote that the shift between readers and traditional news media began long before the booming growth of social channels. Twenty-five years before the rise of citizen journalism, readers still depended on reporters to gather and disseminate the news. Advertisers paid into news publications because everyone was watching the 6:00 p.m. newscast or reading the Sunday paper, and it was a surefire way to get the advertisers' messages in front of a captive audience. Traditional media became less reliable in the eyes of citizens, however, as large conglomerates bought out local papers. Big media companies cut costs by parsing back original reporting in favor of syndicated stories, restricting allocation of column inches and airtime for local coverage to reports on fires and storms. Traditional media lost its relevance with local audiences when it expanded its scope and devoted less time to the news important to small communities.[41]

Unluckily for news media, audience discontent converged with the rise of social media's popularity. Suddenly citizens no longer had to accept roles as passive recipients and had options for digging deeper into the subjects that mattered to them. Citizens took it upon themselves to cover what they found important in their communities. Anyone could shoot photos and videos, blog about local issues, or even break major news stories if they happened to be in the right place at the right time with a smartphone in hand. Local coverage became the dominion of these citizen journalists who took the time to tell stories as they unfolded close to home.[42]

It didn't take long for Facebook, Twitter, and Instagram to become staple sources of news for most people offering diverse perspectives and real-time updates that pressured traditional news providers to adapt and find ways to become relevant once again and survive. The pressure was on for reporters and editors to find a way to leverage social platforms to reconnect with their audiences and win back their loyalty. The two-way nature of social media, however, meant that for news media it wouldn't be as simple as meeting their readers where they were. Plus the additional power social media gave citizens to filter out all but the content they wanted to consume meant the traditional packaging of stories wouldn't translate to new channels.

Skoler explained a cultural shift in how citizens view the role of journalism. People now had an expectation that information is to be shared, and not simply fed to them. Social media set a new standard requiring news media to listen when the audience has questions or commentary to add to a story. The discontentment with how the news industry functioned before the Internet cost traditional media organizations some of the trust of their audiences. Today, audiences expect the news media to build and maintain relationships with them by honing in on and prioritizing the stories audiences care about.[43]

FAMOUS FOR BEING FAMOUS: MEDIA OVEREXPOSURE AND FAMEBALLS

Sometimes news media take exposure to the extreme, as explained by "Gawker," a popular celebrity gossip and news blog based in New York City.

"Gawker" credits the definition of the term "fameball" to Vimeo and CollegeHumor founder Jakob Lodwick, who described fameballs as "individuals whose fame snowballs because journalists cover what they think other people want them to cover."[44] The idea is that coverage of Kim Kardashian's photo shoot with *Paper* magazine,[45] Rihanna's battle with Instagram over topless photos,[46] Justin Bieber's arrests,[47] and Amanda Bynes' erratic online behavior[48] all get people talking due to the shock value.

Controversial and provocative subjects elicit an emotional response of astonishment—and even ridicule—from online communities due to the high-profile subjects of the stories. But the fameball phenomenon perpetuates a cycle of content burnout. Third-party media continue to cover the stories because people keep talking about them—even if that conversation advances to discussions of how tired people are getting seeing the same folks in the headlines day after day. It should be noted that all media institutions fall susceptible to fameball coverage because it drives "clicks." The trend isn't exclusive to institutions that began as traditional news outlets. Perez Hilton's blog and numerous other sites are known to specialize in celebrity gossip and fameball coverage, advancing the visibility of even those who are most famous just for being famous.

Regardless, the response of traditional media companies to society's adoption of social platforms has been evident across the board. Today, televised newscasts include hashtags and Twitter handles of the anchors encouraging viewers to discuss stories in real time. Videos are packaged and posted to YouTube for later review and commentary. Local newscasts have found success with Facebook as an online forum to discuss current events.

Additionally, although migration into the social sphere has given traditional news institutions an opportunity to win back and build relationships with audiences, it has also paved the way for new unique media institutions to enter the fold. The merging of citizen journalism and traditional reporting has given birth to news aggregation blogs such as *The Huffington Post* and *Mashable*. In addition to providing original content these sites benefit from packaging and sharing other content from all across the Web. Born in the digital space, it's something such blogs do well.

The Huffington Post, founded in 2005 as an alternative to more conservative news and commentary sites,[49] covers everything from politics to entertainment, business, tech, media, world news, health, comedy, sports, science, and more. Original blogs and columns elevating voices from minority communities, parents, and columnists across all areas of interest give *The Post* an interesting position as "the Internet's newspaper," with a sense of community stemming from the vast diversity of contributors. That same year, Pete Cashmore founded *Mashable*, covering tech, social media, and news pertinent to what the site calls "the connected generation."

Being native to the Web, news aggregation and online culture blogs are adept at catering to the carefully pruned niche audiences of the Web. Many sites separate their content into narrow verticals, making it easy for users to browse only news about topics of interest. Their social media presence mirrors this organization, and institutions manage multiple social media profiles dedicated to each segment. This structure optimizes these sites for the choosy reader, giving them the opportunity to subscribe and follow only the verticals that interest them.

The massive audiences these sites have earned prove that generation of original content in tandem with digital curation is a recipe for success when working in the social media space, but no matter where the content originates, the relevance and the high-arousal emotional factor must be present. The third-party media institutions that have come out on top have been those whose editors have learned to tell stories in the digital space in a way that lifts up the voices of online communities and elicits those emotional responses.

These principles are the same driving force behind successful content on YouTube. Creators tell their own stories on the Web and share their passions, which elicits that emotional reaction from viewers. It could be argued that vlogging as a genre owes much of its success to the effect it has on an audience. Those raw, less edited clips make content creators more human to their viewers. They help YouTubers build deeper relationships with their fans by pulling them a little more down to earth and giving viewers the opportunity to draw parallels between themselves and content creators as ordinary people. It's those relationships that allow content creation to flourish on the Web, and it's those relationships that are redefining what it means to be oneself on the Web.

SOME CONCLUDING THOUGHTS

Huge changes across the digital media landscape mean that the concept of personal branding never has been more important than it is today. The personal brand has come to symbolize the pursuit of micro-celebrity—commercialization of our self-expression online and the change we can make through the use of our newfound voices in the global conversation. That isn't exactly a negative thing.

Thanks to social media and the numerous outlets for content creation across the Web virtually anyone is enabled to pursue and make a career of their passions. The entire concept of what it means to be a celebrity is changing, and a louder voice for the everyday user means an equal playing field when it comes to opportunities across all content genres—digital and social media have stripped away the barriers that once barred entry into the view of the public eye.

Today it's easier than ever for people to get in front of a camera and find their calling on the Web. We can connect with those who share our quirks—the people who "get" us. Even the most obscure interests can flourish into entire communities of creators and fans, and advertising and marketing intersect with the endeavors of those creators to support them and connect with those communities. Today, it's easier than ever to make a career out of simply being oneself on the Web. All it takes is pulling out a smartphone or turning on a camera and telling the world, "Hello. This is me."

NOTES

1. "Parmigianino, Self-Portrait in a Convex Mirror, 1523–24," narrated by Dr. Beth Harris and Dr. Steven Zucker, Khan Academy, November 5, 2012, https://www.khanacademy.org/humanities/renaissance-reformation/mannerism1/v /parmigianino-self-portrait-in-a-convex-mirror-1523-24.

2. " 'Selfie' Is Named Oxford Dictionaries' Word of the Year 2013," *OxfordWords Blog*, November 19, 2013, http://blog.oxforddictionaries.com/press-releases/oxford -dictionaries-word-of-the-year-2013/.

3. Ibid.

4. Ibid.

5. "Selfie: Definition (U.S. English)," *Oxford Dictionaries*, August 2013, http:// www.oxforddictionaries.com/us/definition/american_english/selfie.

6. Warfield, Katie, "Making Selfies/Making Self: Digital Subjectivit[i]es in the Selfie" (2014), 4.

7. Ibid.

8. Jill Walker Rettberg, *Seeing Ourselves through Technology: How We Use Selfies, Blogs and Wearable Devices to See and Shape Ourselves* (Basingstoke: Palgrave Macmillan, 2014).

9. Megan Rose Dickey, "The 22 Key Turning Points in the History of YouTube," *Business Insider*, February 15, 2013, http://www.businessinsider.com/key-turning -points-history-of-youtube-2013-2?op=1.

10. "Statistics," YouTube, https://www.youtube.com/yt/press/statistics.html, ac-cessed December 12, 2014.

11. "Smosh Exclusive Interview: The Partners Project Episode 13," YouTube, March 10, 2011, https://www.youtube.com/watch?v=HJ_yUutK4pQ.

12. Lev Grossman, "Smosh," *Time*, December 16, 2006, http://content.time.com /time/magazine/article/0,9171,1570729,00.html.

13. Ibid.

14. Kirsten Acuna, "Lionsgate Is Making a Movie with Two of YouTube's Biggest Stars," *Business Insider*, September 18, 2014, http://www.businessinsider.com/smosh -the-movie-announced-2014-9.

15. Felix Kjellberg, "Goodbye Forever Comments," YouTube, August 29, 2014, https://www.youtube.com/watch?v=4_hHKlEZ9Go&list=UU-lHJZR3Gqxm24 _Vd_AJ5Yw.

16. Starcount, "Social Star Awards Live May 23rd," YouTube, May 23, 2013, https://www.youtube.com/watch?v=S7QjrK9vrqk&feature=youtu.be&t=2h17m17s.

17. Jason Schreier, "Pewdiepie Starred Again on Last Night's South Park," *Kotaku*, December 11, 2014, http://kotaku.com/pewdiepie-starred-again-on-last-nights-south -park-1669826680.

18. Stuart Dredge, "YouTube Star PewDiePie Launches $250k Save the Children Fundraiser," *The Guardian*, March 25, 2014, http://www.theguardian.com/technology /2014/mar/25/pewdiepie-youtube-crowdfunding-save-the-children-indiegogo.

19. "PewDiePie Celebrates His 25 Million YouTube Subscriber Milestone by Sup- porting Save the Children," Save the Children, http://www.savethechildren.org/site /c.8rKLIXMGIpI4E/b.9052337/k.ECCD/PewDiePie_Celebrates_his_25_Million _YouTube_Subscriber_Milestone_by_Supporting_Save_the_Children.htm?msource =weklppdp0314.

20. Kamala Kirk, "YouTube Star Grace Helbig Is Coming to Your TV," *E! Online*, January 5, 2015, http://www.eonline.com/news/607365/grace-helbig-is-coming-to-e -with-new-show-premiering-in-april-get-all-the-details.

21. Troye Sivan, "Happy Little Pill—(OFFICIAL AUDIO)—Troye Sivan," You- Tube, July 25, 2014, https://www.youtube.com/watch?v=eXeKoGx9zoM.

22. Bradley Stern, "Troye Sivan Shoots to #1 on iTunes (in 55 Countries!) with Debut EP, 'TRXYE'," *Idolator.com*, August 15, 2014, http://www.idolator.com /7531054/troye-sivan-shoots-to-1-on-itunes-with-debut-ep-trxye.

23. Keith Caulfield, "Troye Sivan Set for Top 10 Debut on Billboard 200," *Billboard*, August 15, 2014, http://www.billboard.com/biz/articles/news/chart-alert /6221699/troye-sivan-set-for-top-10-debut-on-billboard-200.

24. "ARIA Charts—Accreditations—2014 Singles," Australian Recording Indus- try Association, accessed December 14, 2014.

25. Jawed Karim, "Me at the Zoo," YouTube, April 23, 2005, http://youtu.be /jNQXAC9IVRw.

26. Nicole Powers, "Philip DeFranco Is Sxephil," *Suicide Girls*, December 31, 2008, https://suicidegirls.com/girls/nicole_powers/blog/2680080/philip-defranco-is-sxephil/.

27. Michael Humphrey, "ShayCarl's Epic Journey to YouTube Stardom," *Forbes*, May 31, 2011, http://www.forbes.com/sites/michaelhumphrey/2011/05/31/shaycarls -epic-journey-to-youtube-stardom/3/.

28. Shay C. Butler, "SHAYTARDS BEGIN!" YouTube, October 2, 2008, https:// www.youtube.com/watch?v=pLvKz2LMRs8.

29. Apprentice A Productions, *I'm Vlogging Here Documentary*, http://imvlogging here.com/, accessed January 2, 2015.

30. Maker Studios, Inc., "Our Story," Maker, http://www.makerstudios.com /about#/history, accessed January 1, 2015.

31. Tanya De Grunwald, "Meet the YouTube Big Hitters: The Bright Young Vloggers Who Have More Fans than 1D," *Mail Online*, September 18, 2014, http://www.dailymail .co.uk/home/you/article-2656209/The-teen-phenomenon-thats-taking-Youtube.html.

32. Amy-Mae Elliott, "The Million Dollar Question: Why Does the Web Love Cats?," *Mashable*, October 21, 2010, http://mashable.com/2010/10/21/why-does-the -web-love-cats/.

33. Ibid.

34. Jessica Contrera, "Being Bad Luck Brian: When the Meme that Made You Famous Starts to Fade Away," *Washington Post*, January 5, 2015, http://www .washingtonpost.com/lifestyle/style/being-bad-luck-brian-when-the-meme-that -made-you-famous-starts-to-fade-away/2015/01/05/07cbf6ac-907c-11e4-a412 -4b735edc7175_story.html.

35. Laina Morris, "JB Fanvideo," YouTube, June 6, 2012, https://www.youtube .com/watch?v=Yh0AhrY9GjA.

36. Ibid.

37. Michelle Castillo, "From Humor to Heartbreak, BuzzFeed Thrives by Stirring Emotions," *AdWeek*, January 5, 2015, http://www.adweek.com/news/technology /buzzfeeds-new-publisher-says-data-shows-emotional-content-shareability-key-viral -success-162095.

38. K.C. Ifeanyi, "Mom Spins Her 4-Year-Old's Harshest Burns into Hilarious Marketing Gold," *Fast Company*, October 1, 2014, http://www.fastcompany .com/3036533/the-recommender/mom-spins-her-4-year-olds-harshest-burns-into -hilarious-marketing-gold.

39. "Watch this Adorable Disney 'Frozen' Car Sing-a-Long," *FOX2 Now St. Louis*, March 13, 2014, http://fox2now.com/2014/03/13/watch-this-adorable-disney-frozen -car-sing-a-long/.

40. Brian Galindo, "23 Times Rachel from 'Friends' Perfectly Summarized What It's Like to Be in Your Twenties," *BuzzFeed*, January 6, 2015, http://www.buzzfeed. com/briangalindo/23-times-rachel-from-friends-perfectly-summarized-what-its-l# .eeygWB5Vg, accessed June 8, 2015.

41. Michael Skoler, "Why the News Media Became Irrelevant and How Social Media Can Help," *Nieman Reports*, September 16, 2009, http://niemanreports .org/articles/why-the-news-media-became-irrelevant-and-how-social-media-can -help/.

42. Ibid.

43. Ibid.

44. Sheila McClear, "Look, We Made You a Gawker Glossary!" *Gawker*, July 9, 2008, http://gawker.com/5022007/look-we-made-you-a-gawker-glossary.

45. Jolie Lee, "Kardashian Photo Plays off Controversial Black Imagery," *USA Today*, November 13, 2014, http://www.usatoday.com/story/news/nation-now/2014/11/13 /kim-kardashian-photo-black-female-bodies-grio/18962603/.

46. Alyssa Toomey, "Rihanna: I Do Not Have an IG Account!" *E! Online*, May 19, 2014, http://www.eonline.com/news/543294/bye-bye-badgalriri-rihanna-blasts -instagram-imposter-and-confirms-i-do-not-have-an-ig-account.

47. Lesley Messer, "Justin Bieber Arrested for Dangerous Driving and Assault in Canada, Authorities Say," *ABC News*, September 2, 2014, http://abcnews.go.com /Entertainment/justin-bieber-arrested-dangerous-driving-assault-canada-authorities /story?id=25214851.

48. Jeff Nelson, "Amanda Bynes Returns to Twitter, Blasts Her Family and Sam Lutfi," *People*, October 31, 2014, http://www.people.com/article/amanda-bynes -twitter-return-leaves-mental-health-facility.

49. "The Huffington Post | Web Site," *Encyclopedia Britannica Online*, http://www.britannica.com/EBchecked/topic/1192975/The-Huffington-Post. October 29, 2013.

BIBLIOGRAPHY

Acuna, Kirsten. "Lionsgate Is Making a Movie with Two of YouTube's Biggest Stars." *Business Insider*. September 18, 2014. http://www.businessinsider.com/smosh-the-movie-announced-2014-9. Accessed June 7, 2015.

Apprentice A Productions. *I'm Vlogging Here Documentary*. http://imvlogginghere.com/. Accessed January 2, 2015.

"ARIA Charts—Accreditations—2014 Singles." Australian Recording Industry Association. Accessed December 14, 2014.

Butler, Shay C. "SHAYTARDS BEGIN!" YouTube. October 2, 2008. https://www.youtube.com/watch?v=pLvKz2LMRs8. Accessed June 7, 2015.

Castillo, Michelle. "From Humor to Heartbreak, BuzzFeed Thrives by Stirring Emotions." *AdWeek*. January 5, 2015. http://www.adweek.com/news/technology/buzzfeeds-new-publisher-says-data-shows-emotional-content-shareability-key-viral-success-162095.

Caulfield, Keith. "Troye Sivan Set for Top 10 Debut on Billboard 200." *Billboard*. August 15, 2014. http://www.billboard.com/biz/articles/news/chart-alert/6221699/troye-sivan-set-for-top-10-debut-on-billboard-200.

Contrera, Jessica. "Being Bad Luck Brian: When the Meme that Made You Famous Starts to Fade Away." *Washington Post*. January 5, 2015. http://www.washingtonpost.com/lifestyle/style/being-bad-luck-brian-when-the-meme-that-made-you-famous-starts-to-fade-away/2015/01/05/07cbf6ac-907c-11e4-a412-4b735edc7175_story.html. Accessed June 7, 2015.

De Grunwald, Tanya. "Meet the YouTube Big Hitters: The Bright Young Vloggers Who Have More Fans than 1D." *Mail Online*. September 18, 2014. http://www.dailymail.co.uk/home/you/article-2656209/The-teen-phenomenon-thats-taking-Youtube.html.

Dickey, Megan Rose. "The 22 Key Turning Points in the History of YouTube." *Business Insider*. February 15, 2013. http://www.businessinsider.com/key-turning-points-history-of-youtube-2013-2?op=1.

Dredge, Stuart. "YouTube Star PewDiePie Launches $250k Save the Children Fundraiser." *The Guardian*. March 25, 2014. http://www.theguardian.com/technology/2014/mar/25/pewdiepie-youtube-crowdfunding-save-the-children-indiegogo.

Elliott, Amy-Mae. "The Million Dollar Question: Why Does the Web Love Cats?" *Mashable*. October 21, 2010. http://mashable.com/2010/10/21/why-does-the-web-love-cats/.

Galindo, Brian. "23 Times Rachel from 'Friends' Perfectly Summarized What It's Like to Be in Your Twenties." *BuzzFeed*. January 6, 2015. http://www.buzzfeed.com/briangalindo/23-times-rachel-from-friends-perfectly-summarized-what-its-l#.eeygWB5Vg.

Grossman, Lev. "Smosh." *Time*. December 16, 2006. http://content.time.com/time/magazine/article/0,9171,1570729,00.html.

"(The) Huffington Post | Web Site." *Encyclopedia Britannica Online*. October 29, 2013. http://www.britannica.com/EBchecked/topic/1192975/The-Huffington-Post.

Humphrey, Michael. "ShayCarl's Epic Journey to YouTube Stardom." *Forbes*. May 31, 2011. http://www.forbes.com/sites/michaelhumphrey/2011/05/31/shaycarls-epic-journey-to-youtube-stardom/3/.

Ifeanyi, K. C. "Mom Spins Her 4-Year-Old's Harshest Burns into Hilarious Marketing Gold." *Fast Company*. October 1, 2014. http://www.fastcompany.com/3036533/the-recommender/mom-spins-her-4-year-olds-harshest-burns-into-hilarious-marketing-gold. Accessed June 7, 2015.

Karim, Jawed. "Me at the Zoo." YouTube. April 23, 2005. http://youtu.be/jNQXAC9IVRw. Accessed June 7, 2015.

Kirk, Kamala. "YouTube Star Grace Helbig Is Coming to Your TV." *E! Online*. January 5, 2015. http://www.eonline.com/news/607365/grace-helbig-is-coming-to-e-with-new-show-premiering-in-april-get-all-the-details.

Kjellberg, Felix. "Goodbye Forever Comments." YouTube. August 29, 2014. https://www.youtube.com/watch?v=4_hHKlEZ9Go&list=UU-lHJZR3Gqxm24_Vd_AJ5Yw.

Lee, Jolie. "Kardashian Photo Plays Off Controversial Black Imagery." *USA Today*. November 13, 2014. http://www.usatoday.com/story/news/nation-now/2014/11/13/kim-kardashian-photo-black-female-bodies-grio/18962603/.

Maker Studios, Inc. "Our Story." Maker. http://www.makerstudios.com/about#/history. Accessed January 1, 2015.

McClear, Sheila. "Look, We Made You a Gawker Glossary!" *Gawker*. July 9, 2008. http://gawker.com/5022007/look-we-made-you-a-gawker-glossary.

Messer, Lesley. "Justin Bieber Arrested for Dangerous Driving and Assault in Canada, Authorities Say." *ABC News*. September 2, 2014.

Morris, Laina. "JB Fanvideo." YouTube. June 6, 2012. https://www.youtube.com/watch?v=Yh0AhrY9GjA. Accessed June 8, 2015.

Nelson, Jeff. "Amanda Bynes Returns to Twitter, Blasts Her Family and Sam Lutfi." *People*. October 31, 2014. http://www.people.com/article/amanda-bynes-twitter-return-leaves-mental-health-facility.

"Parmigianino, Self-Portrait in a Convex Mirror, 1523-24." Narrated by Dr. Beth Harris and Dr. Steven Zucker. Khan Academy. November 5, 2012. https://www.khanacademy.org/humanities/renaissance-reformation/mannerism1/v/parmigianino-self-portrait-in-a-convex-mirror-1523-24.

"PewDiePie Celebrates His 25 Million YouTube Subscriber Milestone by Supporting Save the Children." Save the Children. http://www.savethechildren.org/site/c.8rKLIXMGIpI4E/b.9052337/k.ECCD/PewDiePie_Celebrates_his_25_Million_YouTube_Subscriber_Milestone_by_Supporting_Save_the_Children.htm?msource=weklppdp0314.

Powers, Nicole. "Philip DeFranco Is Sxephil." *Suicide Girls*. December 31, 2008. https://suicidegirls.com/girls/nicole_powers/blog/2680080/philip-defranco-is-sxephil/.

Rettberg, Jill Walker. *Seeing Ourselves through Technology: How We Use Selfies, Blogs and Wearable Devices to See and Shape Ourselves*. Basingstoke: Palgrave Macmillan, 2014.

Schreier, Jason. "Pewdiepie Starred Again on Last Night's South Park." *Kotaku.* December 11, 2014. http://kotaku.com/pewdiepie-starred-again-on-last-nights -south-park-1669826680.

"Selfie: Definition (U.S. English)." *Oxford Dictionaries.* August 2013. http://www .oxforddictionaries.com/us/definition/american_english/selfie.

" 'Selfie' Is Named Oxford Dictionaries' Word of the Year 2013." *OxfordWords Blog.* November 19, 2013. http://blog.oxforddictionaries.com/press-releases/oxford -dictionaries-word-of-the-year-2013/.

Sivan, Troye. "Happy Little Pill—(OFFICIAL AUDIO)—Troye Sivan." YouTube. July 25, 2014. https://www.youtube.com/watch?v=eXeKoGx9zoM.

Skoler, Michael. "Why the News Media Became Irrelevant and How Social Media Can Help." *Nieman Reports.* September 16, 2009. http://niemanreports.org/articles /why-the-news-media-became-irrelevant-and-how-social-media-can-help/.

"Smosh Exclusive Interview: The Partners Project Episode 13." YouTube. March 10, 2011. https://www.youtube.com/watch?v=HJ_yUutK4pQ.

Starcount. "Social Star Awards Live May 23rd." YouTube. May 23, 2013. https:// www.youtube.com/watch?v=S7QjrK9vrqk&feature=youtu.be&t=2h17m17s.

"Statistics" YouTube. Accessed December 12, 2014. https://www.youtube.com/yt /press/statistics.html.

Stern, Bradley. "Troye Sivan Shoots to #1 on iTunes (in 55 Countries!) with Debut EP, 'TRXYE'." *Idolator.com.* August 15, 2014. http://www.idolator.com/7531054 /troye-sivan-shoots-to-1-on-itunes-with-debut-ep-trxye.

Toomey, Alyssa. "Rihanna: I Do Not Have an IG Account!" *E! Online.* May 19, 2014. http://www.eonline.com/news/543294/bye-bye-badgalriri-rihanna-blasts-instagram -imposter-and-confirms-i-do-not-have-an-ig-account.

Warfield, Katie. 2014. *Making Selfies/Making Self: Digital Subjectivit[i]es in the Selfie,* 4.

"Watch This Adorable Disney 'Frozen' Car Sing-a-Long." *FOX2 Now St. Louis.* March 13, 2014. http://fox2now.com/2014/03/13/watch-this-adorable-disney-frozen -car-sing-a-long/. Accessed June 8, 2014.

13

The Rise of Journalism Accountability

Zac Gershberg

Public trust in journalism has never been lower. According to a 2014 nation-wide poll conducted by Gallup, only 40 percent of Americans expressed confidence in the ability of the mass media—newspapers, television, and radio—to report the news "fully, accurately, and fairly."[1] Another 2014 Gallup poll asked Americans whether they had a "great deal" or "quite a lot" of confidence in the news media, and reported that only 22 percent held such views for newspapers, 19 percent for online news, and 18 percent for television news.[2] Aside from trust, a majority of Americans—63 percent—finds the news media to be biased in its reporting.[3] Complicating matters even more, a host of economic forces beset contemporary journalism, ranging from owner-ship issues such as media consolidation to an overall loss of advertising reve-nue. From the vantage point of the public and of journalists themselves, the outlook seems grim, yet journalism, both as a practice and topic of public concern, is a highly visible issue. The current digital media landscape, filled as it is with bloggers, citizen groups, and professional media watchdogs such as the Media Research Center (on the political right) and Media Matters (on the left), serve a twofold function—to identify bias, fabrications, and inac-curacies; and to provoke a sustained public discussion of the proper role of journalism and journalistic practices.

There is no mistaking the structural problems facing the business and prac-tices of journalism. The question is whether it is worth acknowledging that digital media also has ushered in an age of unprecedented journalism ac-countability, which leads to greater participation in journalism from the public—both as a forum for democratic discourse and as an independent monitor of those who supply society with information. Two key elements drive this change. One is that the news media no longer possess their long-standing gatekeeping function as an elite site of power that assesses what in-formation and issues are worthy of public dissemination.

As Bill Kovach and Tom Rosenstiel describe it, the gatekeeper metaphor referred to how editors, producers, and journalists "stood by an imagined village guardhouse and determined which facts were publicly significant and sufficiently vetted to be made public."[4] This no longer is the case, as online news sites and social media networks can simultaneously cover a multiplicity of subjects unencumbered by the daily confines of newspaper space or a 30-minute television broadcast. As a result, the information available to the public is more fluid and open because of digital media, even if many consumers are self-selecting in choosing where they obtain news. Second, digital media has demonstrated a more immediate ability to assess and discuss the practices of contemporary journalism. The exposure of numerous ethical lapses, fabrications, and plagiarism in the news media over the past two decades certainly has fed into the strong public mistrust of journalists and the mass media, but this newfound propensity of the public to hold journalists and media outlets responsible for such transgressions ultimately serves and reflects a healthy democratic society. That which constitutes effective and ethical journalism is now regularly a part of public discourse, and this self-reflexive function requires media outlets and journalists alike to diligently consider the accuracy of their reporting lest they be cited as biased, unethical, or having fabricated material.

Scandals certainly embarrass and devalue the respect the public has for journalists and the media outlets that publish or broadcast their work, but the ensuing public discourse generated about the practices of journalism, broadly construed, is a relatively new and worthy phenomenon. The tension between the public's perceptions of journalism and the media itself marks an index of democratic alertness in American society. Digital media have engendered a highly discursive forum whereby journalism truly is valued as the fourth estate—or branch—of government. Americans now can more adequately check the news media and exercise the democratic right to voice dissatisfaction with the media. Although media critiques at times could be motivated by cynical political gain, journalists, on the whole, now are held more accountable for their work—which is a positive development. Besides, a perfect press is an illusory goal for a democracy: Although journalists can and should seek an ideal of excellence in reporting, the methods of newsgathering and the dissemination of information never can achieve unqualified neutrality, objectivity, or transparency. A democratic public, moreover, would be wise to distrust a national media infrastructure working in such monolithic lockstep.

Thus, although the economic and structural problems facing the practices of journalism certainly are very real and merit the appropriate concern, the current state of affairs deserves some degree of appreciation. Journalism no longer merely is a repository of information shoveled to a passive public by elite gatekeeping forces. Journalism now represents both diffuse forms of public information and a dialogic clearing space within which citizens, institutions, organizations, and politicians—in addition to media professionals—openly assess how information is procured, published, and broadcast. Both the fierce, contemporary debates about journalism and the recent spike in exposing journalistic malpractice help solidify the separation of powers between the government, the press, and the public. This newfound ability to immediately identify journalistic scandal and then engage in public discourse about it ultimately reflects a more vigorous and democratic public sphere. The public, it is true, exhibits more feelings of distrust in the news media than ever before, but American journalism—owing to the profusion of digital media—never has been held more accountable for its work. To establish this, the following sections review examples of journalistic malfeasance that went unchecked in previous eras of media, and then pivot to consider how—because of digital media—contemporary journalistic misconduct is swiftly exposed and initiates an engaging public discourse about journalism.

(ALMOST) GETTING AWAY WITH IT: A HISTORICAL LACK OF JOURNALISTIC ACCOUNTABILITY

The two key factors contributing to the public's distrust of contemporary journalism are accusations and perceptions of media bias and high-profile scandals—ranging from fabrication and plagiarism to the ethical violations of norms. Digital media platforms have called attention to and in some cases have exacerbated these tensions, but these issues are not new. Bias, for instance, is no stranger to American media. According to Si Sheppard, "For most of American history there was no news as we understand it, in terms of the reporting of the facts; there was only opinion, and highly partisan opinion at that."[5] Bias thus has been the rule and not the exception in American journalism—beginning with the colonial press, which supported independence en masse,[6] and the early partisan press, which soon after independence was fully funded by the Federalist and Democratic-Republican parties as explicit political instruments that served to galvanize public opinion.

Although the private letters of founding fathers such as Thomas Jefferson and passages in Alexis de Tocqueville's *Democracy in America* provide engaging discourses about the constitutional protections and evolving nature of the press, there was little public discussion—from newspapers themselves or citizen-led groups—about journalistic integrity or bias. When the elite, subsidized partisan press eventually yielded to the more open, independent penny press newspapers and magazines of the 1830s, bias by no means disappeared. Many of the news outlets focused exclusively on either promoting the abolition of slavery or opposing women's suffrage.[7] The pioneering journalism of muckraking magazines in the early 20th century, characterized by the investigative work of journalists such as Upton Sinclair and Ida Tarbell, openly championed explicit political reform. There is no doubting the ubiquitous presence of biased political commentary in the current media landscape, but it is hard to accept that the reporting of the news media is now more biased than ever before, given the history of American journalism.

Fabricating material—a gross violation of journalistic ethics—regularly occurred in American history as well. Colonial papers such as the *Boston Journal of Occurrences*, led by Samuel Adams, propagandized in favor of independence through scurrilous accounts of the behavior of British soldiers.[8] Adams is celebrated as a true patriot of the revolution, but it was historians and not his contemporaries who identified and considered the depth with which his deceptions affected the media's relationship to the public. At the same time, the partisan press exhibited a great deal of specious behavior in publishing accounts about its political opposition—from insisting that Jefferson was an atheist[9] to inventing tales of Aaron Burr's sexual licentiousness.[10]

The yellow press of the late 19th century, typified by the newspapers of William Randolph Hearst and Joseph Pulitzer, is remembered for its sensational headlines and lurid crime stories. The media drumbeat for war after the explosion of the U.S.S. *Maine* off the coast of Cuba in 1898, however, relied on willfully ignoring available evidence that exculpated Spain of direct responsibility for the destruction of the naval ship.[11] Hearst is reported to have told an illustrator of his *New York Journal*, whom he sent to Cuba to investigate, in what is now a legendary turn of phrase, "You furnish the pictures and I'll furnish the war."[12]

Although this is just a sampling of journalistic malfeasance in the 18th and 19th centuries,[13] it is worth observing that the fabrications recounted here were neither discovered at the time nor seen to provoke any substantive public dialogue about journalism itself among citizens or within the media. Until recently, there were few if any media watchdogs monitoring the credibility of journalism. As such, the contemporary public distrust of the news media can be seen as a function of the current digital media landscape, which, in contrast to previous eras, features an ability to root out bias, fabrications, and inaccuracies.

The case of Janet Cooke, a reporter for *The Washington Post*, is instructive in this regard. In September of 1980 Cooke published a front-page story, "Jimmy's World," about an eight-year-old heroin addict who lived among drug dealers and had been using drugs since the age of five. It was a heart-wrenching news item that garnered political attention from then Washington, DC, mayor Marion Barry, who initiated a police search for the boy in question. Though Jimmy was never found and, according to Bill Green, the newspaper's ombudsman at the time, there were some staff members who were skeptical of the piece, the newspaper nominated Cooke's article for a Pulitzer Prize—one of the highest honors in the profession of journalism—which she won in April of 1981.[14]

It was, no less, a respected authority of journalism Bob Woodward—the legendary investigative reporter famous for breaking the Watergate scandal with Carl Bernstein—who, as the *Post*'s metro editor, initially defended Cooke's reporting and supported the article's nomination for the award.[15] Not until after the prestigious award was announced and the Associated Press published biographical information on Cooke that contradicted her application materials to the *Post* and to the *Toledo Blade*, where she previously worked, did suspicions about the story grow in earnest. When confronted, Cooke admitted to falsifying her personal history and inventing the child drug user, prompting the Pulitzer committee to withdraw its award. Green, the *Post* ombudsman, reported a flood of calls and letters to the paper expressing outrage about the endangerment of the child soon after the article's publication and the mayor expressed doubt in public about whether such a child existed, but there was no highly visible media or citizen pressure either applied or sustained to ascertain the veracity of the story.[16] The question remains, then, as to whether the story's fabrications would have gone unnoticed if Cooke had not misled the Pulitzer committee about her academic credentials.

For most of American history, unfortunately, fact-checking was difficult. People calling newsrooms or drafting letters to the editor pale in comparison to the digital infrastructure in place now, in which a story such as "Jimmy's World" would draw public attention from social media until it could be confirmed or disproved. Although historians could later piece together contradictory or fabulist renditions of media content, the lag between publication and exposure was substantial. One of the benefits of the digital age—which is characterized by a rapid velocity of communication both about and within the media—is that fact-checking and calling attention to fabrication and ethical transgressions are more immediate.

The first high-profile example demonstrating the power of digital media to root out journalistic fabrications took place in the 1998, when the prolific magazine writer Stephen Glass, who wrote for *The New Republic* among other publications, was identified as a fabulist. According to Buzz Bissinger, Glass fabricated material in 27 of the 31 articles he published in *The New Republic* and covered them by "the production of the false backup materials which he methodically used to deceive legions of editors and fact checkers."[17] Editors initially stood by his work, Bissinger writes, but Glass was exposed when editors from *Forbes Digital Tool*, an early online publication, confronted him in a conference call with *The New Republic* staff about the e-mail addresses and Web sites of companies written about in an article about hacking.[18] Glass, they alleged, had invented the e-mail addresses and Web sites himself, and Bissinger reports this prompted Charles Lane, the editor of The New Republic, to search Glass's computer and eventually discover other anomalies in his work, including having his brother pose as an executive at Jukt Micronics, a fictitious company prominently featured in one of Glass's articles.[19] The editors of Forbes Digital Tool, a forerunner to the news Web site forbes.com, conducted a Lexis-Nexis search and discovered that Jukt had no other citations or mentions.[20] The duplicity and fabulist writings of Glass were thus exposed, yet it is worth remembering that, for a few years of the 1990s, Glass was a highly respected journalist publishing work in some of the more elite magazines in the country. Before the creation of algorithmic search engines available to the public, which might have exposed Glass sooner, it took an article he wrote on hacking, which alarmed editors from a competitor publication familiar with the tech industry and what were then sophisticated digital search tools, to begin the process of sifting through the hoaxes that Glass perpetrated. How long, one could ask, would Glass have continued his journalistic fraudulence had he not chosen subject matter monitored by those with the technical wherewithal to spot a fake Web site? Glass's fabrications ran so deep that his undoing perhaps was inevitable, but now that search tools are so prevalent and accessible, to both editors and the public alike, it is doubtful that Glass would have enjoyed the run of publishing success he experienced a decade later.

Two other scandals, which involved high-profile reporters Jayson Blair and Jack Kelly, broke in 2003 and 2004, respectively, and provide a sense of the exposure lag prior to the current robust digital media landscape. Blair wrote more than 600 articles for The New York Times over a four-year period, at least 36 of which were proven to be fabricated or plagiarized in his last seven months, until he was forced to resign in May 2003.[21] According to the Times' own internal report following his resignation, Blair, like Glass, used e-mail and the Internet to cover his tracks. For Blair, this enabled him to both lie about his whereabouts to create false datelines and plagiarize copy from the work of other reporters around the country.[22] This was in addition to writing that plainly fabricated stories, sources, and quotes. Although some of Blair's editors over the years expressed concern about inaccuracies in his reporting, he continued to be promoted. Blair's unraveling arrived in late April 2003 when Macarena Hernandez, a former colleague of Blair's and reporter for the San Antonio Express-News, read a front-page article in the Times that copied her own work verbatim—without acknowledgment—from the previous week.[23] Hernandez then placed a call to her contact at the Times, where she was once an intern, which set off a chain reaction that ultimately led to Blair's resignation.[24]

Blair was certainly caught and held responsible for his fraudulent reporting. Although, it is worth observing that in addition to the exposure lag,

Blair, like Glass and Cooke before him, gained prominence despite the concerns of some colleagues and flourished professionally—at least for a time—before an unwitting public because at that time there was no access to social media or advanced search engines. Manuel Roig-Franzia, who at the time was Southern bureau chief of *The Washington Post*, later recalled visiting San Antonio in April 2003 and reading the stories authored by Hernandez and Blair. He consulted with his editor about covering "what appeared to be an egregious case of plagiarism," but Roig-Franzia eventually decided against publishing what could be considered "a gotcha piece about another journalist."[25] Reflecting on the incident in May 2014, he wrote, "If all this had taken place 11 days ago, instead of 11 years ago, I'm sure I would have tweeted about the whole mess right there on the side of the road."[26] It is a telling statement by a newspaper editor, one that reflects the innovative developments happening during the past decade—in traditional and new media—that make journalism more accountable. The immediacy and interconnectedness of the communications technologies available now could have halted the fabulist works of Glass and Blair much sooner, but fact-checking was not a social enterprise as it is now.

The case of Jack Kelley, a foreign correspondent for *USA Today* who had written for the paper from its 1982 inception until his resignation in January 2004,[27] makes this clear as well and occurs right at the tipping point of the digital media era. Kelley, who had been nominated for a Pulitzer Prize in 2001, was forced to step down due to "conspiring with a translator to mislead editors overseeing an inquiry into his work." The newspaper subsequently investigated Kelley's journalistic practices and discovered that he "fabricated substantial portions of at least eight major stories, lifted nearly two dozen quotes or other materials from competing publications, [and] lied in speeches he gave for the newspaper."[28] To probe the extent of Kelley's malfeasance, *USA Today* reporters spent seven weeks sifting through more than 700 of his articles and sent fact-checkers to Kelley datelines such as Cuba, Israel, and Jordan to interview his sources, a process that "uncovered evidence that found Kelley's journalistic sins were sweeping and substantial."[29] Additionally, an independent report commissioned and published by the newspaper found that despite concerns expressed by both his professional colleagues as well as government officials objecting to his work, editors consistently "gave him good performance marks."[30] The depth of Kelley's infractions ran spectacularly deep, but it should not be forgotten that he maintained a high-profile position at a national newspaper for more than two decades, despite the independent report finding that his "dishonest reporting date[d] back at least as far as 1991."[31]

Of course, there is no way to prove that Cooke would not have won her Pulitzer or that the careers of Glass, Blair, and Kelley would have been stopped short in the digital media age. Yet the combined forces of innovative communications technology, media watchdogs, and social media serve to bring about journalistic accountability in a much more immediate and public manner. In each instance of journalism misconduct in the quarter-century leading up to the digital era cited above, suspicions were raised by colleagues and government officials. The cover of friendly editors, however, resulted in there being no recourse for those who were suspicious to prove such misgivings. As the following examples drawn from late 2004 to the present make clear, however, the tools available in the current digital landscape supply just such an environment.

JOURNALISM ACCOUNTABILITY IN THE DIGITAL ERA

The year 2004 is used as the point of departure for the digital era in journalism for several reasons. Although the Internet and blogosphere certainly predate that year, 2004 featured a series of events that make it stand out as a shift in the abilities and practices of digital media: Google had its initial public offering, Facebook was created, and, according to the Pew Research Center, for the first time more Americans began to access the Web via broadband connections rather than dial-up connections.[32] Politically speaking, going back as far as to Vice President Spiro Agnew of the Nixon administration, conservatives had labelled the mainstream media as "liberal," yet the 2004 reelection campaign of President George W. Bush focused sharply on running against the biases of the news—most prominently in the speeches delivered at the Republican National Convention that year.[33] Significantly for journalism, 2004 marks the beginning of the swift and public reaction to media wrongdoing, beginning with the now infamous 60 Minutes Wednesday television news story, aired on September 8, 2004. That TV program released documents that purported to reveal how President Bush received preferential treatment during his service in the Air National Guard at the time of the Vietnam War.

The story—which fed into, if not accelerating, preexisting charges of media bias—resulted in the firing of CBS executives and producers, as well as the stepping-down of long-time news anchor Dan Rather. Most notably for the purposes here, is the time frame within which the documents were discredited.

According to Howard Kurtz, The Washington Post media critic at the time, a flurry of published activity on the conservative discussion forum "Free Republic" that night, and blogs Power Line and Little Green Footballs early the next morning analyzed and challenged the authenticity of the documents, which had been attached to the CBS Web site and, under scrutiny, were revealed to have been created using Microsoft Word software.[34] Three of the nation's top daily newspapers covered the story the next day without mentioning any possibility that the documents might be forged.[35] National media eventually questioned the documents by Friday of that week[36] and conducted forensic analyses. The media outlets finally also discredited the documents as forgeries, but they were scooped by the blogosphere, which was both accurate and immediate. In this instance, blogs—which tend to be focused on opinion, not reporting—spurred the national media to vigorously investigate the story. Within a week of the story airing on television, CNN had interviewed the three experts hired by CBS to authenticate the documents before running the piece, each of whom expressed ambivalence about their provenance but whose reservations ultimately were ignored.[37] To this day, Dan Rather continues to stand by the overall reporting in the story,[38] if not the documents themselves.

It is worth noting how the digital media landscape played a role in exposing the forgery. Digital tools enabled both the fabrication of the documents and their subsequent exposure. The mainstream press initially reported the conclusions of the story without authenticating the documents until pressured by the blogosphere to do so. The lag time between initial media fraud and its exposure—which previously required journalistic malfeasance to be discovered through research by later historians or as fortuitous coincidence by other media members suspecting something amiss—disappeared. By late 2004, then, journalists could be seen as operating in a complex media environment providing a check on their reporting practices through a highly visible and public feedback loop of concern and verification.

Blogs perhaps rightly are criticized for injecting political cynicism and polarization into the public sphere, but in this case their ability to monitor the news media served an important role in identifying journalistic malpractice. Daily newspapers took note of this when, in October 2013, 60 Minutes was again called out for airing a discredited story. In this case the producers and reporters did not broadcast fabricated documents but rather based their reporting on a disreputable source who had given conflicting accounts of the 2012 attack on the American consulate in Benghazi, Libya. Soon after the television piece aired, The Washington Post's Karen DeYoung reported that the source in question, using a pseudonym, had written a separate report admitting that he had not been present at the consulate the night of the attack, which conflicted with what he told 60 Minutes; the source was thus proven to be an unreliable witness.[39]

These scandals by no means suggest that 60 Minutes, as a television newsmagazine program with a respected history of investigative journalism, and CBS as a network, are fraudulent perpetrators of news. The stories in question are far less egregious than the examples recounted above by fabulists in print journalism prior to the age of digital media. If anything, the 60 Minutes news staff—executives, producers, and reporters alike—were, in their collective eagerness to break a story, too trusting in their sources and lacked the diligence necessary to verify evidence. As such, the digital age is demanding more accountability of journalists and news media outlets, from an insistence on fact-checking, as in the case of 60 *Minutes*, to original reporting and ethical newsgathering practices.

Two recent plagiarism scandals in journalism bear this out in that each of the reporters in question had, in fact, stolen material from their own previous work. In 2012, Jonah Lehrer—a scientific reporter for newspapers including *The Wall Street Journal* and magazines including *Wired* and *The New Yorker*—was found to have recycled his own previous material. In 2014, CNN's news editor in London, Marie-Louise Gumuchian, cribbed from her earlier written work for *Reuters*.

Lehrer's self-plagiarism was first discovered by Jim Romenesko, who operates *A Blog About Media and Other Things I'm Interested In* at jimromenesko. com, and posted Lehrer's near-identical content to his site a week after the offense appeared.[40] Numerous other instances of self-plagiarism would soon be detailed around the Web. Edward Champion of the blog site *Reluctant Habits* discovered that Lehrer had plagiarized from his *New Yorker* colleague Malcolm Gladwell.[41] Michael Moynihan of *Tablet*, a daily Jewish online magazine, found that Lehrer fabricated material in his book *Imagine: How Creativity Works*.[42]

Marie-Louise Gumuchian lasted only six months in her position writing for CNN.com before "a routine editing check" found an instance of self-plagiarism, which led CNN to run her work through "plagiarism-flagging software," begetting a more problematic, "deeper examination" of her reporting.[43] All told, CNN's internal investigation found more than 50 instances of self-plagiarism in Gumuchian's articles.[44] From a journalistic perspective, the case of Lehrer—who was found guilty of more than mere self-plagiarism—is more troubling than Gumuchian's. Taken together, the plagiarism scandals evince a discerning and media-literate public that monitors journalism in a watchdog capacity as well as a professional news media more sensitive and proactive to plagiarism.

Craig Silverman of the Poynter Institute suggests that the Lehrer example is the latest instance of what he considers "The Google Game," wherein a more active, crowdsourced public, including fellow media professionals, can immediately assess journalism that is suspected of plagiarism by searching for cribbed reporting.[45] Silverman compares the downfall of Lehrer to that of Tim Goeglein, an aide to President George W. Bush, who was forced to resign in 2008 for plagiarizing columns he wrote for the *Fort Wayne (IN) News-Sentinel*. Goeglein's plagiarism was first spotted by a former reporter for the paper, Nancy Derringer who, according to Silverman, used Google's search engine to discover plagiarism and posted the findings to her blog.[46] Within 12 hours of her initial posting, according to Derringer, 20 instances of Goeglein's plagiarism were uncovered by other bloggers, and Goeglein had resigned from the White House.[47]

Though the digital age might be best known for disseminating a prolific quantity of information, these more recent examples of journalistic plagiarism demonstrate how the current media infrastructure possesses an effective toolkit to also evaluate the originality and veracity of content. Beyond just verifying information, digital media also provide an environment for public discourse about the ethical practices of journalism—a new phenomenon that is encouraging both for journalists themselves as well as for the public in general.

A healthy, interactive, and accountable journalistic dialogue ensued rather quickly after *Grantland*, an ESPN-owned webzine about sports and pop culture, published an investigative article into a new and dynamic golf putter by Caleb Hannan. The piece revolved around the putter's inventor, Dr. Essay Anne Vanderbilt, who told Hannan that she was a MIT-trained scientist who used to work on top secret projects for the U.S. Department of Defense.[48] As Hannan investigated further he discovered, among other things, that Vanderbilt's academic credentials were not what she claimed and that she was transsexual. During the course of another interview, Hannan revealed Vanderbilt's gender identity to a business associate who had not known, and then published the sequence including that Vanderbilt had committed suicide before the article went to press.[49]

Bill Simmons, the editor-in-chief at *Grantland*, later reported that although the story initially received praise on Twitter, within 56 hours there was outrage against the site on social media.[50] In addition to writing an article apologizing for never consulting anyone within the transgender community before running the article, Simmons invited a critical response on *Grantland* from Christina Kahrl, a sportswriter for ESPN.com and a member of the board of directors for GLAAD (a media organization devoted to issues affecting the lesbian, gay, bisexual, and transgendered community). Kahrl accused Hannan and his editors of treating Dr. Vanderbilt poorly and ignoring important issues, such as a high rate of suicide among transgendered persons. Additionally, she offered a more ethical way in which the story should have been reported, "[*Grantland*] really should have simply stuck with debunking those claims to education and professional expertise relevant to the putter itself, dropped the element of [Vanderbilt's] gender identity if she didn't want that to be public information—as she very clearly did not—and left it at that."[51]

Given the responses of Simmons and Kahrl, it is clear that the process by which the story was reported, edited, and published was unfortunate, if not unethical and outright tragic. *Grantland*, however, as a Web-based media outlet, was within five days able to publicly apologize for the story, run a critical response to it, and provoke a larger discussion of journalism ethics for the public. Although the ensuing apology and dialogue by no means attenuate the initial grievous errors of judgment committed by the reporter and his editors, the episode speaks to a more self-conscious news media that is able to more quickly identify unethical newsgathering practices and published work that adversely affects vulnerable populations.

A different set of ethical concerns beset *Rolling Stone* magazine in late 2014 but indicate the necessity of responsible reporting and editing in the digital age of journalism accountability. In the opening narrative of an article about the problems of sexual assault on college campuses, Sabrina Rubin Erdely featured a visceral, traumatic story about Jackie, a student at the University of Virginia who alleged that she had been gang-raped at one of the school's fraternity houses.[52] The article went on to discuss a host of issues plaguing college campuses, from sexual assault itself to reporting practices and Title IX (a federal statute requiring gender equity). Jackie's story and the newsgathering practices that led to its publication began undergoing intense scrutiny just a week later. Erdely discussed her reporting on a *Slate* podcast, "DoubleX Gabfest," and admitted that, out of respect for Jackie's concerns, she did not speak to the alleged perpetrators of the act.[53] Although Erdely apparently did contact the local chapter president and a national representative of the fraternity, both she and her editor said they were unable to reach the accused.[54] From there, more challenging questions surfaced about the veracity of Jackie's account, leading *Rolling Stone*, two weeks after the story was published, to issue an apologetic disclaimer that, given the subsequent reporting on the story, conceded errors in judgment.[55] The magazine rightly earned a great deal of criticism for its handling of the story, but a broader examination of the practices of journalism also was provoked. Allison Benedikt and Hannah Rosin wrote of the lessons learned,

> You try very, very hard to reach anyone you're accusing of something. You use any method you can think of. . . . If you fail to reach the person, you write a sentence explaining that you tried—and explaining *how* you tried—as a way to assure your readers that you gave the person a chance to defend themselves. We're not sure why *Rolling Stone* didn't think that was necessary.[56]

And from a legal perspective, law professor Eugene Volokh, whose blog *The Volokh Conspiracy* was acquired by *The Washington Post* Web site, assessed the libel implications if the claims made in the story were materially false.[57] The newsgathering practices of the story thus were quickly and publicly identified as flawed, unethical, and potentially liable to litigation. Though a public parsing of the problems of this case does not mitigate the damage wrought by the initial reporting, it is encouraging that unscrupulous journalistic practices are held accountable through a combination of traditional and online media platforms.

CONCLUDING THOUGHTS

Although the author has sought to trace the positive developments in reaction to journalism scandals over the past decade, some qualifications are necessary. First, the foregoing examples of journalism misconduct, although broadly representative, do not constitute an exhaustive account of media scandals in American history. It is hoped, in this regard, that more in-depth analyses of journalism accountability are spurred moving forward. Second, there are worthy topics of debate about journalism that, it must be admitted, the present essay omits a good deal of academic research,[58] and popular opinion-making[59] attempt to establish (or rebut the existence of) liberal media bias and a lack of objectivity among journalists; identify the harmful consequences of corporate, often conservative, control of news media organizations;[60] and trace the rise of commentary and sensationalism over robust public affairs and investigative reporting.[61] My purpose has thus been decidedly narrow—to establish how journalism has entered into an era of unprecedented accountability due to digital media's swift exposure of plagiarism, fabrication, and unethical behavior. By no means does the present essay suggest that the rise of such accountability can mask or overcome the economic and structural problems that plague the news media or the cynicism that exploits the public's mistrust of journalists. These are harsh realities that media businesses and professionals must openly face.

But for all of the criticisms of traditional journalism, to say nothing of the blogosphere and social media, it is ultimately a bonus for a democratic society to suspiciously monitor and discuss the news. This rise in journalism accountability reflects a digital environment wherein the media share a symbiotic relationship with the consumer-citizen public it reports to. Besides, journalists should never assume that their work will be accepted without any skepticism—irrespective of what factors inform the public's perceptions. There is a possibility that such a state of affairs might intimidate journalists or media outlets from aggressively pursuing important stories, but effective reporting requires following ethical practices of newsgathering, which the public is rightly more sensitive to now. Requiring that journalism be thoroughly fact-checked and original, and sources be properly vetted and respected, are simple truisms that have circulated among newsrooms, professional media codes of ethical conduct, and schools of journalism for at least a century. Now, however, the public is part of the discussion as to what constitutes effective and ethical reporting.

In the digital age, the media is itself a robust subject of reporting, and although this sheds inordinate attention on instances of fabrication, plagiarism, and unethical behavior, journalism deserves such scrutiny—even at the expense of its public image. If anything, this increased monitoring function demands that the news media exercise a due diligence commensurate with the awesome responsibility required for journalism. As such, editors and producers as well as print, online, and television reporters must perforce thoroughly scrutinize their own work lest they be exposed as biased, fraudulent, plagiarized, unethical or as having been deceived by a source. The twin forces of social media and search engines dictate that journalists handle information with proper care because the exposure lag that previously existed in rooting out journalistic misconduct has effectively vanished. It is plain that the scandals of the past decade do not reflect well on journalists, yet digital media has created a landscape wherein journalism is held more immediately accountable to the public and become a site of discussion. Just as the news media provides an important check on the powerful interests of government and industry, the public now is keenly aware of, concerned with, and even participating in making sure the practices of reporting and journalistic content are credible sources of information; and that, for whatever else is said about the media, is an encouraging sign befitting a democratic society.

NOTES

1. Justin McCarthy, "Trust in Mass Media Returns to All-Time Low," *Gallup* (September 14, 2014), http://www.gallup.com/poll/176042/trust-mass-media-returns-time-low.aspx.

2. Andrew Dugan, "Americans' Confidence in News Media Remains Low," *Gallup* (June 19, 2014), http://www.gallup.com/poll/171740/americans-confidence-news-media-remains-low.aspx.

3. McCarthy, "Trust in Mass Media."

4. Bill Kovach and Tom Rosenstiel, *The Elements of Journalism* (New York: Three Rivers Press, 2014), 25.

5. Si Sheppard, *The Partisan Press: A History of Media Bias in the United States* (Jefferson, NC: McFarland & Company, 2008), 19.

6. Philip Davidson, *Propaganda and the American Revolution* (Chapel Hill: University of North Carolina Press, 1941), 226.

7. Roger Streitmatter, *Mightier than the Sword* (Boulder, CO: Westview Press, 2012), 17–46.

8. Davidson, *Propaganda and the American Revolution*, 236.

9. Edward J. Larson, *A Magnificent Catastrophe: The Tumultuous Election of 1800, America's First Presidential Campaign* (New York: Free Press, 2007), 171–73.

10. Nancy Isenberg, *Fallen Founder: The Life of Aaron Burr* (New York: Viking, 2007), 144, 259.

11. Joseph E. Wisan, *The Cuban Crisis As Reflected in the New York Press 1895–1898* (New York: Octagon Books, 1965), 385–438.

12. Streitmatter, *Mightier than the Sword*, 69–70.

13. *See* Marcus Daniel, *Scandal and Civility: Journalism and the Birth of American Democracy* (New York: Oxford University Press, 2009).

14. Bill Green, "The Players: It Wasn't a Game," *Washington Post* (April 16, 1981), http://www.public.asu.edu/~lthornt1/413/jimmy.htm.

15. Ibid.

16. Ibid.

17. Buzz Bissinger, "Shattered Glass," *Vanity Fair* (September 1998), http://www.vanityfair.com/ magazine/archive/1998/09/bissinger199809?src=longreads.

18. Ibid.

19. Ibid.

20. Adam L. Penenberg, "Lies, Damn Lies and Fiction," *Forbes Digital Tool* (May 11, 1998), http://www.forbes.com/sites/michaelnoer/2014/11/12/read-the-original-forbes-takedown-of-stephen-glass/

21. Dan Barry, David Barstow, Jonathan D. Glater, Adam Liptak, and Jacques Steinberg, "Times Reporter Who Resigned Leaves Long Trail of Deception," *New York Times* (May 11, 2003), http://www.nytimes.com/2003/05/11/national/11PAPE.html/?pagewanted=all&src=longreads.

22. Ibid.

23. Seth Mnookin, *Hard News: Twenty-one Brutal Months at* The New York Times *and How They Changed American Media* (New York: Random House, 2005), 103.

24. Ibid., 104.

25. Manuel Roig-Franzia, " 'A Fragile Trust' Exhibits Irresponsibility behind Jayson Blair Plagiarism Scandal," *Washington Post* (May 5, 2014), http://www.washingtonpost.com/lifestyle/style/a-fragile-trust-exhibits-irresponsibility-behind-jayson-blair-plagiarism-scandal/2014/05/05/596aca28-d45d-11e3-95d3-3bcd77cd4e11_story.html.

26. Ibid.

27. Jill Rosen, "The Rise and Fall of Jack Kelley," *American Journalism Review* (April/May 2004), http://ajrarchive.org/article.asp?id=3619.

28. Blake Morrison, "Ex-USA TODAY Reporter Faked Major Stories," *USA Today* (March 19, 2004), http://usatoday30.usatoday.com/news/2004-03-18-2004-03-18_kelleymain_x.htm?src=longreads.

29. Ibid.

30. Bill Hilliard, Bill Kovach, and John Seigenthaler, "The Problems of Jack Kelley and USA Today," *USA Today* (April 22, 2014), http://usatoday30.usatoday.com/news/2004-04-22-report-one_x.htm

31. Ibid.

32. Kathryn Zickuhr, and Aaron Smith, "Home Broadband 2013," *Pew Research Internet Center* (August 26, 2013), http://www.pewinternet.org/2013/08/26/home-broadband-2013/.

33. Sheppard, *The Partisan Press* 5–7, 286–90.

34. Howard Kurtz, "After Blogs Got Hits, CBS Got a Black Eye," *Washington Post* (September 20, 2004), http://www.washingtonpost.com/wp-dyn/articles/A34153-2004Sep19.html.

35. Mark Memmott, "Scoops and Skepticism: How the Story Unfolded," *USA Today* (September 21, 2004), http://usatoday30.usatoday.com/news/politicselections/nation/president/2004-09-21-guard-scoops-skepticism_x.htm.

36. Michael Dobbs and Mike Allen, "Some Question Authenticity of Papers on Bush," *Washington Post* (September 10, 2004), http://www.washingtonpost.com/wp-dyn/articles/A9967-2004Sep9.html.

37. "CBS' Experts Say They Didn't Authenticate Bush Memos," *CNN* (September 15, 2004), http://edition.cnn.com/2004/ALLPOLITICS/09/15/bush.guard.memos/index.html.

38. Dan Rather with Digby Diehl, *Rather Outspoken: My Life in the News* (New York: Grand Central Publishing, 2012), 5.

39. Karen DeYoung, " '60 Minutes' Broadcast Helps Propel New Round of Back-and-Forth on Benghazi," *Washington Post* (October 31, 2013), http://www.washingtonpost.com/world/national-security/60-minutes-broadcast-helps-propel-new-round-of-back-and-forth-on-benghazi/2013/10/31/fbfcad66-4258-11e3-a751-f032898f2dbc_story.html.

40. Jim Romenesko, "Jonah Lehrer's NewYorker.com 'Smart People' Post Borrows from Earlier WSJ Piece" *Jim Romenesko* (June 20, 2012), http://jimromenesko.com/2012/06/19/jonah-lehrers-newyorker-com-smart-people-post-look-familiar/.

41. Edward Champion, "How Jonah Lehrer Recycled His Own Material for *Imagine*," *Reluctant Habits* (June 22, 2012), http://www.edrants.com/how-jonah-lehrer-recycled-his-own-material-for-imagine/.

42. Michael Moynihan, "Jonah Lehrer's Deceptions," *Tablet* (July 30, 2012), http://tabletmag.com/jewish-news-and-politics/107779/jonah-lehrers-deceptions.

43. Erik Wemple, "CNN Fires News Editor Marie-Louise Gumuchian for Plagiarism," *Washington Post* (May 16, 2014), http://www.washingtonpost.com/blogs/erik-wemple/wp/2014/05/16/cnn-fires-news-editor-marie-louise-gumuchian-for-plagiarism/. Accessed June 8, 2015.

44. Meredith Artley, Manuel Perez, and Richard T. Griffiths, "Editors' Note," *CNN.com* (May 16, 2014), http://edition.cnn.com/2014/05/16/world/editors-note/.

45. Craig Silverman, "Jonah Lehrer Is the Latest Target of Google Game, Crowdsourced Investigation," *Poynter*, http://www.poynter.org/news/mediawire/177917/jonah-lehrer-is-the-latest-target-of-google-game-crowdsourced-investigation/. Last updated November 25, 2014.

46. Ibid.

47. Nancy Derringer, "Gone in 60 Seconds," *Slate* (March 3, 2008), http://www.slate.com/articles/news_and_politics/politics/2008/03/gone_in_60_seconds.html.

48. Caleb Hannan, "Dr. V's Magical Putter," *Grantland* (January 15, 2014), http://grantland.com/features/a-mysterious-physicist-golf-club-dr-v/.

49. Ibid.

50. Bill Simmons, "The Dr. V Story: A Letter from the Editor," *Grantland* (January 20, 2014), http://grantland.com/features/the-dr-v-story-a-letter-from-the-editor/.

51. Christina Kahrl, "What Grantland Got Wrong," *Grantland* (January 20, 2014), http://grantland.com/features/what-grantland-got-wrong/.

52. Sabrina Rubin Erdely, "A Rape on Campus: A Brutal Assault and Struggle for Justice at UVA," *Rolling Stone* (November 19, 2014), http://www.rollingstone.com /culture/features/a-rape-on-campus-20141119.

53. Hanna Rosin, June Thomas, and Katy Waldman, "*DoubleX* Gabfest: The Butch Goddess Edition," *Slate* (November, 27, 2014), http://www.slate.com/articles /podcasts/doublex_gabfest/2014/11/the_double_x_gabfest_on_uva_frats_and_rape_in _rolling_stone_husbands_hurting.html.

54. Paul Farhi, "Author on Rolling Stone Article on Alleged U-Va. Rape Didn't Talk to Accused Perpetrators," *Washington Post* (December 1, 2014), http://www .washingtonpost.com/lifestyle/style/   author-of-rolling-stone-story-on-alleged-u-va -rape-didnt-talk-to-accused-perpetrators/2014/12/01/e4c19408-7999-11e4-84d4 -7c896b90abdc_story.html.

55. Will Dana, "A Note to Our Readers," *Rolling Stone* (December 5, 2014), http:// www.rollingstone.com/culture/news/a-note-to-our-readers-20141205.

56. Allison Benedikt and Hannah Rosin, "The Missing Men," *Slate.com* (December 2, 2014), http://www.slate.com/articles/double_x/doublex/2014/12/sabrina _rubin_erdely_uva_why_didn_t_a_rolling_stone_writer_talk_to_the_alleged.html.

57. Eugene Volokh, "Libel Law and the Rolling Stone / UVA Alleged Gang Rape Story," *Washington Post* (December 6, 2014), http://www.washingtonpost.com/news /volokh-conspiracy/wp/2014/12/06/libel-law-and-the-rolling-stone-uva-alleged-gang -rape-story/.

58. For a statistical study confirming the media's liberal bias, see Tim Groseclose and Jeffrey Milyo, "A Measure of Media Bias," *Quarterly Journal of Economics* 120, no. 4 (2005): 1191–237; for a statistical analysis refuting liberal bias, see Dave D'Alessio and Mike Allen, "Media Bias in Presidential Elections: A Meta-Analysis," *Journal of Communication* 50, no. 4 (2000): 133–56.

59. See, from the political right, Bernard Goldberg, *Bias: A CBS Insider Exposes How the Media Distort the News* (Washington, DC: Regnery Publishing, 2002), and from the left, Eric Alterman, *What Liberal Media? The Truth about Bias and the News* (New York: Basic Books, 2003).

60. *See* Robert W. McChesney, *The Political Economy of Media* (New York: Monthly Review Press, 2008); Edward S. Herman and Noam Chomsky, *Manufacturing Consent: The Political Economy of Mass Media* (New York: Pantheon Books, 2002).

61. *See* Kovach and Rosenstiel, *The Elements of Journalism*.

BIBLIOGRAPHY

Alterman, Eric. *What Liberal Media? The Truth about Bias and the News* (New York: Basic Books, 2003) (from the political left).

Artley, Meredith, Manuel Perez, and Richard T. Griffiths. "Editors' Note." *CNN.com* (May 16, 2014). http://edition.cnn.com/2014/05/16/world/editors-note/.

Barry, Dan, David Barstow, Jonathan D. Glater, Adam Liptak, and Jacques Steinberg. "Times Reporter Who Resigned Leaves Long Trail of Deception." *New York Times*

(May 11, 2003). http://www.nytimes.com/2003/05/11/national/11PAPE.html/?pa gewanted=all&src=longreads.

Benedikt, Allison, and Hannah Rosin. "The Missing Men." *Slate.com* (December 2, 2014). http://www.slate.com/articles/double_x/doublex/2014/12/sabrina_rubin _erdely_uva_why_didn_t_a_rolling_stone_writer_talk_to_the_alleged.html.

Bissinger, Buzz. "Shattered Glass." *Vanity Fair* (September 1998). http://www.vanity fair.com/magazine/archive/1998/09/bissinger199809?src=longreads.

"CBS' Experts Say They Didn't Authenticate Bush Memos." *CNN* (September 15, 2004). http://edition.cnn.com/2004/ALLPOLITICS/09/15/bush.guard.memos/index .html.

Champion, Edward. "How Jonah Lehrer Recycled His Own Material for *Imagine*." *Reluctant Habits* (June 22, 2012). http://www.edrants.com/how-jonah-lehrer -recycled-his-own-material-for-imagine/.

D'Alessio Dave, and Mike Allen. "Media Bias in Presidential Elections: A Meta-Analysis." *Journal of Communication* 50: 4 (2000) 133–56. Statistical analysis refuting liberal bias.

Dana, Will. "A Note to Our Readers." *Rolling Stone* (December 5, 2014). http://www .rollingstone.com/culture/news/a-note-to-our-readers-20141205.

Daniel, Marcus. *Scandal and Civility: Journalism and the Birth of American Democracy* (New York: Oxford University Press, 2009).

Davidson, Philip. *Propaganda and the American Revolution* (Chapel Hill: University of North Carolina Press, 1941), 226.

Derringer, Nancy. "Gone in 60 Seconds." *Slate* (March 3, 2008). http://www.slate .com/articles/news_and_politics/politics/2008/03/gone_in_60_seconds.html

DeYoung, Karen. " '60 Minutes' Broadcast Helps Propel New Round of Back-and-Forth on Benghazi." *Washington Post* (October 31, 2013). http://www.wash ingtonpost.com/world/national-security/60-minutes-broadcast-helps-propel-new -round-of-back-and-forth-on-benghazi/2013/10/31/fbfcad66-4258-11e3-a751 -f032898f2dbc_story.html.

Dobbs, Michael, and Mike Allen. "Some Question Authenticity of Papers on Bush." *Washington Post* (September 10, 2004). http://www.washingtonpost.com/wp-dyn /articles/A9967-2004Sep9.html.

Dugan, Andrew. "Americans' Confidence in News Media Remains Low." *Gallup* (June 19, 2014). http://www.gallup.com/poll/171740/americans-confidence-news -media-remains-low.aspx.

Erdely, Sabrina Rubin. "A Rape on Campus: A Brutal Assault and Struggle for Justice at UVA." *Rolling Stone* (November 19, 2014). http://www.rollingstone.com /culture/features/a-rape-on-campus-20141119.

Farhi, Paul. "Author on Rolling Stone Article on Alleged U-Va. Rape Didn't Talk to Accused Perpetrators." *Washington Post* (December 1, 2014) http://www .washingtonpost.com/lifestyle/style/author-of-rolling-stone-story-on-alleged-u-va -rape-didnt-talk-to-accused-perpetrators/2014/12/01/e4c19408-7999-11e4-84d4 -7c896b90abdc_story.html.

Goldberg, Bernard. *Bias: A CBS Insider Exposes How the Media Distort the News* (Washington, DC: Regnery Publishing, 2002) (from the political right).

Green, Bill. "The Players: It Wasn't a Game." *Washington Post* (April 16, 1981). http://www.public.asu.edu/~lthornt1/413/jimmy.htm.

Groseclose, Tim, and Jeffrey Milyo. "A Measure of Media Bias." *Quarterly Journal of Economics* 120, no. 4 (2005): 1191–237. Statistical study confirming the media's liberal bias.

Hannan, Caleb. "Dr. V's Magical Putter." *Grantland* (January 15, 2014). http://grantland .com/features/a-mysterious-physicist-golf-club-dr-v/.

Herman, Edward S., and Noam Chomsky. *Manufacturing Consent: The Political Economy of Mass Media* (New York: Pantheon Books, 2002).

Hilliard, Bill, Bill Kovach, and John Seigenthaler. "The Problems of Jack Kelley and USA Today." *USA Today* (April 22, 2014). http://usatoday30.usatoday.com /news/2004-04-22-report-one_x.htm.

Isenberg, Nancy. *Fallen Founder: The Life of Aaron Burr* (New York: Viking, 2007), 144, 259.

Kahrl, Christina. "What Grantland Got Wrong." *Grantland* (January 20, 2014). http://grantland.com/features/what-grantland-got-wrong/.

Kovach, Bill, and Tom Rosenstiel. *The Elements of Journalism* (New York: Three Rivers Press, 2014), 25.

Kurtz, Howard. "After Blogs Got Hits, CBS Got a Black Eye." *Washington Post* (September 20, 2004). http://www.washingtonpost.com/wp-dyn/articles/A34153 -2004Sep19.html.

Larson, Edward J. *A Magnificent Catastrophe: The Tumultuous Election of 1800, America's First Presidential Campaign* (New York: Free Press, 2007), 171–173.

McCarthy, Justin. "Trust in Mass Media Returns to All-Time Low." *Gallup* (September 14, 2014). http://www.gallup.com/poll/176042/trust-mass-media-returns-time -low.aspx.

McChesney, Robert W. *The Political Economy of Media* (New York: Monthly Review Press, 2008).

Memmott, Mark. "Scoops and Skepticism: How the Story Unfolded." *USA Today* (September 21, 2004). http://usatoday30.usatoday.com/news/politicselections /nation/president/2004-09-21-guard-scoops-skepticism_x.htm.

Mnookin, Seth. *Hard News: Twenty-One Brutal Months at* The New York Times *and How They Changed American Media* (New York: Random House, 2005), 103.

Morrison, Blake. "Ex-USA TODAY Reporter Faked Major Stories." *USA Today* (March 19, 2004). http://usatoday30.usatoday.com/news/2004-03-18-2004-03-18 _kelleymain_x.htm?src=longreads.

Moynihan, Michael. "Jonah Lehrer's Deceptions." *Tablet* (July 30, 2012). http:// tabletmag.com/jewish-news-and-politics/107779/jonah-lehrers-deceptions.

Penenberg, Adam L. "Lies, Damn Lies and Fiction." *Forbes Digital Tool* (May 11, 1998). http://www.forbes.com/sites/michaelnoer/2014/11/12/read-the-original-forbes -takedown-of-stephen-glass/.

Rather, Dan, with Digby Diehl. *Rather Outspoken: My Life in the News* (New York: Grand Central Publishing, 2012), 5.

Roig-Franzia, Manuel. "'A Fragile Trust' Exhibits Irresponsibility behind Jayson Blair Plagiarism Scandal." *Washington Post* (May 5, 2014). http://www.washingtonpost

.com/lifestyle/style/a-fragile-trust-exhibits-irresponsibility-behind-jayson-blair
-plagiarism-scandal/2014/05/05/596aca28-d45d-11e3-95d3-3bcd77cd4e11_story
.html.

Romenesko, Jim. "Jonah Lehrer's NewYorker.com 'Smart People' Post Borrows from Earlier WSJ Piece." *Jim Romenesko* (June 20, 2012). http://jimromenesko .com/2012/06/19/jonah-lehrers-newyorker-com-smart-people-post-look-familiar/.

Rosen, Jill. "The Rise and Fall of Jack Kelley." *American Journalism Review* (April/May 2004). http://ajrarchive.org/article.asp?id=3619.

Rosin, Hanna, June Thomas, and Katy Waldman, "*DoubleX* Gabfest: The Butch Goddess Edition." *Slate* (November, 27, 2014). http://www.slate.com/articles /podcasts/doublex_gabfest/2014/11/the_double_x_gabfest_on_uva_frats_and _rape_in_rolling_stone_husbands_hurting.html.

Sheppard, Si. *The Partisan Press: A History of Media Bias in the United States* (Jefferson, NC: McFarland & Company, 2008), 19.

Silverman, Craig. "Jonah Lehrer Is the Latest Target of Google Game, Crowdsourced Investigation." *Poynter.* http://www.poynter.org/news/mediawire/177917/jonah -lehrer-is-the-latest-target-of-google-game-crowdsourced-investigation/. Last updated November 25, 2014.

Simmons, Bill. "The Dr. V Story: A Letter from the Editor." *Grantland* (January 20, 2014). http://grantland.com/features/the-dr-v-story-a-letter-from-the-editor/.

Streitmatter, Roger. *Mightier than the Sword* (Boulder, CO: Westview Press, 2012), 17–46.

Volokh, Eugene. "Libel Law and the Rolling Stone / UVA Alleged Gang Rape Story." *Washington Post* (December 6, 2014). http://www.washingtonpost.com/news /volokh-conspiracy/wp/2014/12/06/libel-law-and-the-rolling-stone-uva-alleged -gang-rape-story/.

Wemple, Erik. "CNN Fires News Editor for Marie-Louise Gumuchian for Plagiarism." *Washington Post* (May 16, 2014). http://www.washingtonpost.com/blogs/erik-wemple /wp/2014/05/16/cnn-fires-news-editor-marie-louise-gumuchian-for-plagiarism/.

Wisan, Joseph E. *The Cuban Crisis as Reflected in the New York Press 1895–1898* (New York: Octagon Books, 1965), 385–438.

Zickuhr, Kathryn, and Aaron Smith. "Home Broadband 2013." *Pew Research Internet Center* (August 26, 2013). http://www.pewinternet.org/2013/08/26/home-broadband -2013/.

Part III

Community and Globalization

14

Social Media Mechanisms: A Change Agent

Kiran Samuel

On March 27, 2014, a 23-year-old writer and activist Suey Park coined a hashtag so controversial that it sent the Internet into a rabid frenzy within hours of its inception. That hashtag was #CancelColbert.

The term was created by Park in response to a tweet made by @TheColbertReport, the official account for Stephen Colbert's highly popular comedy show of the same name. "I am willing to show #Asian community I care by introducing the Ching-Chong Ding-Dong Foundation for Sensitivity to Orientals or Whatever," it said.[1] The tweet was meant as satire; it was mocking the Washington Redskins Original Americans Foundation, a charitable organization set up by Redskins' owner Daniel Snyder to benefit Native Americans while non-ironically using an offensive Native American slur as part of the foundation's name.

Colbert first made the joke on his show,[2] but @TheColbertReport published the tweet on Thursday night—independent of the aired segment's original context. Park, who didn't see the original segment but caught wind of the tweet,[3] wrote the following in response: "The Ching-Chong Ding-Dong Foundation for Sensitivity to Orientals has decided to call for #CancelColbert. Trend it."[4]

Once she hit send, the Internet roared. Park already had amassed a considerable following on Twitter as a result of previous activist efforts she performed via hashtags. Among them was #POC4CulturalEnrichment, a conversation started in September 2013 about whitewashed (censored) media.[5] It gained traction on Twitter among activists and writers who contributed their own frustrations about mainstream media excluding minority voices.[6] Park gained Twitter fame in December 2013, however, with #NotYourAsianSidekick, which aimed to foster discussion in support of Asian American feminism and to encourage Asian American women to join in to discuss their frustrations with mainstream feminism.[7] #NotYourAsianSidekick landed her into *The*

Guardian's list of "Top 30 Young People in Digital Media."[8] So when Park tweeted #CancelColbert, she already had an audience.

Thousands upon thousands of Colbert supporters flocked to defend the joke, clarify Colbert's context, and attack Park. She, meanwhile, volleyed for her followers to use #CancelColbert enough to make it a "trending topic," Twitter's list of terms or phrases that indicate what is popular in conversation. It did trend, and it did so nationwide. Comedy Central quickly deleted the tweet and released a statement to ensure the awareness that @TheColbertReport was a corporate account and was not actually manned by the comedian.[9] News publications and blogs picked up the story and scrambled to cover it as it evolved, in both news and opinion pieces that ranged in topic from how the event was indicative of a larger cultural conversation about racial sensitivity, to the appropriateness of constant political correctness, to the overall effectiveness of hashtags.[10] The following Monday, Colbert himself addressed #CancelColbert on his show, which aired during its usual time slot. Colbert echoed Comedy Central's statement that he was not the one who posted the tweet, and responded to the hashtag in his famed faux-conservative character, saying, "[T]he interweb tried to swallow me whole but I am proud to say that I got lodged in its throat and it hacked me back up like a hastily chewed chicken wing. I'm still here."[11]

So, did Park's effort fail? Not really. Park never actually wanted to cancel The Colbert Report, as she revealed to news editor Jay Caspian Kang in a follow-up article for *The New Yorker*. Park "saw the hashtag as a way to critique white liberals who use forms of racial humor to mock more blatant forms of racism."[12] But if she didn't want to actually cancel The Colbert Report, why use such extreme language? Kang wrote,

> Park says that the point of the "movement" was to argue that white liberals who routinely condemn what she called 'worse racism' will often turn a blind eye to, or even defend, more tacit forms of prejudice, especially when they come from someone who shares their basic political beliefs.[13]

In essence, Park's aim was to point out the hypocrisy of so-called allies of racial progress who were okay with using racism toward one group (Asian Americans) to condemn racism against another group (Native Americans), especially when coming from a beloved star such as Stephen Colbert.[14]

Through her efforts, Park clearly learned what to say to ignite conversation. She used that knowledge to generate buzz specifically around her issue with Colbert's joke, but also (and perhaps not directly her intention), contributed something intrinsically important in the larger fight for racial progress: Access to her perspective. Even Colbert admitted as much on his show, saying, "To recap: A web editor I've never met posts a tweet in my name on an account I don't control, outrages a hashtag activist, and the news media gets 72 hours of content. The system worked."[15]

As coverage accrued across the Internet in the days that followed, widely read and highly regarded publications including *The New Yorker* and *Salon* called Park for interviews. Park's narrative moved beyond social media to other communication channels, providing exposure to a larger, broader audience that otherwise might not have been privy to the conversation. In the comments section of a *Salon* interview with Park, a reader registered as "thinkbeforeyouspeak" wrote,

> By pointing out her oppression as an Asian woman, she is racist?? She also clearly states that she wishes no harm or ill-will on white people, but she bravely stands up to say what no one is comfortable saying. As a white person, this was tough for me to swallow, too, but y'all really need to pull your heads out of your asses.[16]

Regardless of whether Park persuaded the public with her rationale is an issue independent from the fact that people were exposed to it. The hashtag #CancelColbert generated 350 million impressions,[17] which included direct engagement such as retweets and mentions, but also views—and that's just for the tweet. That type of access is instrumental to amplifying minority voices in their expression of cultural criticism and shared perspectives, especially when those same voices historically have been subjugated, ignored, or deemed less important in the larger media landscape.

This chapter explores how social and digital media provide minority voices with platforms for sharing experiences and enacting change. The chapter makes the following arguments:

- Social media places consumers in a participatory role, which shapes how we perceive the importance of our own perspective and thus encourages us to speak up.
- Social media democratizes media coverage, thus diminishing bias and increasing accountability.
- Social media provides tools for groups to self-protect when traditional justice models fail.
- Social media empowers people to create rallying points around which they can build communities and movements. It thus diminishes the power of gatekeepers to police movements, which further promotes democracy, accountability, and education.

- Social media transforms the relationship between authors and readers from a monologue to a dialogue. This enables authors to become better educators, but also gives them the opportunity to learn from their readers, both through comments and through collaborative writing models.

ACTIVATING THE CONSUMER

Arguably, the differentiator of digital media from traditional channels is the interactivity available to creators and readers online. A commercial broadcast on television, for example, places viewers in a different role than for the same commercial when it is posted on YouTube. Television treats viewers as passive consumers, whereas a social network such as YouTube provides viewers with a range of actions to react or interact with media. In addition to watching, viewers can rewind, fast-forward, "like," "dislike," share, and comment upon that specific piece of content. Viewers also can interact with one another and show their approval for others' comments by casting an up or down vote, and by replying to comments with their own.

The relation between content creator and consumer, however, is not purely reactive: consumers also can upload their own videos to the platform themselves at no cost, corralling the public to interact with their content the same way a big brand would. The same activity is possible across Facebook, Twitter, Instagram, and other sites. Whereas a commercial treats viewers as receptacles for a message, YouTube and other social media networks expect viewers to react and participate.

In general, digital media mechanisms encourage various types of actions from users, but these mechanisms all share a similar function—they all provide users with a role. The ability to participate, conjoined with the increasing ubiquity of social media, builds into consumers an expectation that they can participate. An expectation to participate translates into a right to participate. Social media thus bolsters consumers' belief in their own power, which increases those consumers' actual power.

That kind of expectation and empowerment not only has shaped the ways people interact and react with media from established powers, it also has revolutionized the way they interact and regard one another. One of the consistent functionalities across digital media platforms is the option to share. This certainly isn't a recent development—e-mail arguably was the first advancement that enabled widespread digital sharing—but functionalities such as social plugins on Web sites, which allow people to share content directly to their social networks, and native share options on social media platforms have morphed the public's native behavior to share more often.[18]

Social media thus provides the tools for individuals to amplify each other's voices. On Twitter, users can "retweet" another user's 140-character tweet to their own respective followers, thereby broadening the scope of the original poster's reach. Through this process, a user with 150 followers might amass quadruple the amount of impressions. Similarly, a public Facebook post made by an individual with 500 friends could be shared with just a click of a button to increase visibility by an order of magnitude. The same goes with other platforms, such as Tumblr and Pinterest. These share functionalities serve to echo content through "digital word of mouth."

The result is that the Internet enables consumers to participate in public discussions. They can engage with the people they know in person, the people they follow, or even strangers that they encounter for the first time in the course of engaging in a communal dialogue. These participants are able to extend the scope of their reach when communicating online, uniting with others who might have nothing in common except one common goal: conversation.

CORRECTING MEDIA BIAS

Noninstitutional actors now can participate in producing original content. This directly impacts social justice by creating widespread visibility of content that otherwise might go unnoticed, be distorted, or be underrepresented. Social media played a substantial role, for example, in the coverage of the Arab Spring (the 2011 wave of uprisings in the Arab world fighting for democracy). Protestors used collective activism—action taken by a group of individuals in the advancement of a particular ideology or idea[19]—and cataloged the events that unfolded in front of them using social media, particularly Twitter. This grassroots news coverage provided accurate depictions of the revolutions to circumvent traditional media outlets' biased and lacking media coverage.[20]

That same spirit was mirrored in recent events on U.S. soil after the deaths of Michael Brown and Eric Garner, unarmed black men who died at the hands of police in 2014. Their deaths came just two years after that of Trayvon Martin, an unarmed teenager shot by neighborhood watch volunteer, George Zimmerman. After a long trial, Zimmerman was acquitted of second-degree murder and manslaughter. Martin's death and Zimmerman's subsequent acquittal enraged many people around the country who thought it a miscarriage of justice for Zimmerman to walk free. Brown and Garner's deaths, both of which resulted in no grand jury indictment of the police officers involved, tipped the scales even further into a national outrage.[21]

When 18-year-old Michael Brown was fatally shot by white police officer Darren Wilson in August 2014, Brown's death rippled not only locally in Ferguson, Missouri, amid mounting racial tensions between majority white cops and majority black residents, but also nationwide.[22] There were conflicting reports about what actually happened between Brown and Wilson. Within hours of the shooting, social media teemed with posts about the incident, sharing what limited information was available as well as memorial and protest details.[23] Soon, Ferguson became the central news story on every network and in every medium.

Although we cannot ascertain for sure whether news coverage of the unfolding events would have been as prevalent in the absence of social media, the hundreds of thousands of mentions in the hours that followed are evidence that it played a significant role. As in the case of the Arab Spring, bystanders captured the protests in Ferguson in large part on social media. People used their proximity and access to be the eyes and ears for people listening and watching from afar, and provided on-the-ground reporting.[24] Not only did these eyewitness accounts bypass television broadcasts with their public, first-person accounts of the events, they also provided coverage expeditiously—far faster than traditional channels, because there were no editors with whom to contend.[25] The result was raw, real, emotional, and sometimes untethered from traditional journalistic guidelines, but coverage was plentiful.[26] Social media provided an unrivaled realm of access to what was happening in Ferguson by facilitating a direct connection from witness to reader, reporter to recipient, creator to consumer.

Antonio French and DeRay McKessen were two of the most widely shared and appreciated participants of the protests because of their prolific, on-the-ground reporting, done primarily through their Twitter and Vine accounts. Many heralded French, an alderman for the district of St. Louis, as calm and judicious in his documentation of events through text, photos, and videos.[27] French captured the marches, the chants of "Hands up, don't shoot," and confrontations with the police, which at one point escalated to violence, with tear gas and rubber bullets being shot into the large crowds.[28] McKessen, an activist who drove nine hours from Minneapolis to Ferguson when he caught word of the shooting, was similarly prolific. He also used his social media accounts to get word out about events and protest information, as well as emotional expressions to rally the public. On November 24, 2014, a grand jury eventually chose not to indict Darren Wilson—thus igniting more protests, discussion, and debate. The movement raged on. French and McKessen were just two of thousands who propelled the national debate about racial tensions forward through their use of social media. Among them were trained journalists, academics, politicians, celebrities, and laypersons—all of whom played a part in feeding a public hungry for information and discussion about what was happening in the United States.

Farther north, in Staten Island, NY, a similar yet very different case was also unfolding. Whereas Michael Brown's shooting saw conflicting accounts of the shooting, Eric Garner's death was caught on video. On July 17, plainclothes police officers confronted Garner outside a beauty supply store on suspicion that he was selling loose cigarettes. According to witnesses, Garner had just broken up a fight, which is what might have drawn police attention to him. Garner's friend, Ramsey Orta, recorded the exchange on his phone as police officers surrounded Garner. Officer Daniel Pantaleo put Garner in a chokehold and pushed Garner down to the ground while Garner expressed his inability to breathe. The following is the exchange, including Garner's last words.

> Get away [garbled] for what? Every time you see me, you want to mess with me. I'm tired of it. It stops today. Why would you. . . ? Everyone standing here will tell you I didn't do nothing. I did not sell nothing. Because every time you see me, you want to harass me. You want to stop me [garbled] selling cigarettes. I'm minding my business, officer, I'm minding my business. Please just leave me alone. I told you the last time, please just leave me alone. I can't breathe. I can't breathe. I can't breathe. I can't breathe. I can't breathe. I can't breathe. I can't breathe. I can't breathe. I can't breathe. I can't breathe. I can't breathe.[29]

Orta's footage echoed across the Internet in screenshots and clips. It circulated across Web sites and news publications. It served as visible evidence of excessive force.

On December 3, 2014—just nine days after a grand jury refused to indict Officer Darren Wilson in Michael Brown's death—a grand jury reached a decision to not indict Officer Pantaleo in Eric Garner's death. The news compounded the public's distress and intensified its rage.[30] The country's focus zoomed in even further on police brutality and what many perceived as a broken judicial system. Protests grew, now with added flame under an already stoked fire. "I can't breathe" joined "Hands up, don't shoot" among the protests' messages.[31] Friends, acquaintances, and strangers shared these words in conversations with one another, athletes emblazoned the words on their pre-game t-shirts, and protesters staged die-ins across the country.

The national debate about police brutality—which at the time of this writing is ongoing—is not a result of the mainstream media delegating enough airtime to make it a primary national issue—the conversation is omnipresent as a result of constant participation from people everywhere, using what mediums they have at their disposal as conduits for expressing themselves.

The effect was undeniably large—people from around the globe retweeted, shared, and reblogged content centered around Ferguson, Michael Brown, Eric Garner, and a number of other deaths fueling a nationwide debate.[32] One of McKessen's tweets from Christmas Day, 2014, sums up the power of social sharing in these events: "The fact that you know about the protests means that

they 'are working.' This conversation about racism is now front and center. . . ."
Minute by minute, users flocked to channels to document what they saw,
heard, and felt. The Internet became a vessel for grieving, organizing, and collective action. People of color and allies congregated together—unconfined by
their respective geographies—echoing each other's thoughts and validating
each other's narratives like friends chatting across a dining room table.

SUPPLEMENTING JUSTICE SYSTEMS

Social media also can promote accountability. Once content ends up on
the Internet, it becomes documented by the larger public that has access to
that content. The First Amendment protects all Americans' right of free
speech—which absolves them of prosecution by the legal system, but doesn't
necessarily let them off the hook when it comes to public scrutiny. Not only
are people's words often preserved on the Internet for all to see but often in
cases of derogatory statements those words also can be used against the
"speakers" by a public that wants to hold them accountable and find another
way to impose justice.

The founders of Racists Getting Fired, a Tumblr site, created their Web
site for this express purpose. The readership of Racists Getting Fired documents the abuse of privilege and the racist remarks posted on the Internet by
screenshotting cases and sending the shots to the site's curator, who in turn
posts the wrongdoing, as well as any available employment information.[33] The
site implores its readers to contact the perpetrators' employers and ask that
they take action against them. It's simple, yet effective. The site further documents any resulting action that employers take. In this way, social media can
function as a tool for individuals to supplement already existing justice systems
by facilitating collective action that otherwise would be difficult to organize.

Another way that people bypass traditional justice is by posting customer
testimonials for businesses on a review site. Yelp.com's creators built the site
on the idea that customer reviews fuel new business.[34] Yelp's model is based
on customer testimonials and a 5-star rating system, used to rank everything
from real estate to restaurants. Reviews vary in type, from complimenting
service and pointing out highlights, to voicing gripes and critiquing how the
business is run. Although the implication is that commenters who leave reviews on businesses' Yelp page are customers, however, that's not a requirement—which means anyone who decides to make an account and post a
review can do so.[35] When news that a business treats customers poorly circu-
lates around the Web, that's exactly what people do.

When Big Earl's Bait House and Galley Café in Pittsburg, Texas, refused
service and hurled homophobic slurs at a gay couple, the owners probably
didn't expect it to become national news. They also probably didn't expect

that, upon hearing the news, users would go to Yelp and leave testimonials, knocking their rating down to 1 star, and reviewing the business as the best gay club in town. "[T]his place is great . . . you can really let your freak flag fly here," said one Yelp review.[36] Another said, "On the positive side, it says 'Bait House' but the place was more like a Bath House with all the horny Texas cowboys flirting and slapping each other's behinds. Very gay-friendly atmosphere!"[37]

Big Earl's managed to get the fake reviews taken down, likely because they were disingenuous to the business' standing as a restaurant and not a gay club (a violation of Yelp's content guidelines).[38] It didn't happen immediately, however, and reviews still can be found in articles and blog posts across the Internet that covered the story. Big Earl's mistake was a lesson for all businesses—big and small—to recognize the power of consumers and the power of the tools at their disposal. Big Earl's employees' freedom of speech is protected under the law, but people found justice another way. They most likely always will.

Many wonder whether such actions constitute vigilante justice and are unreasonably harsh, in that people should not be held accountable in such ways for the things they express online in their profiles or offline as personal opinions.[39] Many people, however, disagree. Ayesha Siddiqi, writer and editor-in-chief of *The New Inquiry*, tweeted,

> So many people of color have had their livelihoods ruined by the racist white people around them but we should feel bad about outing bigots? Finally being a bigot carries some threat of consequence and we're supposed to regret this as an era of outrage?[40]

It is not the purpose of this chapter to interrogate the relative merit of using social media to exact noninstitutional justice. What is important to recognize is just that the power is there, and that it is used by communities that traditional justice models tend to marginalize.

CREATING RALLYING POINTS—HASHTAG ACTIVISM

Any discussion about the relationship between social media and social movements would be incomplete without addressing the importance of hashtag activism. Hashtag activism is the act of creating a movement online through a hashtagged word or phrase. Park is one of many who have seen success using hashtags to ignite conversation around a social issue. With its use, people across the world are able to join together in discussion or debate. Critics of hashtag activism frequently refer to it as "slacktivism" because participation in a hashtag does not require any action outside of online spaces.[41]

However, hashtag activism can spread widespread awareness of an issue or cause. Social progress requires awareness. Therefore, hashtag activism can play an important role.

Hashtags by design encourage behavior. They do this in two ways, by categorizing conversations and by inviting others to participate using the same hashtagged word or phrase. The former is a facet of their functionality, in that their use on social media and Web sites enables similar content to show up in conjunction when explored. A user might type 20 different hashtags to accompany a photo of a Maine landscape, among them: #NewEngland, #trees, #ilovefall. Once posted these hashtags become links. On Instagram, clicking on a hashtag expands other content from users whose link to each other is that they all used the same hashtag. As a result, someone from Finland who snapped a photo of a pretty scene in Helsinki and tagged his photo with #trees would show up in the same content stream.

The other effect of hashtags—encouraging others to participate using the same hashtag—is based upon Twitter users' familiarity with the categorizing function. Participating with one another through hashtags in this way is a learned behavior. So, yes, Twitter users who tag #myworstfirstdate categorize their tweet and enter a conversation with other Twitter users who previously contributed with that same phrase, but they also recruit people in their network to participate by virtue of the fact that their tweet is specifically shared with them, those people who follow the original poster. A hashtag is an entry phrase into a conversation, and by using it a participant provides a shared umbrella under which responses—both among the same immediate network and outside of it—can string together.

With that said, there are no rigid guidelines for their implementation. The process of tagging content or integrating hashtags into colloquial digital language is an extremely frequent occurrence. People use hashtags as the punch lines to jokes, invent hashtags that start conversations in which no one else participates, or employ them in unconventional contexts to layer on new meaning.

The Importance of Understanding Your Audience

There are no guidelines for participants, therefore hashtags can be co-opted by whoever decides to contribute—an occurrence that sometimes yields disastrous results.

On April 22, 2014, the NYPD coined and pushed out #MyNYPD with the express purpose of generating support for their police force through kind anecdotes sourced from New York City residents. Their account, @NYPD News, tweeted, "Do you have a photo w/ a member of the NYPD? Tweet us & tag it #myNYPD. It may be featured on our Facebook."[42]

Submissions poured in, but instead of seeing heartwarming stories about the NYPD, stories about racial profiling, photos of police brutality, and general grievances with the police force filled up the NYPD's mentions immediately.[43]

"#MyNYPD makes me feel paranoid, sweaty, and nervous when I'm doing absolutely nothing wrong. 'Justice' shouldn't make me fear for my life." said @williamcson.

"#MyNYPD rolled up on me demanding I show them the 'weapon I had concealed in my pocket.' It was my iPod. I felt so safe!" said @Are0h.

The NYPD, in turn, could do nothing about it; control was out of their hands. Anyone who came across the hashtag could click on it to expand the content it housed. It was a prime illustration of crowdsourcing the Internet as akin to opening Pandora's Box.

That type of influx of conversation is precisely the point of hashtag activism. Its effectiveness in offline activism aside, what hashtag activism seeks to accomplish is more narrow and specific in its purpose: It creates visibility for a topic and opens a discussion for both sides of a debate. Suey Park, through #CancelColbert, was able to raise attention through her choice in hashtag— controversial enough to spark attention amid people who were ally or were perceived enemy, those familiar with Park or, more likely, passersby. The beauty of the hashtag is that it begged explanation—which perusal of the hashtag might not have provided. Nevertheless, the hashtag certainly added to the controversy, propelling it into a national conversation with diverse participants.

Subverting Mainstream Awareness through Minority Opinion

Although #CancelColbert is unique in phrasing, the phenomenon it produced is not an anomaly. Hashtag activism has led to a number of conversations propelled forward by activists, academics, comics, and everyday people who use social media to generate interest or awareness of the multifaceted discussion of a particular topic.

Take Mikki Kendall's #solidarityisforwhitewomen, a conversation she started to highlight the divisiveness of mainstream feminism in only uplifting white feminists and highlighting their voices, but actively ignoring the experiences of WOC, trans women, and other minority feminists.[44] Kendall coined the hashtag as a direct response to the Twitter meltdown of self-proclaimed male feminist and oft called-upon "expert" Hugo Schwyzer.[45] In his extensive barrage of tweets, he admitted to stalking, targeting, and stamping out the voices of women of color through backchannel politics. "So to the many people I hurt I am sorry. To the women of color I trashed in 2008, I am so sorry," said one tweet. Schwyzer continued, "And WOC, yes you

@amaditalks and @Blackamazon, you were right. I was awful to you because you were in the way."[46] Schwyzer's semi-famous presence in digital feminism despite these documented occurrences against minority voices spoke to the larger problem regarding so-called feminist sites not actually representing all women. They consistently afforded people like Schwyzer a platform, but Kendall and others were shut out.[47]

It's a controversial problem. Although the Internet affords the potential for visibility by giving everyone a voice, gaining mainstream visibility for those voices still depends on media publications that are mostly curated and operated by white editors.[48] For nonwhite feminists, mass coverage of their message depends on white feminists empowering and making room for others to be heard. Not only did Kendall see that not happening, but the opposite was occurring—minority voices were being actively silenced.[49]

> [I]f you're publishing on gender issues and all of the people you choose to reach out to are white and/or cis, then you've contributed to the erasure of trans people, or people of color, and someone is eventually going to notice that maybe you're not really committed to the advancement of all women.[50]

Issues like racism and transphobia were seen as separate from feminism, which alienated the core identities of many women. A refusal to acknowledge these experiences, not only by covering these issues but by not hiring qualified people who are living examples of those experiences, is antithetical to inclusive female empowerment.[51]

Kendall's hashtag struck a chord because so many women shared in her personal experience. #Solidarityisforwhitewomen soon made its way into a trending topic nationally, with global contributions from feminists around the world who shared in the experience.

"#SolidarityIsForWhiteWomen when convos about gender pay gap ignore that white women earn higher wages than black, Latino and Native men," said @raniakhalek.[52]

"#solidarityisforwhitewomen when pink hair, tattoos, and piercings are "quirky" or "alt" on a white woman but "ghetto" on a black one," said @zblay.[53]

"#SolidarityIsForWhiteWomen is when Femen gets to decide the Muslim women's attire," said @roadtopalestine.[54]

Nobody from mainstream networks had to give license to Kendall and others who were allies in #solidarityisforwhitewomen—the mere act of so many women congregating under the hashtag was subversive and overwhelming enough to cause people to take notice on social media, and carry the conversation to other platforms. What side people were on was secondary to the fact that the discussion was public, and the discourse that ensued was rich, intimate, and necessary for moving feminism forward.

National Events Become Personal

After footage was released of Baltimore Ravens football player Ray Rice in an elevator punching his then-fiancée Janay and knocking her out cold, a number of questions arose. What was to become of Rice? What does it mean that Janay still married him? How could she be so weak to stay? Bev Gooden, an HR manager in North Carolina, wasn't an academic or an activist, but Janay's story triggered a reflection of her own experiences in an abusive relationship that turned her into a powerful spokesperson.[55]

As the tweets came in, mostly victim-blaming Janay for staying and then for marrying Ray, Gooden was fed up.

> When I saw those tweets, my first reaction was shame. The same shame that I felt back when I was in a violent marriage. It's a sort of guilt that would make me crawl into a shell and remain silent. But today, for a reason I can't explain, I'd had enough. I knew I had an answer to everyone's question of why victims of violence stay. I can't speak for Janay Rice, I can only speak for me.[56]

Gooden recounted her experience on Twitter. "I tried to leave the house once after an abusive episode, and he blocked me. He slept in front of the door that entire night—#WhyIStayed," she tweeted.[57] "I stayed because my pastor told me that God hates divorce. It didn't cross my mind that God might hate abuse, too. #WhyIStayed."[58] As she continued voicing her own experience, #WhyIStayed immediately took off as other women began to contribute their own.

"#WhyIStayed—Because he told me that no one would love me after him, and I was insecure enough to believe him," said @BBZaftig.[59]

"I was determined to make it work, wanted kids to have their dad, convinced myself that what he did to me wasn't affecting them #WhyIStayed" said @ReIgniteRomance.[60]

"#WhyIStayed: because my word was the only evidence," said @RachelMcKibbens.[61]

Within hours, the discussion evolved into including a new hashtag: #WhyILeft.

"#WhyIStayed—I thought my baby boys needed their dad. #WhyILeft? I didn't want them to learn to hit any woman ever," said @notokmaybeok.[62]

"#whyistayed: He told me 'no one will ever love you like I do' #whyileft: I realized that no one should ever 'love' me like he did," said @kirin_rosemary.[63]

The social media sphere shifted the conversation from a place of negativity and victim blaming to a space in which women felt supported and empowered as survivors with important stories to tell—all via a hashtag.

Educating through Widespread Awareness

What do these mentions and contributions amount to? Well, the mere fact that nobody had to grant Kendall or allies of #solidarityisforwhitewomen a place to speak is, in itself, huge. Women were able to shift the national dialogue about Janay Rice from victim blaming to survivorship also is significant. The conversations that ensue from joining all kinds of perspectives together are subversive to mainstream coverage, and necessary in including all voices.

Education is a critical byproduct of hashtag activism, and with specifically that goal in mind, it's tremendously successful. The mechanisms of social networks allow for connectivity to thrive among people who do not know each other and might not share the same views, yet coexist in a space where their messages are seen, acknowledged, discussed, and amplified. One of the necessary byproducts of #solidarityisforwhitewomen was to be heard by white feminists and compel them to be introspective about the ways in which they are complicit in the media sphere.[64] On August 12, 2013, feminist organizer Shelby Knox tweeted, "Fellow white feminists: #SolidarityIsForWhite-Women is not for us to defend, explain, protest. It's time for us to take a damn seat & listen."

The prevalence of these conversations—whether they are about feminism, domestic violence, racism, LGBTQ issues, or something else—is helped by social media mechanisms. The connectivity, visibility, and amplification inherent in social media propels forward otherwise subjugated voices into a space where they cannot be ignored by mainstream media systems. The access this affords is tremendous in pushing the envelope toward inclusivity and progress.

BUILDING COMMUNICATION CHANNELS BETWEEN AUTHOR AND READER

One of the compelling and dynamic ways that people take advantage of their individual power is through commenting systems. Whether on Facebook or YouTube, news Web sites or blogs, digital media has provided a way for people to express themselves and respond to what's being presented to them. In turn, the infrastructure of digital media systems has welcomed this behavior or, at the very least, made accommodations for its inevitability. The empowered reader's newfound capability as collaborator and public dissenter has changed the way in which the public consumes information, at least online. The author is no longer infallible—the model shifts from broadcast messaging to conversation, joining together different perspectives to flesh out the whole story.

It's not only on social media that people are given a role. Web sites such as blogs and digital publications have comment sections that empower people in similar ways. In both forms, digital media has allowed ample space for people's opinions to sit side by side one another. It's democratized the Web in that everyone has the right to speak.

Many publications not only embrace but also encourage commenters. Gawker Media, the parent company of affiliate sites including Jezebel, Deadspin, and ValleyWag, developed a platform called "Kinja," which treats comments as necessary components of the articles themselves. As a result, comment sections on Gawker Media sites are thriving—readers act as copyeditors, supporters, dissenters, collaborators, and authors.[65] Gawker writers often engage with readers to clarify, thank, validate, and argue. Commenters can create their own pages, hosted by Kinja, which allows readers to write their own articles; some of these are reposted to the main Gawker page.[66]

The Atlantic rewards commenters, too. Ta-Nehisi Coates, senior editor, arguably the most widely read journalist on the site and definitely the most celebrated on the topic of race and culture, has built a strong community of intellectual discourse through his comment section.[67] He expects a lot from participants—demands it, really. Coates himself is the second-highest-ranked commenter site-wide.[68] He has no problem banning trolls (Internet slang referring to people who try to create discord, start arguments, or upset people by posting in an online community and deliberately intend to provoke others into an emotional response or otherwise disrupt normal on-topic discussion) or ripping into people who stoop to making slurs. Coates acts not only as a writer, but as an educator, ally, and student as well.[69] Watching an article he's written about racial politics evolve and deepen from robust commentary within a few hours demonstrates the benefits of new media in real time.

THE TROUBLE WITH ANONYMITY ONLINE

Commenting systems are not perfect and rarely if ever function without flaw. Not every site sees the type of well-formed discussions typical of Gawker and *The Atlantic*—these examples are closer to anomalies than to the norm.[70] One of the most valid criticisms against commenting systems is the veil of anonymity it affords people when they aren't forced to affiliate their identities with their comments, which happens on Web sites. That, along with the lack of interpersonal connection across a computer screen, can turn discussions destructive, dangerous, and wholly unhelpful.[71] Of people online, 73 percent have witnessed harassment online.[72] Studies have shown that people holding predisposed opinions are likely to further polarize their views based on their exposure to uncivil comments—they also defend their views publicly in those forums to outdo or outwit the opposing side for the sake of their reputations or their opinions.[73]

As a result of rampant vitriol and spam, some sites have shut down their comment sections and rerouted discussions to social networks. Popular Science and Mic are among the sites that phased out their comment sections. Reuters only permits commenting on its opinion pieces. Medium and Quartz launched without a comment section entirely.[74] The sentiment is that participation is still valued, but not on that forum.

That might not be the optimal route, especially when the visibility of informed dissenters can be educational and add alternate opinions[75]— inherently important for social progress, especially for the visibility of minority opinions. Even if conversation thrives on platforms such as Twitter, the visibility depends on who you follow. Also, although journalists and editors usually have Twitter accounts, not every reader has one—only 23 percent of online adults use Twitter.[76] That puts a big limit on the conversation as well as the participants involved.

So what's the solution? There is no clear-cut answer, but certain publications are definitely trying to champion great contributions. There is a segment of sites that instead of redirecting conversation have started curating it. *The New York Times* and *Washington Post* have hired community managers to pick out comment highlights. Gawker Media founder Nick Denton flat out refuses to give up on that potential.

It's not just about leveling the playing field between commenters or readers and writers—we want sources as well, we want them to be able to participate in these discussions. And the principle is that in order to be able to achieve the potential of the Internet we need to harness the collective intelligence of the readership.[77]

CONCLUSION

However strong the cultural appetite is for enriching conversation to yield progress for representation of minority voices, there are many situations in which social and digital media can result in a negative and even ugly state of affairs.

The ease in which people are able to disseminate information and talk to and about each other doesn't mean what's being shared among them is necessarily true. Things can be taken out of context, sensationalized, and misreported. The veil of anonymity (discussed previously) also can relieve people of accountability and interpersonal sensitivity,[78] making them mean, gossipy, and altogether cruel to one another.

Monica Lewinsky, in an essay for *Vanity Fair*, talks about her own experiences as a part of what she refers to as "humiliation culture." Lewinsky, whose catapult into infamy happened almost two decades ago, in 1997, didn't have to contend with overexposure that would have been inevitable had her affair with President Bill Clinton happened today. As much as it played out online and in publications, in living rooms, and at watercoolers, those sentiments weren't expressed as publicly and rampantly as they would have through social media. Lewinsky still feels the aftereffects, however, even today.

> Yes, we're all connected now. We can tweet a revolution in the streets or chronicle achievements large and small. But we're also caught in a feedback loop of defame and shame, one in which we have become both perps and victims. We may not have become a crueler society—although it sure feels as if we have—but the Internet has seismically shifted the tone of our interactions. The ease, the speed, and the distance that our electronic devices afford us can also make us colder, more glib, and less concerned about the consequences of our pranks and prejudice. Having lived humiliation in the most intimate possible way, I marvel at how willingly we have all signed on to this new way of being.[79]

As truthful as her words are, they don't mitigate progress made on other fronts. It can be true that the democratization of the Internet will yield progress for minority voices as a result of accessibility and connectivity, and that people will have more opportunity to do damage to one another, too. These results can and do happen simultaneously; they are not mutually exclusive.

But in terms of net benefit to our society, the long-term effects of education through exposure to new ideas is an undeniable positive, especially regarding creating a socially progressive, more forward-thinking world. Embrace it or not, it is inevitable—and we are changed as a result of it.

NOTES

1. Jay Caspian Kang, "The Campaign to 'Cancel' Colbert," *New Yorker* (March 30, 2014), http://www.newyorker.com/news/news-desk/the-campaign-to-cancel-colbert.

2. "Sport Report—Professional Soccer Toddler, Golf Innovations & Washington Redskins Charm Offensive," *The Colbert Report*, March 26, 2014 (New York: Comedy Central).

3. Prachi Gupta, "#CancelColbert Activist Suey Park: 'This Is Not Reform, This Is Revolution'," *Salon* (April 3, 2014), http://www.salon.com/2014/04/03/cancelcolbert_activist_suey_park_this_is_not_reform_this_is_revolution/.

4. Kang, "The Campaign to Cancel Colbert."

5. Muna Mire, "Tweeting for Racial Justice: Millennials Take to Organizing Online," *Youngist* (September 9, 2013), http://youngist.org/tweeting-for-racial-justice-millennials-take-to/#.VLnVzWTF93V.

6. Ibid.

7. Yoonj Kim, "#NotYourAsianSidekick Is a Civil Rights Movement for Asian American Women," *The Guardian* (December 17, 2013), http://www.theguardian.com/commentisfree/2013/dec/17/not-your-asian-sidekick-asian-women-feminism.

8. Kang, "The Campaign to 'Cancel' Colbert."

9. Nellie Andreeva, " 'The Colbert Report' Embroiled in Racial Controversy," *Deadline* (March 27, 2014), http://deadline.com/2014/03/the-colbert-report-embroiled-in-racial-controversy-705803/.

10. Gupta, "#CancelColbert Activist Suey Park: 'This Is Not Reform, This Is Revolution'."

11. Linette Rice, "Stephen Colbert on #CancelColbert: 'We Almost Lost Me'—VIDEO," *Entertainment Weekly* (April 1, 2014), http://insidetv.ew.com/2014/04/01/stephen-colbert-cancelcolbert/.

12. Kang, "The Campaign to 'Cancel' Colbert."

13. Ibid.

14. Suey Park and Eunsong Kim, "We Want To #CancelColbert," *Time* (March 28, 2014), http://time.com/42174/we-want-to-cancelcolbert/.

15. Bill Chappell, "When the Twit Hit the Fan: 'I'm Still Here,' Colbert Says," NPR (April 1, 2014), http://www.npr.org/blogs/thetwo-way/2014/04/01/297683346/when-the-twit-hit-the-fan-i-m-still-here-colbert-says.

16. thinkbeforeyouspeak, June 17, 2014, comments on "#CancelColbert Activist Suey Park: 'This Is Not Reform, This Is Revolution'," *Salon* (April 3, 2014), http://www.salon.com/2014/04/03/cancelcolbert_activist_suey_park_this_is_not_reform_this_is_revolution/.

17. TweetArchivist, March 28, 2014–May 31, 2014, last updated May 31, 2014, http://www.tweetarchivist.com/becmarydunn/9/source.

18. Jenny Price, "Study Shows Role of Media in Sharing Life Events," *University of Wisconsin-Madison News* (July 24, 2014), http://www.news.wisc.edu/23018.

19. Tom Postmes and Suzanne Brunsting, "Collective Action in the Age of the Internet: Mass Communication and Online Mobilization," *Social Science Computer Review* 20 (3) (2002): 290–301.

20. Alexandra Petri, "Ferguson, MO, FOX News, and the Difference between Looking at and Seeing," *Washington Post* (August 14, 2014), http://www.washingtonpost.com/blogs/compost/wp/2014/08/14/ferguson-mo-fox-news-and-the-difference-between-looking-at-and-seeing/.

21. "Timeline: Eric Garner Death," *NBC New York* (December 4, 2014), http://www.nbcnewyork.com/news/local/Timeline-Eric-Garner-Chokehold-Death-Arrest-NYPD-Grand-Jury-No-Indictment-284657081.html.

22. Ibid.

23. Rubina Madan Fillion, "How Ferguson Protesters Use Social Media to Organize," *Wall Street Journal* (November 24, 2014), http://blogs.wsj.com/dispatch/2014/11/24/how-ferguson-protesters-use-social-media-to-organize/.

24. Ibid.

25. Deron Lee, "In Ferguson, Local News Coverage Shines," *Columbia Journalism Review* (August 20, 2014), http://www.cjr.org/united_states_project/local_coverage_ferguson_michae.php?page=all.

26. Ibid.

27. Laura Mandaro, "300 Ferguson Tweets: A Day's Work for Antonio French," *USA Today* (August 26, 2014), http://www.usatoday.com/story/news/nation-now/2014/08/25/antonio-french-twitter-ferguson/14457633/.

28. Ibid.

29. The Advise Show TV, "NYPD Publicly Executes Eric Garner for Illegal Cigarettes" (*YouTube* video) (July 22, 2014), http://youtu.be/g-xHqf1BVE4.

30. "Timeline: Eric Garner Death"

31. Ibid.

32. Fillion, "How Ferguson Protesters Use Social Media to Organize."

33. Racistsgettingfired.tumblr.com.

34. Yelp, 10 Things You Should Know about Yelp (December 13, 2014), http://www.yelp.com/about.

35. Yelp, "Responding to Reviews" (December 13, 2014), https://biz.yelp.com/support/responding_to_reviews.

36. Khushbu Shah, "Yelpers Slam Texas Restaurant for Asking Gay Couple to Not Return," *Eater* (May 29, 2014), http://www.eater.com/2014/5/29/6216059/yelpers-slam-texas-restaurant-for-asking-gay-couple-to-not-return#4141546.

37. Cavan Sieczkowski, "Texas Restaurant that Refused 'Fag' Customers Touted As Gay Bar Online," The Huffington Post (May 31, 2014), http://www.huffingtonpost.com/2014/05/31/texas-restaurant-fag-customers-gay-bar_n_5423727.html.

38. Yelp, "Responding to Reviews."

39. Hessie Jones, "Racists Getting Fired: Vigilante Justice or Civic Duty?" ArCompany (December 19, 2014), http://arcompany.co/racists-getting-fired-vigilante-justice-or-civic-duty/.

40. Ayesha Siddiqi, Twitter post (December 17, 2014; 3:08 PM), https://twitter.com/pushinghoops/status/546079768903696385.

41. Caitlin Dewey, "#Bringbackourgirls, #Kony2012, and the Complete, Divisive History of 'Hashtag Activism'," *Washington Post* (May 8, 2014), http://www

.washingtonpost.com/ news/the-intersect/wp/2014/05/08/bringbackourgirls-kony2012
-and-the-complete-divisive-history-of-hashtag-activism/.

42. NYPD News, Twitter post (April 22, 2014; 10:55 PM), https://twitter.com
/NYPDnews/status/458665477409996800.

43. Ryan Broderick, "The NYPD Learned A Very Valuable Lesson about Asking
the Internet to Use a Twitter Hashtag," *Buzzfeed* (April 22, 2014), http://www
.buzzfeed.com/ryanhatesthis/the-nypd-just-learned-a-very-valuable-lesson-about
-asking-th#.voGQ2LAaN.

44. Mikki Kendall, "#SolidarityIsForWhiteWomen: Women of Color's Issue with
Digital Feminism," *The Guardian* (August 14, 2013), http://www.theguardian.com
/commentisfree/2013/aug/14/solidarityisforwhitewomen-hashtag-feminism.

45. Ibid.

46. Ibid.

47. Ibid.

48. Jia Tolentino, "A Chat with Mikki Kendall and Flavia Dzodan about
#SolidarityIsForWhiteWomen," *The Hairpin* (August 16, 2013), http://thehairpin
.com/2013/08/solidarity-is-for-hairpin.

49. Ibid.

50. Ibid.

51. Ibid.

52. Erin Gloria Ryan, "Our Favorite #SolidarityIsForWhiteWomen Tweets
[Updated]," *Jezebel* (August 13, 2013), http://jezebel.com/our-favorite-solidarityisfor
whitewomen-tweets-1125272401.

53. Ibid.

54. Ibid.

55. Audie Cornish, and Bev Gooden, "Hashtag Activism in 2014: Tweeting 'Why
I Stayed,' " NPR (December 23, 2014), http://www.npr.org/2014/12/23/372729058
/hashtag-activism-in-2014-tweeting-why-i-stayed.

56. Ibid.

57. Ibid.

58. Ibid.

59. Ibid.

60. Ibid.

61. Ibid.

62. Jared Keller, "19 #WhyIStayed Tweets that Everyone Needs to See," *Mic*
(September 8, 2014), http://mic.com/articles/98326/19-why-istayed-tweets-that
-everyone-needs-to-see.

63. Ibid.

64. Tolentino, "A Chat with Mikki Kendall and Flavia Dzodan about
#SolidarityIsForWhiteWomen."

65. Bob Cohn, "Comments on the Web: Engaging Readers or Swamping
Journalism?" *The Atlantic* (August 2, 2013), http://www.theatlantic.com/technology
/archive/2013/08/comments-on-the-web-engaging-readers-or-swamping-journalism
/278311/.

66. Derek Thompson, "The Case for Banning Internet Commenters" *The Atlantic* (September 23, 2013), http://www.theatlantic.com/business/archive/2013/09/the -case-for-banning-internet-commenters/279960/.

67. Cohn, "Comments from the Web. . . ."

68. Chris Ip, "Ta-Nehisi Coates Defines a New Race Beat," *Columbia Journalism Review* (October 29, 2014), http://www.cjr.org/feature/ta-nehisi_coates_defines_a _new.php?page=all.

69. Ibid.

70. Thompson, "The Case for Banning Internet Commenters."

71. Elizabeth Suhay, "Comment Threads Are Messy, but So Is Democracy," *Washington Post* (December 18, 2014), http://www.washingtonpost.com/blogs/monkey -cage/wp/2014/12/18/comment-threads-are-messy-but-so-is-democracy/.

72. Maeve Duggan, "Part 5: Witnessing Harassment Online," *Pew Internet Project* (October 22, 2014), http://www.pewinternet.org/2014/10/22/part-5-witnessing -harassment-online/.

73. Suhay, "Common Threads Are Messy, but So Is Democracy."

74. Thompson, "The Case for Banning Internet Commenters."

75. Monica Anderson, and Andrea Caumont, "How Social Media Is Reshaping News," *Pew Research Center* (September 24, 2014), http://www.pewresearch.org/fact -tank/2014/09/24/how-social-media-is-reshaping-news/.

76. Duggan, "Part 5: Witnessing Harassment Online."

77. Mathew Ingram, "Gawker Founder Nick Denton Is Still Trying to Reinvent Reader Comments—and It's Working," *Gigaom* (September 23, 2013), https:// gigaom.com/2013/09/23/gawker-founder-nick-denton-is-still-trying-to-reinvent -reader-comments-and-its-working/.

78. Seth Fiegerman, "No Comment: Why News Websites Are Ditching Comment Sections," *Mashable* (December 17, 2014), http://mashable.com/2014/12/17/no -comment/.

79. Monica Lewinsky, "Shame and Survival," *Vanity Fair* (June 2014), http://www .vanityfair.com/society/2014/06/monica-lewinsky-humiliation-culture.

BIBLIOGRAPHY

The Advise Show TV. "NYPD Publicly Executes Eric Garner for Illegal Cigarettes" (*YouTube* video) (July 22, 2014). http://youtu.be/g-xHqf1BVE4.

Anderson, Monica, and Andrea Caumont. "How Social Media Is Reshaping News." *Pew Research Center* (September 24, 2014). http://www.pewresearch.org/fact -tank/2014/09/24/how-social-media-is-reshaping-news/.

Andreeva, Nellie. " 'The Colbert Report' Embroiled in Racial Controversy." *Deadline* (March 27, 2014). http://deadline.com/2014/03/the-colbert-report-embroiled -in-racial-controversy-705803/.

Broderick, Ryan. "The NYPD Learned a Very Valuable Lesson about Asking the In- ternet to Use a Twitter Hashtag." *Buzzfeed* (April 22, 2014). http://www.buzzfeed .com/ryanhatesthis/the-nypd-just-learned-a-very-valuable-lesson-about-asking -th#.voGQ2LAaN.

Brunsting, Suzanne, and Tom Postmes, "Collective Action in the Age of the Internet: Mass Communication and Online Mobilization." *Social Science Computer Review* 20 (3) (2002): 290–301.

Chappell, Bill. "When the Twit Hit the Fan: 'I'm Still Here,' Colbert Says." NPR (April 1, 2014). http://www.npr.org/blogs/thetwo-way/2014/04/01/297683346 /when-the-twit-hit-the-fan-i-m-still-here-colbert-says.

Cohn, Bob. "Comments on the Web: Engaging Readers or Swamping Journalism?" *The Atlantic*, August 2, 2013. http://www.theatlantic.com/technology/archive /2013/08/comments-on-the-web-engaging-readers-or-swamping-journalism/278311/

Colbert, Stephen. "Sport Report—Professional Soccer Toddler, Golf Innovations & Washington Redskins Charm Offensive" (television program). Performed by Stephen Colbert (2014). New York, Comedy Central.

Cornish, Audie, and Bev Gooden, "Hashtag Activism in 2014: Tweeting 'Why I Stayed'." NPR (December 23, 2014). http://www.npr.org/2014/12/23/372729058 /hashtag-activism-in-2014-tweeting-why-i-stayed.

Dewey, Caitlin. "#Bringbackourgirls, #Kony2012, and the Complete, Divisive History of 'Hashtag Activism'." *Washington Post* (May 8, 2014). http://www .washingtonpost.com/ news/the-intersect/wp/2014/05/08/bringbackourgirls-kony 2012-and-the-complete-divisive-history-of-hashtag-activism/.

Duggan, Maeve. "Part 5: Witnessing Harassment Online." *Pew Internet Project* (October 22, 2014). http://www.pewinternet.org/2014/10/22/part-5-witnessing -harassment-online/.

Fiegerman, Seth. "No Comment: Why News Websites Are Ditching Comment Sections." *Mashable* (December 17, 2014). http://mashable.com/2014/12/17 /no-comment/.

Fillion, Rubina Madan. "How Ferguson Protesters Use Social Media to Organize." *Wall Street Journal* (November 24, 2014). http://blogs.wsj.com/dispatch/2014/11 /24/how-ferguson-protesters-use-social-media-to-organize/.

Gupta, Prachi. "#CancelColbert Activist Suey Park: 'This Is Not Reform, This Is Revolution'." *Salon* (April 3, 2014). http://www.salon.com/2014/04/03 / cancelcolbert_activist_suey_park_this_is_not_reform_this_is_revolution/.

The Huffington Post (April 1, 2014). http://www.huffingtonpost.com/2014/04/01 /cancel-colbert-stephen-colbert_n_5068592.html.

Ingram, Mathew. "Gawker Founder Nick Denton Is Still Trying to Reinvent Reader Comments—and It's Working." *Gigaom* (September 23, 2013). https://gigaom .com/2013/09/23/gawker-founder-nick-denton-is-still-trying-to-reinvent-reader -comments-and-its-working/.

Ip, Chris. "Ta-Nehisi Coates Defines a New Race Beat." *Columbia Journalism Review* (October 29, 2014). http://www.cjr.org/feature/ta-nehisi_coates_defines_a_new .php?page=all.

Jones, Hessie. "Racists Getting Fired: Vigilante Justice or Civic Duty?" ArCompany (December 19, 2014). http://arcompany.co/racists-getting-fired-vigilante-justice -or-civic-duty/.

Kang, Jay Caspian. "The Campaign to 'Cancel Colbert'." *New Yorker* (March 30, 2014) . http://www.newyorker.com/news/news-desk/the-campaign-to-cancel-colbert.

Keller, Jared. "19 #WhyIStayed Tweets That Everyone Needs to See." *Mic* (September 8, 2014). http://mic.com/articles/98326/19-why-istayed-tweets-that -everyone-needs-to-see.

Kendall, Mikki. "#SolidarityIsForWhiteWomen: Women of Color's Issue with Digital Feminism." *The Guardian* (August 14, 2013). http://www.theguardian.com /commentisfree/ 2013/aug/14/solidarityisforwhitewomen-hashtag-feminism.

Kim, Yoonj. "#NotYourAsianSidekick Is a Civil Rights Movement for Asian American Women." *The Guardian* (December 17, 2013). http://www.theguardian .com/commentisfree/2013/dec/17/not-your-asian-sidekick-asian-women -feminism.

Lee, Deron. "In Ferguson, Local News Coverage Shines," *Columbia Journalism Review* (August 20, 2014). http://www.cjr.org/united_states_project/local_coverage _ferguson_michae.php?page=all.

Lewinsky, Monica. "Shame and Survival." *Vanity Fair* (June 2014). http://www.vani tyfair.com/society/2014/06/monica-lewinsky-humiliation-culture.

Mandaro, Laura. "300 Ferguson Tweets: A Day's Work for Antonio French," *USA Today*, August 26, 2014. http://www.usatoday.com/story/news/nation-now/2014/08/25 /antonio-french-twitter-ferguson/14457633/

Mire, Muna. "Tweeting for Racial Justice: Millennials Take to Organizing Online." *Youngist* (September 9, 2013). http://youngist.org/tweeting-for-racial-justice -millennials-take-to/#.VKXpomTF93U.

NYPD News. Twitter post (April 22, 2014; 10:55 PM). https://twitter.com /NYPDnews/status/458665477409996800.

Park, Suey, and Eunsong Kim. "We Want to #CancelColbert" *Time* (March 28, 2014). http://time.com/42174/we-want-to-cancelcolbert/

Petri, Alexandra. "Ferguson, MO, FOX News, and the Difference Between Looking at and Seeing." *Washington Post* (August 14, 2014). http://www.washingtonpost .com/blogs/compost/wp/2014/08/14/ferguson-mo-fox-news-and-the-difference -between-looking-at-and-seeing/.

Postmes, Tom, and Suzanne Brunsting. "Collective Action in the Age of the Internet: Mass Communication and Online Mobilization." *Social Science Computer Review* 20 (3) (2002): 290–301.

Price, Jenny. "Study Shows Role of Media in Sharing Life Events." *University of Wisconsin–Madison News* (July 24, 2014). http://www.news.wisc.edu/23018.

Racists Getting Fired. Tumblr site. racistsgettingfired.tumblr.com. Last accessed December 15, 2014.

Rice, Linette. "Stephen Colbert on #CancelColbert: 'We Almost Lost Me' VIDEO." *Entertainment Weekly* (April 1, 2014). http://insidetv.ew.com/2014/04/01/stephen -colbert-cancelcolbert/.

Ryan, Erin Gloria. "Our Favorite #SolidarityIsForWhiteWomen Tweets [Updated]" *Jezebel*, August 13, 2013. http://jezebel.com/our-favorite-solidarityisforwhitewomen -tweets-1125272401.

Shah, Khushbu. "Yelpers Slam Texas Restaurant for Asking Gay Couple to Not Return." *Eater* (May 29, 2014). http://www.eater.com/2014/5/29/6216059 /yelpers-slam-texas-restaurant-for-asking-gay-couple-to-not-return#4141546.

Siddiqi, Ayesha. Twitter post (December 17, 2014; 3:08 PM). https://twitter.com /pushinghoops/status/546079768903696385.

Sieczkowski, Cavan. "Texas Restaurant that Refused 'Fag' Customers Touted as Gay Bar Online." *Huffington Post* (May 31, 2014). http://www.huffingtonpost.com /2014/05/31/texas-restaurant-fag-customers-gay-bar_n_5423727.html.

Suhay, Elizabeth. "Comment Threads Are Messy, but So Is Democracy." *Washington Post* (December 18, 2014). http://www.washingtonpost.com/blogs/monkey-cage /wp/2014/12/18/comment-threads-are-messy-but-so-is-democracy/.

"thinkbeforeyouspeak" (June 17, 2014). Comment on "#CancelColbert activist Suey Park: 'This is not reform, this is revolution'." *Salon* (April 3, 2014). http://www .salon.com/2014/04/03/cancelcolbert_activist_suey_park_this_is_not_reform_this _is_revolution/.

Thompson, Derek. "The Case for Banning Internet Commenters." *The Atlantic* (September 23, 2013). http://www.theatlantic.com/business/archive/2013/09/the-case -for-banning-internet-commenters/279960/.

"Timeline: Eric Garner Death." *NBC New York* (December 4, 2014). http://www .nbcnewyork.com/news/local/Timeline-Eric-Garner-Chokehold-Death-Arrest -NYPD-Grand-Jury-No-Indictment-284657081.html.

Tolentino, Jia. "A Chat with Mikki Kendall and Flavia Dzodan about #SolidarityIs ForWhiteWomen." *The Hairpin* (August 16, 2013). http://thehairpin.com/2013/08 /solidarity-is-for-hairpin.

TweetArchivist, March 28, 2014—May 31, 2014. Last updated May 31, 2014. http:// www.tweetarchivist.com/becmarydunn/9/source.

Yelp. 10 Things You Should Know about Yelp. Company Web site. Accessed December 13, 2014. http://www.yelp.com/about.

Yelp. Responding to Reviews. Company Web site. Accessed December 13, 2014. https://biz.yelp.com/support/responding_to_reviews.

15

Habermas in the African E-Village: Deliberative Practices of Diasporan Nigerians on the Internet

Farooq Kperogi

The notion of the public sphere is at the core of the reconstruction of deliberative democracy.[1] In the age of the Internet, deliberative democracy is increasingly instrumentalized through spatially dispersed but nonetheless organic online communities. Mary Chayko calls these "the portability of social connectedness,"[2] and they have been mushrooming exponentially even in "digital backwaters" of the world such as Africa, that Manuel Castells had gloomily characterized as the "black hole of informational capitalism."[3] Thanks to the ubiquity of Internet-ready mobile devices, several African countries—including Nigeria, the continent's most populous nation and biggest economy—are active participants in participatory, many-to-many, user-led online communities.

This chapter examines a vibrant Nigerian online community called the "Nigerian Village Square" (www.NigeriaVillagesquare.com). Over the years, this community has functioned as an arena for the vigorous exchange of ideas among Nigerians both at home and in the diaspora, and as a veritable locus for the initiation of petition drives to change or influence state policies in Nigeria. The author argues that several of the deliberative practices of Nigerianvillagesquare.com resonate with—or at least consciously seek to abide by—some features of Habermas's characterization of the 17th- and 18th-century public spheres in Britain, France, and Germany.

This chapter first gives a brief review of the literature on the public sphere, with a special reference to the conception of the public sphere as popularized by Jürgen Habermas. It next reviews and interrogates the contending theoretical constructions on the deliberative potential and practices—or lack thereof—of the Internet. It also dissects the internal categories of the Habermasian "ideal speech situation" and reviews the trajectory of scholarship that affirms or repudiates the symbolic nexus between the traditional, normative

conception of the public sphere that Habermas theorizes, and the phenome-
nally dizzying democratic marketplace of ideas that, in the author's opinion,
the Internet enables. The author contends that the Internet, although sometimes falling short of the requirements of a normative Habermasian conception of the public sphere, in many respects also creates the opportunity for the
robust flowering of a variety of "public spheres," especially transnational, diasporic spheres of public discourse.

A BRIEF OUTLINE OF THE CLASSICAL PUBLIC SPHERE

Jürgen Habermas, arguably the most notable of the second-generation members of the Frankfurt School,[4] popularized the concept of the public sphere in
the English-speaking world when his book *The Structural Transformation of the
Public Sphere*, first published in 1962, was translated into English in 1989.[5] The
remarkable popularity of the Internet as a new arena for deliberative practices,
coupled with the dramatic corporatization of and disillusionment with both
the content and form of the traditional media of mass communication that
were hitherto construed as embodying the public sphere, have been incentives
in the renewed focus on Habermas' theory of the public sphere.

It is noteworthy, however, that although Habermas's theory of the public
sphere might be the most influential, it by no means stands alone; there exists a
multiplicity of competing conceptions of what constitutes the public sphere.
An influential conception of the public sphere that is coeval with but radically
different from Habermas's, for instance, is that of Reinhart Koselleck. Koselleck's
concept of the public sphere is found in his historicization of the role that secret
societies such as the Freemasons played in challenging totalitarian authority in
Europe.[6] By countervailing the totalitarianism prevalent in Europe in the 18th
century, he argued, the secret societies created moral alternatives that expanded
the range of discourse in much the same way that the bourgeois public sphere
Habermas historicized served as counterweights to feudal absolutism.

For John Dewey, an influential American theorist, the public is called into
being by the concatenation of "indirect, extensive, enduring and serious consequences of conjoint and interacting behavior" of individuals and groups in
the society.[7] What is noteworthy about Dewey's conception of the public is
that, unlike Habermas's and Koselleck's, it is not uncoupled from the state. It
is beyond the scope of this chapter, however, to explore all the conceptions of
the public.[8] Nevertheless, the notion of the public sphere that forms the theoretical bedrock of this chapter is that propounded by Habermas. This is because, as Nancy Fraser argues, "no attempt to understand the limits of actually
existing late-capitalist democracy can succeed without in some way or another making use of [Habermas's early conception of the public sphere]."[9]
Habermas (1991: 398) defines the public sphere as follows.

By "public sphere" we mean first of all a domain of our social life in which such a thing as public opinion can be formed. Access to the public sphere is open to all citizens. A portion of the public sphere is constituted in every conversation in which private persons come together to form a public. When the public is large, this kind of communication requires certain means of dissemination and influence; today newspapers and periodicals, radio and television are the media of the public sphere. We speak of a political public sphere when the public discussions concern objects connected with the practice of the state.

Habermas's theory derives inspirational strength from a historical juncture during the 17th and 18th centuries in Western Europe—particularly in England, France, and Germany—when coffee houses, salons, societies, the town hall, the village church, the tavern, the public square, convenient barns, union halls, parks, factory lunchrooms, and even street corners became the arenas of debate, political discussion, and action. Habermas extends this to a normative model of popular involvement in the public sphere for contemporary times. He explains that in these venues, everyone had an equal right to speak as if they were equals. In England, for example, the coffee house conferred discursive sanctuary not only on the nobility but also on "the wider strata of the middle class, including craftsmen and shopkeepers."[10] The dialogue, he posited, transpired in a profoundly democratic forum where the status differentials and positional hierarchies of participants were bracketed, and issues were discussed without external coercion.

The public sphere, Habermas contends, was governed by a moral-practical discourse, and the apparatus for the mediation of this discourse was rational and critical argumentation. His analysis of communication in the archetypal bourgeois public sphere reveals that every participant who engaged in the moral-practical discourse of the time made recourse to a number of normative conditions, which he later called the "ideal speech situation" in his extension of this theory. The requirements to qualify for the Habermasian "ideal speech situation" are:

- The discourse should be independent from state and corporate interferences;
- The exchange of points of views during a discourse should be amenable to criticism and review, and dogmatism should be eschewed;
- Participants in the discourse should demonstrate reflexivity and a willingness to question both their individual assumptions and those of the social milieu in which they live;

- Participants should show a capacity for tolerance, sympathy, and even a vicarious identification with points of views that are at variance with theirs and also avoid the use of emotive and insulting language;
- Participants must make an effort to be sincere in their search for the truth; and
- There must be discursive inclusivity and equality.[11]

It is important to note that Habermas himself has recognized that these are mere ideals that have never been fully realized even in the classical bourgeois public sphere.

Nonetheless, as persuasive as his historicization and idealization of the bourgeois public sphere is, critics from different theoretical and ideological orientations have criticized it. Specifically, it has been criticized as Eurocentric, biased in favor of the bourgeoisie and against the working class, as patriarchal, and as logocentric.[12] Poststructuralists like Lyotard even questioned the functional relevance of Habermas's model of consensus through rational-critical discourse.[13] He specifically queried the emancipatory and social utility of excessive rationalism in the conduct of discourse for the wider strata of society. The concept is also criticized for instituting a linear, evolutionary, and progressive history of the world that ignores the differential socio-historical experiences of non-Western or, to be sure, non-European societies. It is accused of falsely conferring on the idiosyncrasies of Enlightenment Europe a universality it never possessed, and of consigning the differential temporalities of other societies to the discursive fringes.[14]

Feminist theorists such as Nancy Fraser also point out the exclusion of women in Habermas's public sphere.[15] As Neil McLaughlin observes, a typical participant in the public sphere usually was male, educated, and propertied; with the means and leisure to take part in public discourse.[16] Economically disaffiliated segments of the society that had a need—indeed an obligation—to work hard to survive the vicissitudes and cruelties of the incipient capitalist socioeconomic order did not have the luxury to expend energies and time to participate in the discussions. Certainly, women at that time were either too ensnared by the drudgery of domestic engagements or the suffocating stranglehold of male tyranny and oppression to participate in the discursive indulgence of the public sphere. Several feminist theorists such as Anne Fernald,[17] including a whole host of other critics in an edited volume titled *Feminists Read Habermas: Gendering the Subject of Discourse*,[18] also point out that the putative discursive openness of Habermas' public sphere was premised on practices of deliberative omission not only of women but also of other marginal and subordinate groups in the society. Jodi Dean even suggests that Habermas's account of the bourgeois public sphere was at best apocryphal.[19]

Aside from poststructuralist, postcolonial, and feminist critics, other critics such as Oskar Negt, Alexander Kluge,[20] Nancy Fraser, and Michael Warner[21] also instructively call attention to Habermas's privileging of a hegemonic public sphere, which they argue was structured to be congenial only to the preferences and prejudices of people who occupy the upper end of the social scale. In place of one overarching, dominant public sphere, they postulate the concept of multiple public spheres that are not only oppositional to the hegemonic public sphere but that also incorporate the aspirations of marginal groups in the society such as the working class, women, and racial and sexual minorities. The symbolic utility of their argument in terms of contemporary scholarship in computer-mediated communication is that the landscape of the public sphere has been shifted from a historico-transcendental veneration of Europe during the Enlightenment to a multiplicity of trans-historical loci of discourses. This crucial change in the notion of the public sphere assumes its full consequence when it is seen in relation to the Internet, which, in many ways, defies simplistic spatial and temporal categorizations, and encompasses a robust array of spheres of discourses in ways that are probably unparalleled in human history.

Now, how does the Internet relate with classical notions of the public sphere and how does it depart from it? Do Internet chat rooms, e-zines, news groups, electronic bulletin boards, and cyber salons qualify to be labeled modern-day public spheres? What conditions must they fulfill to approximate the status of public spheres? The next section addresses how some scholars have answered these questions.

THE INTERNET AND THE PUBLIC SPHERE

Although it is obvious that Habermas did not envision the Internet when he formulated his theory of the public sphere, his caution against "Athens envy" in his subsequent work no doubt anticipates the debates about the deliberative capacities of the Internet. Habermas—in an important extension of his theory of the public sphere—argued that if democracy is to be implemented in today's diverse and complex world, then society has to learn to adjust to the reality of the impracticability of a spatially bound agglomeration of mutually consenting members in the public sphere. Instead, he proposes that citizens who are not necessarily physically co-present can develop forms of communication that dispense with the necessity for corporeal presence.[22] Although Habermas did not specifically refer to the Internet when he said this, many communication scholars have interpreted him as affirming the deliberative potential of the Internet. As Heinz Brandenburg instructively observed,

[I]t is not the case that Habermas himself or deliberative theorists went on the search to discover an effective and inclusive forum of public deliberation on a mass scale and came across the Internet. It is rather the other way around: namely that early cyber-enthusiasts quickly embraced Habermas' notion of the public sphere and the theory of deliberative democracy and began to claim that the Internet provides just that: a virtual public sphere.[23]

One of the most notable early cyber-theorists who rhapsodized over the emancipatory and deliberative potential of the Internet was Howard Rheingold.[24] Rheingold popularized a cyber-enthusiastic vision of the electronic agora, or what Robin and Webster have characterized as the vision of an "Athens without slaves."[25] Rheingold believes that Internet technology, "if properly understood and defended by enough citizens, does have democratizing potential in the way that alphabets and printing presses had democratizing potential."[26] He postulates that the formation of virtual communities on the Internet provides the chance to revitalize the public sphere, and that the chance to do this has been put back in the hands of the public in a manner that is unparalleled in the records of human democratic progress. One of the main benefits of the Internet, he said, is the wide latitude it gives its users to find others who share similar interests, concerns, and worries with them. He made the point that although a person cannot, for instance, simply pick up a phone and ask to be connected with someone who wants to talk about Islamic art or Californian wine, but a person can join a newsgroup on any of those topics and converse with the people there, either privately or publicly.

Rheingold, in spite of his faith in the strengths and promises of the Internet, was cautious not to draw a mechanical link between the normative public sphere and the Internet, preferring to use the word "potential." Hubertus Buchstein was less restrained. He was certain that the Internet is the Habermasian public sphere reincarnated in an electronic form. He contends that it "looks like the most ideal speech situation."[27] Douglas Kellner also states that the Internet has "produced new public spheres and spaces for information, debate, and participation that contain the potential to invigorate democracy and to increase the dissemination of critical and progressive ideas."[28] Similarly, Hauben and Hauben, from their empirical inquiry into the discursive practices of Usenet groups and other forms of deliberation on the Web, conclude that the Internet expands the range and diversity of communication and viewpoints "through the freewheeling and rambling discussion" that it enables.[29]

Several other scholars, however, do not share in the optimism of these cyber-enthusiasts. Dahlberg, for instance, on the basis of an empirical research, concludes that the gains of Internet discussion groups in terms of opening up new vistas for advancing the concept of the public sphere are vitiated by the growing commercialization and commoditization of cyberspace by state and corporate concerns.[30] He also notes that the Internet as presently constituted is hallmarked by a dearth of reflexivity, a deficit of mutual tolerance, difficulty in authenticating identity claims and information put forward by participants in Internet discussion groups, the disaffiliation of large sections of people from online political forums, and the monopolization of cyber-discourse by a few individuals and groups. Dahlberg contends therefore that this reality detracts from the ascription of the status of public sphere to the Internet. In their introduction to *Resisting the Virtual Life*, Brook and Boal were even harsher in their critique of the Internet. They say it is "pernicious" when it is deployed as "substitutes for face-to-face interactions."[31]

There are scholars such as Susan O'Donnell, however, who strike a happy balance between the cyber-enthusiasts and the cyber-skeptics. O'Donnell's research applies the public sphere concept to the Internet and argues that although the Internet does have democratizing potential it often fails as a public sphere in practice.[32] How do these views relate to an actual, extant site of discourse on the Internet?

THE NIGERIAN VILLAGE SQUARE AS A COLLABORATIVE, HABERMASIAN ONLINE COMMUNITY

The Nigeriavillagesquare.com was founded in 2003 by a group of immigrant Nigerians based in the United States. The main figure associated with the site is Philip Adekunle, a Chicago-based computer information systems specialist. The Web site was created to serve as a locus for the untrammeled exchange of ideas and opinions about the homeland by Nigerians both in Africa and in the diaspora.[33] It is the reinvention, in an electronic form, of the deliberative content of the "village square" in the precolonial African social formation where "people from all corners [met] at the Village Square after a hard day's work to sip unadulterated palm-wine, share news, gossip, jokes, music, dance, events and opinions."[34] In many respects, the precolonial African village square that the owners of this Web site reference has many resonances with the early European bourgeois public sphere that Habermas historicized; only that the African village square was premodern, prebourgeois, and did not function as a counterweight to the ruling class, nor did it have any purposive, codified, normative ideals that guided its deliberative practices. It was a core cultural institution, however, that was crucial to the intergenerational perpetuation of traditions, customs, and mores, which were disrupted with the advent of colonialism.

The reincarnated village square in electronic form, however, both replicates and transcends the structures and discursive practices of its predecessor. Although the NigeriaVillagesquare.com site has guidelines on the form of articles to be posted on its front pages and in its discussion forums, the site is largely unmoderated. It accepts opinion articles, news commentaries, trivia, and even fictional creative writing from all Nigerians and non-Nigerians interested in Nigerian affairs. Contributors to the site do not have to be subscribers or registered users to submit articles, although they must register to participate in online discussions of materials posted on the site. This speaks to the site's inclusivity.

Although the site is not primarily a news site, it publishes breaking news stories, has a citizen media project called "i-Witness,"[35] which collates reports from citizen reporters from all over Nigeria—especially during periods of conflicts—and it often is the interim medium for many of Nigeria's robust diasporan citizen media projects.[36] Many of the strictly news citizen media sites still share their news stories on the Village Square's message boards, and a great number of politically consequential citizen reports that went on to change state policy started on this site.

The site also has links to the Web addresses of major Nigerian newspapers that have an online presence, it periodically posts high-impact and controversial news stories both from the domestic newspapers and from diasporan online newspapers on its front page, and it invites discussion from subscribers. This feature has made it one of the most popular Internet sites for Nigerians both at home and in the diaspora. The debates in the site's forums are not only robust, and sometimes frenzied, even emotional, but also evince a studious concern for civility in public discourse.

Because Internet deliberation is easily susceptible to degeneration into ad hominem attacks, the site has what it calls the "Nigerian Village Square Publishing Guide," which not only gives instructions on how commentaries should be posted on the site but also provides the ground rules for deliberation. It stresses the importance of eschewing emotive language and embracing what Habermas would call rational-critical debate. One of the mechanisms established to ensure that this rule is observed is the formation of what is called the "village dumpster" where articles, comments, and discussions that are deemed irreverent or overly personal and insulting are consigned.[37] The decision about what posts should be pushed to the "village dumpster" often is arrived at through the votes of registered members of the site. This structural check has imposed self-moderation on many discussions in the forum, but it also occasionally raises allegations of majoritarian tyranny.

It is instructive that, over the years, the site has transformed from being a mere cyber salon for quotidian dialogic disputations to a close-knit, many-to-many, collaborative cyber community where deliberations and decisions about national politics take place. In 2006, for instance, when former Nigerian president Olusegun Obasanjo wanted to manipulate the national legislature to amend the constitution to allow him to run for a third term, "villagers," as members of the discussion forums on the site call themselves, started an Internet petition drive to stop the move. They generated petitions from thousands of Nigerians in the diaspora and contributed money to send a representative to deliver the petition to the president of the Nigerian Senate and the speaker of the House of Representatives. It was also delivered to the president when he visited the United States.[38] This move generated publicity in the Nigerian national media and contributed to the defeat of the president's third-term bid.

Similarly, in the same year, when a Nigerian immigrant by the name of Osamuyia Aikpitanhi was murdered by Spanish immigration officers, the "villagers" not only vigorously deliberated on the issue, they contributed money and sent a delegation to Spain to demand an explanation of the circumstances that led to the death of their compatriot. The delegation also met with the parents of the deceased and gave them a $1,200 check that the "villagers" contributed. It was robustly covered by Nigeria's biggest, state-run television network, the Nigerian Television Authority, more popularly known by its initialism, NTA.[39] The Nigerian Village Square delegation also had a meeting with Nigerian government officials and got the president and the Nigerian legislature to request that Spanish authorities explain the circumstances that surrounded the death of the Nigerian and to pay compensation to the family if need be. Weeks after this visit, what would have been an ignored issue was elevated to a major diplomatic row between Nigeria and Spain.[40] At last, the Spanish government created a commission to investigate the murder of the Nigerian and issued an interim apology to the Nigerian government while the investigation was in progress.[41]

The "villagers," many of whom have never met except through the virtual village square, directly intervene in many other domestic political issues. Another prominent example of their collaborative political activism was the petition drive that they started in the village square in the midpoint of 2007 to force the then Nigerian vice president Goodluck Jonathan (who later became president) to declare his assets. Although the Nigerian constitution does not require public officials to declare their assets publicly, Nigeria's then newly elected president, in a bid to show his seriousness in fighting corruption, publicly declared his assets for the first time in the country's history.[42] This singular act earned him praise, but it also paved the way for citizens to demand the same forthrightness from other elected representatives. The vice president insisted that because he was under no constitutional obligation to declare his assets publicly, he would not be railroaded into doing so. Again, members of the Nigerian Village Square started a petition drive to which hundreds of Nigerians in the diaspora appended their signatures. A representative physically went to Nigeria not only to deliver the petition but to attract wide media attention to this event.[43] This pressure contributed to forcing the vice president to publicly declare his assets.[44] Efforts were doubled to shame other well-placed public officers into also publicly declaring their assets.

The media attention that the interventionist activities of members of the Nigerianvillagesquare.com has generated has given it a lot of visibility and clout in government circles, and many well-known personalities in government are known to be registered members of the forum either anonymously or under their full names. A case in point is that of a Mr. Olusegun Adeniyi, who was the spokesman for the late Nigerian president, Umar Musa Yar'adua. In the wake of withering attacks against him in the village square over a frivolous trip he had taken to the United States,[45] he appeared in the forum under his own name to defend himself.[46] In no way, however, can this be construed as an infiltration of the forum by government. On the contrary, it represents a dialogue with citizens who ordinarily would not have had the privilege to engage in this deliberative exchange had they sought the traditional means of communication with the government.

As most scholars argue, at the heart of most conceptions of the public sphere—especially the Habermasian one—is the idea of conversation, which this Internet forum seems to enhance in more ways than spatially bounded notions of deliberation can. Indeed many scholars contend that conversation in the public sphere is a precondition for democracy. Michael Schudson, for instance, notes that "[d]emocracy is government by discussion."[47] Bruce Ackerman also states that "[d]ialogue is the first obligation of citizenship."[48] This Internet discussion group does certainly make possible the type of categories necessary for most, if not all, of Habermas's "ideal speech situation" to occur.

CONCLUSION

The communicative acts that take place in this Nigerian virtual village square confirm, to a large degree, that the Internet is capable of facilitating discourse that replicates the central construction of rational-critical debate and that, in a variety of ways, approximate the prerequisites of the Habermasian public sphere. It is discursively inclusive, as evidenced in its policy of not moderating its discussions and in not requiring that potential contributors of articles to the site be registered members. Participation also is entirely voluntary and is not the product of coercion by government or corporate interests, and the site's deliberative practices so far have impacted governance in the homeland. What is more, although members of the forum come from different social backgrounds—including professors in United States, United Kingdom, and Nigerian universities; engineers; doctors; students; and Nigeria-based working-class people—there has not been any record of open social discrimination in the discursive enterprise. Although the main preoccupation of the forum is politics in the homeland, however, it also features short-story fiction and trivia and sometimes can get bogged down by petty personality disputes.

Bohman notes that a crucial deficiency of Internet public spaces is that they have no linkages with structures of power, a condition, he says, that divests them of the capacity to "secure the conditions of publicity but also in order to promote the interaction among publics that is required for deliberative democracy."[49] Bohman's arguments are difficult to sustain when applied to the Nigerian Village Square because the activities of this Internet site generate a lot of media attention in Nigeria.

Nancy Fraser's recent exposition of what constitutes a transnational public sphere also strengthens the case for the Nigerian Village Square to be ascribed the status of a transnational online public sphere. According to Fraser, "a public sphere is supposed to be a vehicle for mobilizing public opinion as a political force. It should empower the citizenry vis-à-vis private powers and permit it to exercise influence over the state."[50] This Web site's continually productive engagements with the Nigerian state on domestic policy issues certainly elevate it to a politically consequential public sphere.

Admittedly, however, the relative numerical inferiority of this Internet-based public sphere, and the uncertainty whether it can sustain its activities, potentially detract from its potency. Similarly, although many contributors to the forum are identifiable by their names, others use anonymous handles. Scholars have debated whether the incidence of anonymity on the Internet invalidates the notion of conversation in news groups and chat rooms in the real sense of the word. This concern is perfectly legitimate but, as Dean reminds us, "anxieties around authenticity on the Net function primarily to reassure our trust in the authenticity of other sorts of mediated interactions, indeed, to pathologize our justifiable paranoia."[51] The point, therefore, is that although some of the Internet's shortcomings significantly detract from the Habermasian status of the public sphere that many scholars ascribe to it, there are certainly important respects in which it fulfils this requirement, as the analysis of the Nigerian Village Square suggests.

The foregoing, though, does not seek to institute the relative discursive openness of the Nigerian Village Square as representative of the sort of interaction that takes place on all Nigerian online media forums. Although several such examples abound on the Internet, there is also a multiplicity of Web sites that do not have deliberative democracy—or anything remotely related to such democracy—as their raison d'être. It is incontestable that the Internet meshes with existing and preexisting social functions and extends them in many fresh and new ways, but it does not fit easily in comparison to characteristically modern organizations or stereotypically idyllic early modern social and cultural institutions such as the Habermasian public sphere.

The new peculiarities and idiosyncrasies of the Internet can become intelligible only if a conceptual frame of reference is adopted that does not limit the discussion of the Internet from the outset to predetermined and pre-given patterns of interpretation. More importantly, the Nigerian Village Square has shown the possibility for the coexistence between online communities and citizen journalism. The site provides the platform for citizen reports, for discussions about politics in the homeland, and for direct action. This fits the outlines of the traditional conceptions of alternative journalism. The Village Square, however, is a loose collection of disparate people and interests that are not all united by notions of progressive ideology. In more ways than one, it problematizes the boundaries between alternative and citizen journalism in that it provides an arena for both forms of interaction.

NOTES

1. Deliberative democracy is the reconceptualization of democracy in more participatory and substantive terms. It expands the notion of democracy beyond such institutional rituals as periodic elections to more popular and practical civic and discursive engagement in the process of governance. For a useful introductory discussion on the origin and development of this concept, see Heinz Brandenburg, "Pathologies of the Virtual Public Sphere," in *Civil Society, Politics and the Internet: A Comparative Perspective*, eds. S. Oates, D. Owen, R. Gibson (London: Routledge, 2005). Also see Jon Elster, "Introduction" in *Deliberative Democracy: Essays on Reason and Politics*, eds. J. Bohman, W. Rehg (Cambridge, MA: The MIT Press, 1997): 1; and John Dryzek, *Deliberative Democracy and Beyond. Liberals, Critics, Contestations* (Oxford: Oxford University Press, 2000): 162.

2. *See* Mary Chyko (2008), *Portable Communities: The Social Dynamics of Online and Mobile Connectedness* (Albany: SUNY Press).

3. *See* Manuel Castells (1998), *End of Millennium* (Malden, MA: Blackwell).

4. The Frankfurt School, also known as the Institute of Social Research, was founded in Frankfurt, Germany, in 1923 by German Jewish Marxists Theodor Adorno, Herbert Marcuse, Max Horkheimer, and Erich Fromm. As a result of Nazi persecution, they fled Germany and went into exile first in Geneva and later to California and to Columbia University in New York. For a discussion of the influence they brought to bear on the social sciences and philosophy, see Neil McLaughlin, "Origin Myths in the Social Sciences: Fromm, the Frankfurt School and the Emergence of Critical Theory," *Canadian Journal of Sociology* 24 (1) (1999): 109–39.

5. Habermas (1989), *The Structural Transformation of the Public Sphere*.

6. *See* Reinhart Koselleck, *Critique and Crisis: Enlightenment and the Pathogenesis of Modern Society* (Cambridge: MIT Press, 1988).

7. John Dewey, *The Public and Its Problems* (Athens: University of Ohio Press, 1927): 64.

8. For other influential conceptions of the public sphere, see, for instance, Hanna Arendt's *The Human Condition* (Chicago: University of Chicago Press, 1958), Walter

Lipmann's *Public Opinion* (New York: Free Press, 1922). For an excellent comparison of the different traditions of public sphere theorizing, see Seyla Benhabib, "Models of Public Space: Hannah Arendt, the Liberal Tradition, and Jürgen Habermas," in *Habermas and the Public Sphere*, edited by Craig Calhoun (MIT Press, 1992). Benhabib delineates three models of the public sphere and associates them with the work of particular theorists. She calls the first model, represented by the work of Hanna Arendt, the agonistic model. She characterizes the second model, represented by the work of Ackerman, as the legalistic model, and labels the third model, represented by the work of Habermas, as the discursive model.

9. Nancy Fraser, "Rethinking the Public Sphere: A Contribution to the Critique of Actually Existing Democracy," in *Habermas and the Public Sphere*, edited by Craig Calhoun (MIT Press, 1992): 111.

10. Jürgen Habermas, *The Structural Transformation of the Public Sphere*, 33.

11. This summary of Habermas's requirements for a rational-critical debate in the quintessential public sphere was adapted from the model developed in Lincoln Dahlberg, "Computer-mediated Communication and the Public Sphere: A Critical Analysis," *Journal of Computer-Mediated Communication* 7 (1) (2001): par.4 [journal online], available at http://jcmc.indiana.edu/vol7/issue1/dahlberg.html, accessed October 12, 2007. It encapsulates the whole range of Habermas's theoretical postulations, published in different books, on the public sphere.

12. Habermas responded to the criticisms of his conception of the public sphere in *The Inclusion of the Other: Studies in Political Theory*, edited by Ciaran Cronin and Pablo De Greiff (Cambridge: MIT Press, 1998).

13. Jean-Francois Lyotard, *The Postmodern Condition*, trans. Brian Massumi and others (Minneapolis: University of Minnesota Press, 1984).

14. For a review of the criticism of Habermas' public sphere, see Craig Calhoun, "Introduction," in *Habermas and the Public Sphere*, edited by Craig Calhoun (Cambridge, MA: MIT Press, 1992).

15. Nancy Fraser, "Rethinking the Public Sphere," *Social Text* 25/26 (1990): 80–56.

16. Neil McLaughlin, "Feminism, the Public Sphere, Media and Democracy," *Media Culture and Society* 15 (4) (1993): 599.

17. Anne Fernald E., "A Feminist Public Sphere? Virginia Woolf's Revisions of the Eighteenth Century," *Feminist Studies* 31 (1) (2005): 158–250.

18. Johanna Meehan (ed.), *Feminists Read Habermas: Gendering the Subject of Discourse* (New York: Routledge, 1995).

19. *See* Jodi Dean, "Why the Net Is not a Public Sphere," *Constellations* 10 (1) (2003): 96.

20. Oskar Negt and Alexander Kluge, *Public Sphere and Experience: Toward an Analysis of the Bourgeois and Proletarian Public Sphere*, trans. Peter Labanyi and others (Minneapolis: University of Minnesota Press, 1993).

21. *See* Nancy Fraser, "Rethinking the Public Sphere," in Michael Warner, *Publics and Counterpublics* (New York: Zone Books, 2005).

22. Cited in John Durham Peters, "Distrust of Representation," *Media, Culture and Society* 15 (4) (1993): 564.

23. Heinz Brandenburg, "Pathologies of the Virtual Public Sphere," in *Civil Society, Politics and the Internet: A Comparative Perspective*, edited by S. Oates, D. Owen, R. Gibson (London: Routledge, 2005): 4.

24. Howard Rheingold's seminal text, *The Virtual Community: Homesteading on the Electronic Frontier* (Boston, MA: Addison-Wesley Publishing Company, 1993): 279, provides, in the view of many scholars, the first systematic, scholarly reflection on online sociability and the democratizing promise of the Internet. The book basically is a recounting of Rheingold's experiential encounters with online environments, particularly with the WELL (Whole Earth 'Lectronic Link—one of the Internet's earliest bulletin board systems).

25. Kevin Robins and Frank Webster, "Athens without Slaves . . . or Slaves without Athens? The Neurosis of Technology," in *Science as Culture* vol. 1 (London: Free Association).

26. Rheingold, *The Virtual Community*, 279.

27. Hubert Buchstein, "Bytes that Bite: The Internet and Deliberative Democracy." *Constellations* 4 (October 1997): 250.

28. Douglas Kellner, "Intellectuals, the New Public Spheres, and Techno-politics," In *The Politics of Cyberspace: A New Political Science Reader*, edited by C. Toulouse and T. W. Luke (New York: Routledge, 1998): 167–86.

29. Michael Hauben and Rhonda Hauben (eds.). *Netizens: On the History and Impact of Usenet and the Internet* (Los Alamitos, CA: IEEE Computer Society Press, 1997): 69.

30. Lincoln Dahlberg, "Computer-Mediated Communication and the Public Sphere."

31. James Brook and Iain A. Boal (eds.), *Resisting the Virtual Life: The Culture and Politics of Information* (San Francisco: City Lights, 1995): vii.

32. Susan O'Donnell, "Analyzing the Internet and the Public Sphere: The Case of Womenslink," *Javnost-The Public* 8 (1) (2001): 39–58.

33. *See* The Nigerian Village Square, "About Us," http://www.nigeriavillagesquare .com/about-us.html.

34. Ibid. For another insightful fictional reconstruction of precolonial deliberative practices in what were called village squares in Nigeria and, by extension, Africa, read Chinua Achebe's *Things Fall Apart*.

35. http://iwitness-nigeria.com/index.php/main.

36. See Farooq Kperogi (2008), "Guerillas in Cyberia: The Transnational Alternative Online Journalism of the Nigerian Diasporic Public Sphere," *Journal of Global Mass Communication* 1 (1/2): 72–87; Farooq Kperogi (2011), "Webs of Resistance: The Citizen Online Journalism of the Nigerian Digital Diaspora," Ph.D. Dissertation. Georgia State University. http://scholarworks.gsu.edu/communication_diss/27.

37. Isaac Olawale Albert (2010, April 19), "Whose Deliberative Democracy? A Critique of Online Public Discourses in Africa," Nigerian Village Square. www .nigeriavillagesquare.com.

38. Concerned Nigerians in Diaspora (2006, March 29), "Letter to President obasanjo on Third Term Agenda on His Visit to the White House." Nigerian Village Square. www.nigeriavillagesquare.com.

39. *See* Ahaoma Kanu (2007, August 13), "NVS Members Donate $1200 to the Aikpitanhi Family, NTA Goes Haywire," Nigerian Village Square. http://www.nigeriavillagesquare.com.

40. Shola Adekoya (2007, June 29), "FG Probes Killing of Nigerian Aboard Spanish Plane," *Nigerian Tribune*.

41. El Pais (2007, November 11), "Aikpitanhi's Death Causes Major Policy Change in Spain." Nigeria Village Square. http://nigeriavillagesquare.com. The report was culled from a Spanish newspaper called *El Pais*.

42. Bashir Adigun (2007, June 28), "Nigerian President Declares Assets," The Associated Press.

43. *Guardian* (2007, August 8), "DYA [Declare Your Assets] Campaign: The Vice President's Assets."

44. P.M. News (2007, August 8), "VP Jonathan Declare Assets," *P.M. News*.

45. Olusegun Adeniyi (2007, September 7), "Why the Hell Am I Here?" Nigerian Village Square, http://www.nigeriavillagesquare.com/.

46. Empowered Newswire (2007, September 4), "I am Here to Understudy the Americans–Adeniyi, Yar'Adua's Spokesman, on US Visit." Nigerian Village Square, http://www.nigeriavillagesquare.com.

47. Michael Schudson, "Why Conversation Is Not the Soul of Democracy," *Critical Studies in Mass Communication* 14 (4) (1997): 297–309.

48. Bruce Ackerman, "Why Dialogue?", *Journal of Philosophy* 86 (1989): 6

49. James Bohman, "Expanding Dialogue: The Internet, the Public Sphere and Prospects for Transnational Democracy," in *After Habermas: New Perspectives on the Public Sphere*, edited by Nick Crossley and John M. Roberts (Oxford: Blackwell Publishing, 2004): 146.

50. Nancy Fraser, "Transnationalizing the Public Sphere: On the Legitimacy and Efficacy of Public Opinion in a Post-Westphalian World," *Theory, Culture & Society* 24 (4) (2007): 7.

51. Jodi Dean, "Virtually Citizens," 277–78.

BIBLIOGRAPHY

Achebe, Chinua (1986). *Things Fall Apart* (London: Heinemann, 1986) (originally published in 1958).

Ackerman, Bruce (1989). "Why Dialogue?" *Journal of Philosophy* 86, 6.

Adekola, Shola (2007). "FG Probes Killing of Nigerian Aboard Spanish Plane." *Nigerian Tribune* June 29.

Adeniyi, Olusegun (2007, September 7). Why the Hell Am I Here? *Nigerian Village Square*. http://www.nigeriavillagesquare.com/.

Adigun, Bashir (2007, June 28). "Nigerian President Declares Assets." The Associated Press.

Albert, Isaac Olawale (2010, April 19). Whose Deliberative Democracy? A Critique of Online Public Discourses in Africa. *Nigerian Village Square*. www.nigeriavillagesquare.com.

Arendt, Hanna (1958). *The Human Condition* (Chicago: University of Chicago Press).

Benhabib, Seyla (1992). "Models of Public Space: Hannah Arendt, the Liberal Tradition, and Jürgen Habermas." In Craig Calhoun, *Habermas and the Public Sphere* (MIT Press).

Bohman, James (2004). "Expanding Dialogue: The Internet, the Public Sphere and Prospects for Transnational Democracy." In Nick Crossley and John M. Roberts, *After Habermas: New Perspectives on the Public Sphere* (Oxford: Blackwell Publishing).

Brandenburg, Heinz (2005). *Pathologies of the Virtual Public Sphere*. In S. Oates et al., *Civil Society, Politics and the Internet: A Comparative Perspective* (London: Routledge).

Brook, James, and Boal, Iain (eds.) (1995). *Resisting the Virtual Life: The Culture and Politics of Information* (San Francisco: City Lights).

Buchstein, Hubert (1997). "Bytes that Bite: The Internet and Deliberative Democracy." *Constellations* 4, 250.

Calhoun, Craig (1992). "Introduction." Craig Calhoun, *Habermas and the Public Sphere* (Cambridge, MA: MIT Press).

Castells, Manuel (1998). *End of Millennium* (Malden, MA: Blackwell).

Chyko, Mary (2008). *Portable Communities: The Social Dynamics of Online and Mobile Connectedness* (Albany: SUNY Press).

Concerned Nigerians in Diaspora (2006, March 29). Letter to President Obasanjo on Third Term Agenda—On His Visit to the White House. Nigerian Village Square. www.nigeriavillagesquare.com.

Dahlberg, Lincoln (2001). "Computer-Mediated Communication and the Public Sphere: A Critical Analysis." *Journal of Computer-Mediated Communication* 7 (1) [journal online]. http://onlinelibrary.wiley.com/journal/10.1111/%28ISSN%291083-6101.

Dean, Jodi (2003). "Why the Net Is Not a Public Sphere." *Constellations* 10 (2): 95–112.

Dean, Jodi (2001). "Cybersalons and Civil Society: Rethinking the Public Sphere in Transnational Technoculture." *Public Culture* 13 (2): 243–65.

Dean, Jodi (1997). "Virtually Citizens." *Constellations* 4 (2): 277–78.

Dewey, John (1927). *The Public and Its Problems* (Athens: University of Ohio Press).

Dryzek, John (2000). *Deliberative Democracy and Beyond. Liberals, Critics, Contestations* (Oxford: Oxford University Press).

El Pais (2007, November 11). Aikpitanhi's Death Causes Major Policy Change in Spain. Nigeria Village Square. http://nigeriavillagesquare.com. The report was culled from a Spanish newspaper called *El Pais*.

Elster, Jon (1997). "Introduction." In J. Bohman, and W. Rehg, *Deliberative Democracy: Essays on Reason and Politics* (Cambridge, MA: The MIT Press).

Empowered Newswire (2007, September 4). I Am Here to Understudy the Americans—Adeniyi, Yar'Adua's Spokesman, on US Visit. Nigerian Village Square. http://www.nigeriavillagesquare.com.

Fernald, Anne (2005). "A Feminist Public Sphere? Virginia Woolf's Revisions of the Eighteenth Century." *Feminist Studies* 31 (1): 158–82.

Fraser, Nancy (1992). "Rethinking the Public Sphere: A Contribution to the Critique of Actually Existing Democracy." In Craig Calhoun, *Habermas and the Public Sphere* (Cambridge, MA: MIT Press).

Fraser, Nancy (1990). "Rethinking the Public Sphere." *Social Text* 25/26, 56–80.

Fraser, Nancy (2007). "Transnationalizing the Public Sphere: On the Legitimacy and Efficacy of Public Opinion in a Post-Westphalian World." *Theory, Culture & Society* 24 (4): 7.

Guardian (2007, August 8). "DYA [Declare Your Assets] Campaign: The Vice President's Assets."

Habermas, Jurgen (1989). *The Structural Transformation of the Public Sphere: An Inquiry into a Category of Bourgeois Society*. Trans. T. Burger (London: Polity Press).

Habermas, Jurgen (1991). "The Public Sphere." In C. Mukerji and M. Schudson, *Rethinking Popular Culture: Contemporary Perspectives in Cultural Studies* (Berkeley: University of California Press).

Habermas, Jürgen (1998). *The Inclusion of the Other: Studies in Political Theory*. Edited by Ciaran Cronin and Pablo De Greiff (Cambridge, MA: MIT Press).

Hauben, Michael, and Hauben, Rhondaeds (1997). *Netizens: On the History and Impact of Usenet and the Internet* (Los Alamitos, CA: IEEE Computer Society Press).

Kanu, Ahaoma (2007, August 13). NVS Members Donate $1200 to the Aikpitanhi Family, NTA Goes Haywire. Nigerian Village Square. http://www.nigeriavillagesquare .com.

Katz, Jon (1997). "Birth of a Digital Nation." *Wired* 5 (4): 190.

Kellner, Douglas (1998). "Intellectuals, the New Public Spheres, and Technopolitics." In C. Toulouse and T. W. Luke, *The Politics of Cyberspace: A New Political Science Reader* (New York: Routledge).

Koselleck, Reinhart (1988). *Critique and Crisis: Enlightenment and the Pathogenesis of Modern Society* (Cambridge, MA: MIT Press).

Kperogi, Farooq (2008). "Guerillas in Cyberia: The Transnational Alternative Online Journalism of the Nigerian Diasporic Public Sphere." *Journal of Global Mass Communication* 1 (1/2): 72–87.

Kperogi, Farooq (2011). Webs of Resistance: The Citizen Online Journalism of the Nigerian Digital Diaspora. Ph.D. Dissertation. Georgia State University. http:// scholarworks.gsu.edu/communication_diss/27.

Lipmann, Walter (1922). *Public Opinion* (New York: Free Press).

Lyotard, Jean-François (1984). *The Postmodern Condition*. Brian Massumi et al. trans. (Minneapolis: University of Minnesota Press).

McLaughlin, Neil (1999). "Origin Myths in the Social Sciences: Fromm, the Frankfurt School and the Emergence of Critical Theory." *Canadian Journal of Sociology* 24 (1): 109–39.

McLaughlin, Neil (1993). "Feminism, the Public Sphere, Media and Democracy." *Media Culture and Society* 15 (4): 599.

Meehan, Johanna (ed.) (1995). *Feminists Read Habermas: Gendering the Subject of Discourse* (New York: Routledge).

Negt, Oskar, and Kluge, Alexander (1993). *Public Sphere and Experience: Toward an Analysis of the Bourgeois and Proletarian Public Sphere*. Peter Labanyi et al. trans. (Minneapolis: University of Minnesota Press).

O'Donnell, Susan (2001). "Analyzing the Internet and the Public Sphere: The Case of Womenslink." *Javnost-The Public* 8 (1): 39–58.

Peters, John Durham (1993). "Distrust of Representation." *Media, Culture and Society* 15 (4): 564.

P.M. News (2007, August 8). "VP Jonathan Declare Assets." *P.M. News*.

Rheingold, Howard (1993). *The Virtual Community: Homesteading on the Electronic Frontier* (Massachusetts: Addison-Wesley Publishing Company).

Robins, Kevin, and Webster, Frank (1987). "Athens without Slaves . . . or Slaves without Athens? The Neurosis of Technology." In *Science as Culture* vol. 1 (London: Free Association Books).

Schudson, Michael (1997). "Why Conversation Is Not the Soul of Democracy." *Critical Studies in Mass Communication* 14 (4): 297–309.

Warner, Michael (2005). *Publics and Counterpublics* (New York: Zone Books).

16

When Bad Timing Is Actually Good: Reconceptualizing Response Delays

Stephanie A. Tikkanen and Andrew Frisbie

This chapter is about how time can communicate. Time—although often overlooked as a part of the communication process—plays an integral role in how we understand messages. Have you ever sent an e-mail to a friend or coworker and never received a reply? Where you left wondering whether your friend or coworker received the message? Similar situations are studied by communication researchers. The study of time and its role in the communication process is called "chronemics." Specifically, chronemics is the study of time in nonverbal communication.

Before discussing how chronemics vary in different settings, this chapter first examines how time is understood in general. When it comes to studying time there are two ways that humans understand time, "monochronic" and "polychronic" (also known as "M time" and "P time"). People that have a monochronic view of time like to do one (mono) thing at a time. They tend to focus on the task at hand and will also put an emphasis on punctuality. Conversely, people functioning with a polychronic view of time might focus on several different tasks and do so with ease. This is not to say that people on P time can handle multiple tasks better, but that they might have less rigid guidelines for completing tasks and place more emphasis on building relationships. Also, a P-time person might not find a deep personal commitment to punctuality. These differences typically are understood as cultural. Regardless of orientation, our understanding of time shapes the way we communicate with others and our understanding of how others communicate with us. Consequently, as technology increases the amount of mediated communication in which we engage every day, chronemics begin to play a powerful role in our interpersonal communication processes.

THE ROLE OF TECHNOLOGY IN CHRONEMIC RHYTHMS

The world is quickly becoming a global village. Vast amounts of information are sent from person to person, or from group to group, across long distances. Not only do technologies close the gap between people who are geographically separated, but new technologies also change the way in which we communicate. Social media (e.g., Facebook) and microblogging (i.e., Twitter) are reorganizing norms of communication.

One way to think about time is to think of communication rhythms. During face-to-face (FTF) conversations, patterns or rhythms emerge over the course of the discussion. Typically, FTF conversations are fluid exchanges of dialogue with short intervals of silence. Communication exchanges that are fairly immediate and responsive are considered *synchronous* dialogue. That is to say that two or more individuals have sustained a dialogue that is "in sync."

There also are *asynchronous* modes of communication; dialogue that is sporadic and littered with pauses. It might be easier to think of this type of communication as "punctuated equilibrium," or moments of fast dialogue separated by long durations of silence. Have you ever been texting a good friend and the conversation is quickly developing with messages sending one right after another? Maybe you're both texting so much that the conversation becomes difficult to track because each of you is commenting on multiple things at once. Then, all of a sudden, the texts stop. Maybe your friend went to class and couldn't text anymore or maybe they stepped into a meeting and had to turn their phone off. After about an hour goes by you get a reply and the messages start up right where they left off. Those moments of intense messaging and periods of silence are an example of punctuated equilibrium.

Just like notes on sheet music, each word that we send is surrounded by hosts of symbols that make up sentences (measures) and full conversations (a full song). Depending on the musicians or conversationalists, the rhythms created are different with each song or conversation. Unlike FTF conversations, when we communicate through text-based media (i.e., e-mail, mobile phone text messages, snail mail) there pauses between the message we send and the reply we get. Conversations develop different patterns that are created by both the individual users and the type of technology that is used. Much like music can evoke different emotions, patterns of communication over time can create different feelings that we associate with technology.

The way we communicate varies from medium to medium. Media Richness Theory (MRT) states that when it comes to transferring information not all media are equal.[1] Each medium (e.g., FTF, e-mail, phone call, text message) allows for information to be sent through a variety of cue channels (e.g., sight, sound, visuals). One such difference across media in MRT is the potential for immediate feedback, an essential component of asynchronous communication. There are certain text-based media that limit feedback, and thus are considered strictly asynchronous. Much like how MRT places media forms on a continuum of "richness" based on how many cue systems are available to the user, it's best to think of different types of technology ranging on a scale between really fast and really slow communication. The fastest type of communication happens face to face. Slower (but still quite fast) conversations take place via text message, Snapchat, and in online chat sites such as Match.com. Even slower communication takes place through e-mail, microblogs, and through mail delivered by a postal system. (Importantly, this scale is subjective; some individuals use e-mail in a way similar to text messages and reply very quickly, whereas others might be particularly slow at answering text messages.)

First, let's talk about technologically mediated communication that is generally considered to be fairly fast. Many people have cell phones and use those cell phones to connect with a variety of people each day. Text messaging has become a popular way for people to communicate because it allows brief intermittent exchanges without disrupting someone's schedule. Depending on the phone's network, most texts can be sent and read almost instantaneously. Because cell phone service is fast the authors consider mobile phone texting to be nearly synchronous. Texting enables friends, family, coworkers, and acquaintances to interact with very little delay. The unique thing about text messages, however, is that they can quickly become an asynchronous way of communicating. Mobile phone text messaging lets conversations oscillate from synchronic to asynchronic rhythms at any given moment. As mentioned, sometimes when people are deep in a thumb-tapping typhoon, conversations can go quiet without notice. Texting has the unique affordance for cell phone users to switch back and forth between instantaneous conversation and very sporadic, slow conversation. Texting is discussed further elsewhere in this chapter.

E-mail and microblogs are slightly slower than text messages because users might be using a desktop computer. E-mails can take on a slightly more formal style and require more care for proper grammar and syntax. Accessibility can play a significant role in contacting someone. If a person doesn't have access to a computer then it could take them awhile to respond, let alone know that you tried to contact them. E-mail conversations also can vary by the nature of the context. If you send an e-mail to your coworker it is likely that you will get a quicker response than what you would if you sent an e-mail to your senator. Your senator, or your senator's assistant, might reply, but it most likely won't be immediate or even within the same day. Typically, people with high status are expected to take more time to respond. Friends and peers with the same social status are expected to respond more quickly. One reason for the difference in expectations is the personal relationship that exists outside of the technology. Relationships with people help shape our understanding of a person's communication whether it is face to face or through text messaging. Over time, people in a relationship learn their friend's or partner's style of communication and dispel some of the initial uncertainty that surrounds text-based conversations. Interpersonal communication through a text-based channel can be just as rich, in some instances, and can be more personal than FTF communication.

According to many communication scholars,[2] one vital difference in FTF and mediated communication is the lack of other cues available for interpretation. Reader might ask, "So what? What's the big deal when the person you're talking with can't see your facial expressions and hear your tonal inflections?" The answer is that there is slightly more potential for miscommunication to happen when a person does not have all the signs they are used to having when interpreting another person. To try and compensate for the lack of some of those social cues, new media might incorporate visual aids to help the communication process flow smoothly. Many smartphones have "waiting bubbles" that appear when the person on the other end of a text message is typing a response. Some phones even include receipts to mark the time that a text was delivered and read. Smartphones are not the only devices that employ nonverbal cues; e-mail, Facebook, and some online dating sites (i.e., OK Cupid, Zoosk, Plenty of Fish) are starting to adopt extra cues to let communicators decipher what is happening. All of these cues help to reduce uncertainty in a conversation. Generally, the more information a person has in a conversation the more apt she or he is to understand what is being said.

Thus, text-based communication isn't necessarily a bad way to communicate. But what about the pauses that are inherent with text-based communication? Texting affords the ability to pause a conversation mid-sentence, if one so chooses. The conversational expectations of face-to-face communication do not allow interlocutors to postpone responses. In text conversations, adjusting the rate and speed of replying becomes a way to communicate in itself. Or does it? Is the timing between texts during a conversation a facet that creates meaning? The answer is not clear. How an individual interprets or perceives the behaviors of their partner in text-message conversations remains fairly understudied. What is known, however, is that response delays do indeed hold meaning for communicators, and often are perceived as inherently negative.

ACKNOWLEDGING "THE BAD": NEGATIVE PERCEPTIONS OF RESPONSE DELAYS

Imagine a couple that just went out on a blind date set up by a mutual friend. They met for the first time over lattes and muffins at a local coffee shop. While getting to know one another they exchanged phone numbers. After a couple of days had passed, Jordan, one of the people on the date, decided to ask Cameron out for a second date and sent a text to see if Cameron would like to see a movie. Hoping to hear back, Jordan waited to see what Cameron would say. As time passed, the suspense started to build and Jordan started to wonder. Jordan started to come up with reasons as to why Cameron wasn't responding and even thought about calling. As time passed the stories became more elaborate as Jordan tried to make sense of the silence.

The experience of waiting for a response can be agonizing. In FTF communication, a longer-than-average response time can be uncomfortable for both parties—the sender must mull over possible response options immediately in front of the recipient, who is left to wonder what is going on in the other person's mind. When communication is conducted through some mediated form (e.g., telephone, e-mail, Facebook, other social network site), the sender is freed from "normal" time constraints—but the recipient still is left wondering, only now the recipient lacks any nonverbal clues from the sender. This is perhaps the root of the majority of frustration stemming from time delays: as communicators, people often try to infer meaning from these delays, and commonly do so incorrectly.

Americans, as a monochronic society, value immediate replies. As technology advances to make individuals more accessible, our expectations of others' accessibility also increase.[3] Consequently, response delays—any difference between when a response was anticipated and when one was actually received—can hold a wide array of negative meanings for an expectant recipient.

Each person deals with uncertainty in different ways. Humans do not like cognitive dissonance (mental discomfort) and will go to great lengths to make sense of conversations. In monochronic cultures there is an emphasis on efficiency and immediacy. When those expectations of immediacy are not met, negative inferences can be made. Depending on the individuals involved and on the particulars of the situation, response delays can be interpreted as confusion, anger, carelessness, or indecision, among other things. Studies have shown that response delays hold implications for our perceptions of status differences,[4] credibility,[5] and personal perceptions,[6] such that longer responses indicated higher status but more negative perceptions. Simply put, the longer it takes for someone to get back to us, we often fear the worst. Consider examples such as a job applicant waiting for a decision, or a teenager confessing his feelings to his crush. In both cases, rarely does the waiting party view the delay as anything other than rejection.

This fear likely stems from the established link between uncertainty and anxiety.[7] When expectations for how something should occur are violated, often some level of uncertainty is experienced;[8] thus, when a response takes longer than a person believes it should, the sender is left to decipher the meaning of the pause. Consequently, this uncertainty leads anxiety. Tikkanen, Afifi, and Merrill[9] found that parents who were waiting for a response from their adolescent children experienced increased uncertainty during the delay, and those who perceived their child to potentially be at risk also experienced heightened anxiety. When parents were unsure of their child's safety, many developed negative explanations for the delay (e.g., that their child was in danger), which—unsurprisingly—led to this anxiety. Notably, however, those parents who did not perceive their child to be at risk did not experience any heightened anxiety levels.

Risk perception, however, is not the only possible negative implication of response delays. Individuals in other close relationships outside of the parent-child relationship also can experience anxiety or upset as a result of a partner taking longer than normal to respond to a contact effort. In these moments, many relational qualities and contextual cues can impact the recipient's attributions for the response delay. A sender who knows that the recipient is unavailable due to school or work obligations, for example, might not be alarmed by a slow response. Someone who can reasonably expect a response but does not receive one could experience a more negative reaction. Emotional contexts can impact these explanations, as well; an individual who is fighting with a significant other could attribute silence to anger more often than those who are not fighting.

Frisbie[10] explored the attributions that individuals in close relationships make when they do not receive responses within the expected time frame. He found that many people do not conjure up negative explanations for response delays, instead attributing them to unavailability. Immediate responses convey positive attitudes and are typically appreciated. One might think that long delays are interpreted negatively, but many participants showed that long delays might indicate consideration, appreciation, or preoccupation. In short, although response delays certainly can lead to anxiety or attributions of negative effect, the majority of individuals seem to accept or even ignore the meaning of these delays. Because this text focuses on both the good and the bad of mediated communication forms, it can be argued that—despite the potential for these negative outcomes—the asynchronicity of mediated communication also presents myriad benefits for both the sender and receiver.

FINDING "THE GOOD": MOVING TOWARD POSITIVE INTERPRETATIONS OF RESPONSE DELAYS

As discussed previously in this chapter, text-based communication such as text messages, e-mails, or messages sent via online forums span a continuum of synchronicity, with some being more synchronous than others. Although these time delays certainly can cause some uncertainty or anxiety, they also can prove the old adage "Good things come to those who wait." In terms of response time, there is conflicting research between FTF interactions and mobile-to-mobile interactions. On one hand, it is understood that immediate response to another individual is preferred, especially when potentially private or intimate information is shared.[11] In another study, however, respondents listed "large gaps in conversation" as a sign for a well-thought-out response.[12] Asynchronous channels let interlocutors pause between messages. Pausing gives time for interlocutors to reflect and construct messages as desired. Depending on the social currency available to an individual, channels that afford delays might be preferred for tough conversations that would not be easy to address in person.

It can be difficult for someone to form a good message when under pressure. People also often feel nervous being in front of a crowd or an important individual, which can make a person somewhat tongue-tied. In such situations, a bit of a time delay can be beneficial for all involved; taking time enables a speaker to send messages that fit into the image being portrayed, and the audience receives a more relevant, well-prepared, and thoughtful message. This is the "good" in asynchronicity; such often-dreaded delays frequently translate into richer, clearer, and more supportive communication. The following examines how response delays can be effective in four interpersonal communication contexts—impression management, self-disclosure, social support, and persuasion.

IMPRESSION MANAGEMENT

Early researchers of computer-mediated communication argued that it would never compare to FTF communication in terms of quality and interpersonal benefits. As time progressed and both technology and society's knowledge of technology advanced, however, people began to find that, in some cases, interpersonal communication actually was *better* on the computer. People had richer, more fulfilling relationships—which, of course, was entirely counterintuitive. How could someone have close, personal relationships with people they had never even met?

Joseph Walther, a noted communication researcher, developed a rationale to explain this phenomenon. His hyperpersonal model[13] explored the role of response delays in relationship development. He argued that when individuals are given time to cautiously craft messages, they are able to better manage the impressions they are sending to others. In other words, when we have time to think about what we're saying, we can come off as wittier, smarter, or even kinder online—depending on the image we wish to portray. Because computer-mediated communication is characterized by these delays, it is an environment rich in *impression management,* the act of carefully molding the impressions others form of us.

The model originally was crafted to explain how relationships between two strangers could develop so rapidly through strictly mediated communication, but also can be used to explain how CMC contributes to relationship satisfaction in established relationships, as well. The hyperpersonal model credits the amount of control over messages we have in computer-mediated communication with enabling us to create better impressions. When we are not face to face with someone, we decide what messages we want to send. If a person chooses to communicate via e-mail when nervous, for example, the audience will not hear a quivering voice or see shaky hands. Likewise, the audience will not be privy to unintentional facial expressions or reactions.

Although the lack of nonverbal cues is important, chronemics also plays a special role. The time delays inherent in mediated communication—such as those used to type a message—provide more control over the messages than when in a face-to-face, real-time communication. These delays, however long or short, give us a bit more control over how we choose to portray ourselves. They give us time to think when we are confused about how to respond, time to calm ourselves when we are angry or upset, and time to look up things we do not know so that we can sound informed when replying. The hyperpersonal model argues that when recipients receive only these carefully formed versions of ourselves, they form idealized images of us. (Importantly, note that these images are not inherently deceptive; they still are reflections of who we really are, just often more polished.) In turn, we form idealized images of our audience. Satisfaction in a relationship largely is based on our perceptions of our partner, thus idealized images contribute to interpersonal success. Consequently, the time delays that enable us to manage the impression that others have of us are important in developing successful relationships.

SELF-DISCLOSURE

A second implication of asynchronicity in mediated communication concerns a person's willingness and desire to share personal information, also known as *self-*disclosure. Self-disclosure is a key part of developing and maintaining healthy interpersonal relationships,[14] and for increasing liking[15] and trust.[16] Sharing things about ourselves also can be scary, however, because it can make us vulnerable to ridicule or rejection. In mediated communication, self-disclosure often is increased due to feelings of anonymity; Kim and Raja[17] found that in mediated settings individuals were more likely to engage in "face threatening behaviors," such as disclosure of highly personal information. Presumably, knowing that you will not meet the recipient of your secrets makes you more likely to disclose them.

Time delays in mediated communication can influence our willingness to disclose, as well. By distancing us from the immediate reactions of our audience, asynchronicity shields us from possible negative reactions until we are ready to face them. This buffer allows for what psychologist John Suler[18] refers to as an "emotional hit-and-run." Many gay and lesbian adolescents talk openly on YouTube about their experiences with coming out on sites such as Facebook. They state that changing their sexual preferences from opposite-sex to same-sex on the site enabled them to communicate to all of their family and friends en masse about their sexuality—simultaneously giving themselves some time prepare for the onslaught of responses, both positive and negative. This buffer also is helpful in a variety of other self-disclosure scenarios, such as offering social support.

SOCIAL SUPPORT

One context in particular where self-disclosure is affected by response delays is the online forum, in which individuals discuss sensitive issues in a sort of online support group. In this mediated context, individuals are empowered to seek support and advice from others who are in similar situations and who have relevant knowledge and experiences. Here, users are encouraged to share their stories and can emotionally benefit from the delay between their own statements and others' responses. Additionally, users can share their stories without interruptions, control what details to include, and hide any nonverbal cues such as crying or stammering. These forums offer a sort of cathartic, long-form space to share one's story, and also to receive supportive and informative replies. In these spaces, users are able to creatively express themselves and share sensitive information with minimal levels of embarrassment.[19]

The sender is not the only party who benefits from asynchronicity in this context, however. Because the messages in these forums often contain sensitive and emotional information, recipients could feel confused about how to respond in a supportive way. Often, when face to face with a person in distress, we do not know what to say or how to respond. As a result, we might offer poor support, or struggle to manage our own emotions in conjunction with the discloser's. Because messages online are meant to be delayed an immediate response is not required; this gives recipients time to process the message emotionally, and to determine the best way to respond.[20] This helps us to feel more competent and confident in our reply, and also ensures that the original sender receives the best possible support available. Additionally, Pfiel, Zaphiris, and Wilson[21] found that social support group users stated that asynchronicity allowed them to not only give more honest and frank responses, but also to check for misunderstandings before sending. Because the provision of social support often can be difficult—both informationally and emotionally—extra time to process can be invaluable.

PERSUASION

Lastly, asynchronicity can have persuasive advantages for users. Patrick O'Sullivan[22] argues in his Impression Management Model that individuals strategically choose the best medium to portray messages. Often this is used to manage the information given to a recipient and to uphold a certain impression that the sender is trying to make (much like in the hyperpersonal model), but O'Sullivan also notes that strategic media selection can be useful in conflict scenarios.

There can be many communication goals when people are engaged in conflict. Each might wish to provide their side of the story without being interrupted, for example, or want to make their emotional state exceedingly clear, or even might wish to ensure that those involved in the conflict have equal opportunity to defend themselves. Each of these goals is differentially facilitated by media; asynchronous media is especially useful when individuals wish to share information without interruption. In FTF communication interruptions are nearly inevitable. People talk over one another, or plan their next statement instead of actively listening to the speaker. As a result, messages sent in this format often can be misunderstood or cut off prematurely.

In contrast, messages sent via computer-mediated communications—particularly through less synchronous forms such as e-mail—often flow uninterrupted, and can be fully processed by the recipients. This heightens the persuasive power of these messages, allowing senders to fully express their position completely, and giving recipients time to process both the original message as well as their own response without having to do so simultaneously. This can lead to more productive conflict, as both sides can take time to fully express their opinions in addition to listening more carefully to each other.

Although it is important to acknowledge the power of asynchronicity in developing strong verbal messages, the delay itself also can send an important—and persuasive—nonverbal message. Much like teachers use uncomfortable silences to compel reluctant students to speak up in class, silence can be a powerfully persuasive tool in other situations. Johannesen[23] speaks to the utility of silence in a variety of interpersonal functions. He notes that in political communication voters most often are swayed by a politician's silence on an issue; silence prompts listeners to interpret meaning themselves. As such, the "silence" of a delayed response can serve a persuasive function in and of itself; a sender who does not receive a response might reword the message in a more favorable way, or interpret the silence as a rejection of ideas and begin to negotiate meaning—regardless of whether the original recipient intended for this to occur.

NEGOTIATING THE GOOD AND THE BAD: CHANGING HOW WE INTERPRET RESPONSE DELAYS

Given the examples outlined in his chapter, the perception that response delays are inherently negative is flawed. Certainly, a longer-than-expected delay has the potential to cause anxiety, especially when one is concerned about the safety of the sender. Immediate responses are preferred most of the time, and studies have shown that immediate and frequent responses are signs of relational quality. That means that simply sending a reply quickly tells the other person that they are important. Conversely, long pauses do not necessarily indicate negative feelings with regard to the relationship. Although long pauses might cause some uncertainty, the person waiting for a reply does not always automatically think that a delay is negative. Pauses become part of the conversation between people who are texting, but texters have not been found to use pauses as a way of communicating. Long durations of silence actually can indicate that the receiver is thinking about what was said, and is taking time to develop a response. Thoughtfulness is a process that can take time, and—to a certain point—taking time to think about a response might be appreciated by the other person.

Response delays serve several important interpersonal functions, such as more liberated and idealized self-expression as well as improved listening and support provision. As technology advances and gives users greater freedom in when they choose to communicate, a better understanding of how asynchronicity affects communicative processes can hopefully improve the reputation of response delays.

The root of most negative perceptions surrounding response delays stems from a violation of expectations: A response took longer than expected, and the sender then is left to interpret the meaning of that delay. Just as a response that is quicker than expected can imply positive feelings (e.g., eagerness, excitement, expertise), these delays often are taken to mean the opposite (e.g., apathy, reluctance, lack of knowledge). Rarely, however, are other factors considered. Delays can mean thoughtfulness, caution, or simply having a phone set to silent. The solution, then, is simple: Manage expectations.

Tikkanen and colleagues[24] suggest that parents have discussions with teens in advance of outings to set rules and guidelines surrounding phone use. By making expectations explicit, teens are less likely to violate the rules. If parents expect a call back within five minutes of a missed contact effort, for example, then teens know they should be more diligent in checking their phone.[25] Similarly, setting expectations for a response can help reduce the anxiety associated with waiting. Employers can provide a time line for a job search so that prospective employees can set reasonable expectations for expecting a reply. This does not even need to be an explicit conversation; expectations can be managed technologically, by setting an automated out-of-office reply or by not turning off read receipts if one knows they cannot immediately reply. By managing expectations and explaining reasons for response delays, some of the negative associations and interpretations can be reduced or avoided.

Though the asynchronous nature of mediated communication has long been touted as a reason for its inferiority to FTF communication, it is hoped that this chapter has persuaded readers otherwise. Not only can delayed responses be viewed just as positively as negative responses, but the delays intrinsic to asynchronous communication actually could make it better than more synchronous forms. As mediated communication grows in popularity, there are just as many opportunities to create meaningful relationships with people on the other end of the conversation. As technologies emerge, with them come profoundly original ways to communicate with one another. In many ways, the medium does not limit the capacity to share—it simply alters it.

NOTES

1. Richard Daft and Robert Lengel, "Organizational Information Requirements, Media Richness and Structural Design," *Management Science* 32 (5) (1986): 555.

2. Ibid., 554; Sara Kielser, Jonathan Siegel, and Timothy McGuire, "Social Psychological Aspects of Computer-mediated Communication," *American Psychologist* 39 (1984): 1123; Lee Sproull and Sara Kiesler, "Reducing Social Context Cues: Electronic Mail in Organizational Communication," *Management Science* 32 (1986): 1492.

3. Naomi Baron, *Always On: Language in an Online and Mobile World* (Oxford University Press, 2008).

4. Nicola Döring and Sandra Pöschl, "Nonverbal Cues in Mobile Phone Text Messages: The Effects of Chronemics and Proxemics," in *The Reconstruction of Space and Time*, edited by Rich Ling and Scott Campbell (New Brunswick, NJ: Transaction Publishers, 2008): 109; Joseph Walther and Lisa Tidwell, "Nonverbal Cues in Computer-mediated Communication, and the Effect of Chronemics on Relational Communication," *Journal of Organizational Computing* 5 (1995): 355.

5. Yorman Kalman and Sheizaf Rafaeli, "Chronemic Nonverbal Expectancy Violations in Written Computer-mediated Communication" (presentation, Annual Convention of the International Communication Association, Montreal, Canada, 2008).

6. Oliver Sheldon, Melissa Thomas-Hunt, and Chad Proell, "When Timeliness Matters: The Effect of Status on Reactions to Perceived Time Delay within Distributed Collaboration," *Journal of Applied Psychology* 91 (2006): 1385.

7. Maria Miceli and Cristiano Castelfranchi, "Anxiety As an 'Epistemic' Emotion: An Uncertainty Theory of Anxiety," *Anxiety, Stress, and Coping* 18 (2005): 291.

8. Kathy Kellerman and Rodney Reynolds, "When Ignorance Is Bliss: The Role of Motivation to Reduce Uncertainty in Uncertainty Reduction Theory," *Human Communication Research* 17 (1990): 5.

9. Stephanie Tikkanen, Walid Afifi, and Anne Merrill, "Gr8 Textpectations: Parents' Experiences of Anxiety in Response to Adolescent Mobile Phone Delays," in *Family Communication in an Age of Digital and Social Media*, edited by Carol Bruess (New York: Peter Lang International, 2015).

10. Andrew Frisbie, "Mobile Phone Text Messaging: Implications of Response Time within an Asynchronous Medium" (presentation, Annual Convention of the Organization for the Study of Communication, Language, and Gender, Houghton, MI, October 10–13, 2013).

11. Sandra Petronio, *Boundaries of Privacy: Dialectics of Disclosure* (Albany: State University of New York Press, 2002).

12. Dominick Madell and Stephen Muncer, "Control over Social Interactions: An Important Reason for Young People's Use of the Internet and Mobile Phones for Communication?" *CyberPsychology and Behavior* 10 (2007): 137.

13. Joseph Walther, "Computer-Mediated Communication: Impersonal, Interpersonal, and Hyperpersonal Interaction," *Communication Research* 23 (1996): 20; Joseph Walther, "Group and Interpersonal Effects in International Computer-Mediated Collaboration," *Human Communication Research* 23 (1997): 342.

14. Irwin Altman and Dalmas Taylor, *Social Penetration: The Development of Interpersonal Relationships* (New York: Holt, Rinehart and Winston, 1973).

15. Nancy Collins and Lynn Miller, "Self-Disclosure and Liking: A Meta-Analytic Review," *Psychological Bulletin* 116 (1994): 457.

16. Robert Larzelere and Ted Huston, "The Dyadic Trust Scale: Toward Understanding Interpersonal Trust in Close Relationships," *Journal of Marriage and Family* 42 (1980): 595.

17. Min-Sun Kim and Narayan Raja, "Verbal Aggression and Self-disclosure on Computer Bulletin Boards" (presentation, Annual Conference of the International Communication Association, Chicago, IL, May 1991).

18. John Suler, "The Online Disinhibition Effect," *CyberPsychology & Behavior* 7 (2004): 321.

19. Dawn Braithwaite, Vincent Waldron, and Jerry Finn, "Communication of Social Support in Computer-mediated Groups for People with Disabilities," *Health Communication* 11 (1999): 123.

20. Marsha White and Steve Dorman, "Receiving Social Support Online: Implications for Health Education," *Health Education Research* 16 (6) (2001): 693.

21. Ulrik Pfeil, Panayiotis Zaphiris, and Stephanie Wilson, "Older Adults' Perceptions and Experiences of Online Social Support," *Interacting with Computers* 21 (2009): 159.

22. Patrick O'Sullivan, "What You Don't Know Won't Hurt Me: Impression Management Functions of Communication Channels in Relationships," *Human Communication Research* 26 (2000): 403.

23. Ronald Johannesen, "The Functions of Silence: A Plea for Communication Research," *Western Journal of Communication* 38 (1974): 25.

24. Stephanie Tikkanen, Walid Afifi, and Anne Merrill, "Gr8 Textpectations: Parents' Experiences of Anxiety in Response to Adolescent Mobile Phone Delays," in *Family Communication in an Age of Digital and Social Media*, edited by Carol Bruess (New York: Peter Lang International, 2015).

25. Alyssa Bereznak, "Mom-Made App Allows Parents to Lock Their Kids' Phones Until They Call Back," *Yahoo! Tech*, accessed August 18, 2014, https://www.yahoo.com/tech/mom-made-app-allows-parents-to-lock-their-kids-phones-95121969584.html. In a more extreme version of parental monitoring, mother Sharon Standifird created an app called "Ignore-No-More" in which parents can remotely lock their child's phone until the child calls them for the password, forcing the child to return the parent's calls.

BIBLIOGRAPHY

Altman, Irwin, and Dalmas Taylor. *Social Penetration: The Development of Interpersonal Relationships* (New York: Holt, Rinehart and Winston, 1973).

Baron, Naomi. *Always On: Language in an Online and Mobile World* (Oxford University Press, 2008).

Bereznak, Alyssa. "Mom-Made App Allows Parents to Lock Their Kids' Phones Until They Call Back." *Yahoo! Tech*. Accessed August 18, 2014. https://www.yahoo.com/tech/mom-made-app-allows-parents-to-lock-their-kids-phones-95121969584.html.

Braithwaite, Dawn. O., Vincent Waldron, and Jerry Finn. "Communication of Social Support in Computer-Mediated Groups for People with Disabilities." *Health Communication* 11 (1999): 123–51.

Collins, Nancy, and Lynn Miller. "Self-Disclosure and Liking: A Meta-analytic Review." *Psychological Bulletin* 116 (1994): 457–75.

Daft, Richard, and Robert Lengel. "Organizational Information Requirements, Media Richness and Structural Design." *Management Science* 32 (5) (1986): 554–71.

Döring, Nicola, and Sandra Pöschl. "Nonverbal Cues in Mobile Phone Text Messages: The Effects of Chronemics and Proxemics." In *The Reconstruction of Space and Time*, edited by Rich Ling and Scott Campbell, 109–36 (New Brunswick, NJ: Transaction Publishers, 2008).

Frisbie, Andrew. "Mobile Phone Text Messaging: Implications of Response Time within an Asynchronous Medium." Presentation at the Annual Convention of the Organization for the Study of Communication, Language, and Gender, Houghton, MI, October 10–13, 2013.

Johannesen, Ronald, L. "The Functions of Silence: A Plea for Communication Research." *Western Journal of Communication* 38 (1974): 25–35.

Kalman, Yorman, and Sheizaf Rafaeli. "Chronemic Nonverbal Expectancy Violations in Written Computer-Mediated Communication." Presentation at the Annual Convention of the International Communication Association, Montreal, Canada, 2008.

Kellerman, Kathy, and Rodney Reynolds. "When Ignorance Is Bliss: The Role of Motivation to Reduce Uncertainty in Uncertainty Reduction Theory." *Human Communication Research* 17 (1990): 5–75.

Kielser, Sara, Jonathan Siegel, and Timothy McGuire. "Social Psychological Aspects of Computer-Mediated Communication." *American Psychologist* 39 (1984): 1123–34.

Kim, Min-Sun, and Narayan Raja. "Verbal Aggression and Self-Disclosure on Computer Bulletin Boards." Presentation at the Annual Conference of the International Communication Association, Chicago, IL, May 1991.

Larzelere, Robert, and Ted Huston. "The Dyadic Trust Scale: Toward Understanding Interpersonal Trust in Close Relationships." *Journal of Marriage and Family* 42 (1980): 595–604.

Madell, Dominick, and Stephen J. Muncer. "Control over Social Interactions: An Important Reason for Young People's Use of the Internet and Mobile Phones for Communication?" *CyberPsychology and Behavior* 10 (2007): 137–40.

Miceli, Maria, and Cristiano Castelfranchi. "Anxiety As an 'Epistemic' Emotion: An Uncertainty Theory of Anxiety." *Anxiety, Stress, and Coping* 18 (2005): 291–319.

O'Sullivan, Patrick, B. "What You Don't Know Won't Hurt Me: Impression Management Functions of Communication Channels in Relationships." *Human Communication Research* 26 (2000): 403–31.

Petronio, Sandra. *Boundaries of Privacy: Dialectics of Disclosure* (Albany: State University of New York Press, 2002).

Pfeil, Ulrik, Panayiotis Zaphiris, and Stephanie Wilson. "Older Adults' Perceptions and Experiences of Online Social Support." *Interacting with Computers* 21 (2009): 159–72.

Sheldon, Oliver, Melissa Thomas-Hunt, and Chad Proell. "When Timeliness Matters: The Effect of Status on Reactions to Perceived Time Delay within Distributed Collaboration." *Journal of Applied Psychology* 91 (2006): 1385–95.

Sproull, Lee, and Sara Kiesler. "Reducing Social Context Cues: Electronic Mail in Organizational Communication." *Management Science* 32 (1986): 1492–512.

Suler, John. "The Online Disinhibition Effect." *CyberPsychology & Behavior* 7 (2004): 321–26.

Tikkanen, Stephanie, Walid Afifi, and Anne Merrill. "Gr8 Textpectations: Parents' Experiences of Anxiety in Response to Adolescent Mobile Phone Delays." In *Family Communication in an Age of Digital and Social Media*. Edited by Carol Bruess (New York: Peter Lang International. 2015).

Walther, Joseph. "Group and Interpersonal Effects in International Computer-Mediated Collaboration." *Human Communication Research* 23 (1997): 342–369.

Walther, Joseph. "Computer-Mediated Communication: Impersonal, Interpersonal, and Hyperpersonal Interaction." *Communication Research* 23 (1996): 1–43.

Walther, Joseph, and Lisa Tidwell. "Nonverbal Cues in Computer-Mediated Communication, and the Effect of Chronemics on Relational Communication." *Journal of Organizational Computing* 5 (1995): 355–78.

White, Marsha, and Steve Dorman. "Receiving Social Support Online: Implications for Health Education." *Health Education Research* 16 (6) (2001): 693–707.

In Defense of "Slacktivism": How KONY 2012 Got the Whole World to Watch[1]

Christopher Boulton

In March 2012, the Internet video KONY 2012 swept across Facebook and Twitter racking up more than 100 million views in just six days,[2] making it the most viral video in history.[3] KONY 2012's stated intent was to draw attention to how Joseph Kony's Lord's Resistance Army (LRA) abducts, abuses, and forces children to fight as soldiers in and around Uganda; and to inspire young activists to pressure celebrities and politicians to do whatever it takes to stop Kony.[4] The backlash was swift; critics objected to the facts, foreign policy agenda, and racial politics of the video as well as to the financial priorities of Invisible Children (IC), the organization behind the video.[5] Many of these objections were well founded. The video oversimplified and exaggerated Kony's power (comparing Kony to Hitler when Kony had only a few hundred followers), called for a United States–led military intervention into an oil-rich country, and cast young, mostly white teenagers as Africa's saviors (echoing neocolonial notions).[6] International Children also came under scrutiny for spending more money on making and showing movies than actually directing aid to the affected region.[7]

Other critics questioned the approach of the KONY 2012 campaign, dismissing the sharing of a video on social media as the epitome of "slacktivism,"[8] a term combining the lazy connotations of "slacker" with "activist" to convey how the Internet makes political expression more convenient or, as Snopes founder Barbara Mikkelson first put it to The New York Times back in 2002, "the desire people have to do something good without getting out of their chair."[9] Thus, although this author concedes that KONY 2012 made mistakes—some of them ugly—this chapter argues that the film also did something much more important by turning suburban teens into slacktivists[10]—it made human rights "cool."

Malcolm Gladwell most certainly would beg to differ. Writing in The *New Yorker* two years before the release of *KONY 2012*, Gladwell quipped that "the revolution will not be tweeted" and cited the 1960's lunch counter sit-ins of the Civil Rights Movement as the kind of high-commitment protests necessary for social change.[11] Although various good causes might use Facebook as a platform to increase online participation, Gladwell argues they can only do so by "lessening the level of motivation that participation requires."[12] In other words, it is easy to get a slacktivist to click the "like" button, but it's more difficult to get them to actually show up at a protest occurring offline, much less be willing to accept personal injury or arrest. Indeed, strategic, disciplined, and hierarchical organizations such as the National Association for the Advancement of Colored People (NAACP) and the Congress of Racial Equality (CORE) trained Civil Rights activists ahead of time to expect and even provoke such risks, and to conduct their civil disobedience nonviolently. Though technology enthusiasts were quick to characterize more recent protests in Moldova and Iran as "Twitter revolutions," Gladwell points out that social media did not drive or even organize the protests so much as provide a platform for Westerners to discuss them.[13] Thus, the crux of the "slacktivist" critique: All those retweeted hashtags might have raised awareness, but did they actually change anything on the ground? Does this kind of attention create international solidarity or global rubbernecking? The answer, of course, is a bit of both. History suggests that watching in itself matters, however, even if the audience never gets off the couch.

In August 1968, thousands of students gathered in a park outside of the Democratic National Convention in Chicago to protest the Vietnam War. On the eve of the convention, police invaded the park with fire trucks, launching teargas canisters and swinging batons. The next day, Don Rose, press secretary for the National Mobilization Committee to End the War in Vietnam (MOBE), was asked how to respond to the police brutality and said, "Tell them the whole world is watching, and they'll never get away with it again."[14] Two days later, during what would later become known as "The Battle of Michigan Avenue," demonstrators gathered in front of the Conrad Hilton Hotel, where many of the convention delegates and news media were staying. Police moved in to clear the street, beating and arresting demonstrators and onlookers alike. As the television news cameras rolled, the protestors chanted, "The whole world is watching! The whole world is watching! The whole world is watching!" And they were right. Nearly 83 million Americans saw the horrifying events unfold on their TV screens, and the images quickly traveled around the globe.[15]

Even if the demonstrators could not get the convention to adopt a peace platform, having the whole world watch helped them to publicly shame the Democratic Party and Chicago city officials. Similarly, civil rights protestors knew that nonviolent civil disobedience in the South alone might not be enough to persuade racist shop owners to desegregate their lunch counters, but it would draw the inherent violence of Jim Crow apartheid, typically enforced in the shadows, out into the light for the cameras—and thus people all around the country at home on their couches—to see.[16]

I understand, on a very intuitive level, why KONY 2012 went so viral so fast. I think there are two central reasons for this—inspiration and aspiration. When I was in high school and college, for instance, I began to learn about injustice in the world and longed to be part of a good cause. I often wished I had been alive in the 1960s so I could have joined the antiwar or civil rights movements. They seemed, in retrospect, to be so clear, pressing, and righteous, culminating in bright shining moments like the March on Washington in 1964 or the National Moratorium Antiwar Demonstrations of 1970.[17] These were events full of courageous, young, antiestablishment, and presumably cool people just like me, and I longed to have been part of the scene. I aspired to be like the charismatic leaders and was inspired by their mission to change the world. Where could I find my movement in the 1990s? Who could I follow? And could they connect me with like-minded people? Where was my Woodstock? I looked to my idols, rock bands of course, for guidance—U2 introduced me to Amnesty International, a human rights organization that writes letters requesting the release of political prisoners. REM got me interested in Greenpeace, a radical environmental organization that uses direct action to confront polluters.

The bands' endorsements of these causes were very important to me; they made the issues seem "cool" and, most importantly for an insecure adolescent, popular. So, when I organized the Amnesty International club at my high school, I was trying to change the world to be sure, but was also hoping to make friends with other people like me . . . and meet cute girls who would think that I was cool, and popular, too. In short, if KONY 2012 and the Internet had been around at the time, I'm sure I would have jumped on the bandwagon as another slacktivist ready to share the latest—and coolest—cause with everyone I hoped to impress. This chapter examines how KONY 2012 used inspiration, aspiration, role models, and the lure of friendship to make it the type of cause that so many wanted to spread, and defends slacktivism as an online reboot of the cherished protest march tradition.

KONY 2012 is inspiring. Narrator Jason Russell raises the stakes in the introduction by making historic—even revolutionary—claims. The opening sequence depicts planet Earth as seen from outer space and then, over poignant moments from YouTube and the Arab Spring, Russell declares that social media "is changing the way the world works" such that "governments are trying to keep up," "the older generations are concerned," and "the game has new rules."[18] In this way, Russell winks at his target audience of young Facebook users and challenges them to seize their destiny. Russell then tells a very simple story of good versus evil, casting three characters in starring roles: Joseph Kony, warlord and kidnapper, as the villain; Jacob Acaye, escaped child soldier, as the victim; and himself, along with his audience, as the hero that saves the day. Russell shows us his own son's birth as a symbol of the universal value of human life then introduces us to "another boy" who would change his life entirely: Jacob, his friend from Uganda whom he had met 10 years earlier.[19] We then see a clip of Jacob crying about seeing his brother killed and Russell promising Jacob that he will stop Kony. Russell, back in his narrator role, then invites his audience to join the crusade "because that promise is not just about Jacob or me, it's also about you. And this year, 2012, is the year that we can finally fulfill it."[20] With the main characters clearly established, the introduction closes by boiling down a very complex situation into a simple call to action: Stop Joseph Kony and "change the course of human history."[21] This goal is presented as both feasible and urgent, with the added bonus of narrative closure because the "movie expires on December 31, 2012."[22] Thus, like a summer blockbuster action flick, the fate of the planet hangs in the balance and the hero must avenge the victim by defeating his nemesis once and for all.

Russell, who once described IC as "the Pixar of human rights stories,"[23] is tapping into a long, successful narrative tradition in Hollywood, described by Matthew Hughey in his book *The White Savior Film*.[24] In this genre, the (white) hero acts as a bridge character for white audiences—entering a hostile territory populated by people of color, making a sacrificial rescue, and completing a journey of self-discovery. In *Cry Freedom*, Denzel Washington plays Stephen Biko, the slain antiapartheid activist, but the film is really about Kevin Kline's white journalist who befriends Biko and risks his family's life to liberate Black South Africans from white rule. Willem Dafoe's FBI agent rescues blacks from the KKK in *Mississippi Burning*. Kevin Costner goes native and sacrifices himself for the Sioux in *Dances with Wolves*. Tom Cruise goes to Japan, learns martial arts, joins the rebellion, defeats the ninjas, and ends up as *The Last Samurai* standing. Clint Eastwood defends his Hmong neighbors in *Gran Torino*. *Avatar*, much like *Dances with Wolves*, features a white protagonist who goes native, sacrifices himself for the Navi nation and, to top it all off, gets reincarnated. Sandra Bullock literally picks up a football star on the side of the road in *The Blind Side*. *The Help* stars a white woman who discovers the stories of black maids, publishes a book and goes to New York while the maids stay home. Other examples include *Lawrence of Arabia*, *Glory*, *Dangerous Minds*, *Amistad*, *Blood Diamond*, and *Cool Runnings*.[25] The point here is that Russell's audience has already been trained to identify with bridge characters, so casting himself (and his son) in such prominent roles surely helped inspire his white Western audience to embrace the film and its mission to save other African children like Russell's friend Jacob.

KONY 2012 is aspirational. The video targeted—and was, in turn, endorsed and retweeted by—famous celebrity role models such as Oprah Winfrey, Justin Bieber, Angelina Jolie, Bill Gates, Rihanna, Ryan Seacrest, Nicole Richie, Diddy, and the Kardashian sisters.[26] This exposure helped distribute the video to a wider audience while building its credibility as a popular cause. In addition to fame, the campaign embraced rebellion—another aspirational value of youth culture—by co-opting the street art tactics portrayed in *Exit through the Gift Shop*, a documentary nominated for a 2010 Academy Award.[27] Ostensibly directed by the anonymous and widely celebrated street-artist Banksy, *Exit through the Gift Shop*'s opening credit sequence presents a montage of young people—often wearing hoodies or kerchiefs over their faces—sneaking around at night to glue up posters, spray on stencils, and paint graffiti onto buildings, bridges, tunnels, traffic signs, and other urban public spaces. Most of the sequence occurs after dark—echoing the song's refrain that "[t]onight, the streets are ours"—and culminating in a young man escaping two pursuing police officers by scaling a wall and disappearing into the shadows.[28] As the narrator explains, this footage presents a behind-the-scenes perspective on "an explosive new movement that would become known as street art," a "hybrid form of graffiti . . . driven by a new generation using stickers, stencils, posters, and sculptures to make their mark by any means necessary."[29] Although the earliest street art largely was local and ephemeral, "with the arrival of the [I]nternet, these once temporary works could be shared by an audience of millions; street art was poised to become the biggest countercultural movement since punk."[30] As an example of one of the first artists to cross over into the mainstream, *Exit through the Gift Shop* introduces Shepard Fairey.

> One day, Shepard would be famous for transforming the face of an unknown senator [Barack Obama] into a universally recognized icon. But, even back in 2000, Shepard was the world' most prolific street artist. Shepherd's experiment with the power of repetition went back to 1989 and an image based on 1970s wrestler Andre the Giant. Combining Andre's face with the command to "Obey," Shepherd had already clocked out over a million hits around the world.[31]

Fairey then explains how the "Obey" campaign slowly gained influence over time and space.

> Even though the Andre the Giant sticker was just an inside joke and I was just having fun, I liked the idea of the more stickers that are out there, the more important it seems, the more people want to know what it is, the more they ask each other and it gains real power from perceived power.[32]

Toward the end of *KONY 2012*, Russell adapts Fairey's tactics when he poses the problem that IC's version of street art will solve: Kony is invisible, so the best way to stop him is to make his image ubiquitous so Americans will care and subsequently pressure the U.S. government to continue to support local efforts to locate and arrest him. Russell predicts that this DIY negative publicity campaign will make Kony "world news" by "redefining the propaganda we see all day, every day, that dictates who and what we pay attention to."[33]

The next sequence strings together images of billboards and magazine ads for with a media consumer watching television to set up the video's counter-cultural alternative to passivity, apathy, and conformity, as explained by street art's patron saint. Again, Shepard Fairey:

A lot of people feel powerless to communicate their ideas. They think that, "Okay, I'm not a corporation. I don't own my own magazine or news station. I just don't have any say." But seeing what I've done, I think it's empowered a lot of people to realize that one individual can make an impact. I actually want to demystify and say, "Here are these really simple tools. Go out and rock it!"[34]

Russell immediately follows Fairey's injunction to "rock it" by promising "and that's just what we intend to do" as the video soars over a crowd of young people in red *KONY 2012* T-shirts raising their arms in a synchronized peace symbol salute. As the soundtrack blasts "I can't stop," and another group of teens runs through a parking garage, Russell explains that IC is prepared to distribute "hundreds of thousands of posters, stickers, yard signs, and flyers" with Kony's image so that "all of these efforts will culminate on one day, April 20th, when we cover the night."[35] He promises that young people from all over the world will "meet at sundown and blanket every street in every city till the sun comes up" so that, the next day, everyone will "wake up to hundreds of thousands of posters demanding justice on every corner."[36] This is a big, bold claim, but the imagery backs him up. In a sequence reminiscent of the opening of *Exit through the Gift Shop*, *KONY 2012* pre-enacts[37] what the night of April 20th will look like with a rapid-fire montage of young people running around at night, putting up posters, unfurling banners, throwing flyers off a bridge, dashing across bridges, and ducking through road flair–lit tunnels. Most of these shots are handheld, and thus connote the rough authenticity of amateur camcorder footage. Although there are no depictions of direct confrontations with police, the sense of deviance and transgression through petty vandalism is palpable, especially when a young man stands in front of a mirror and menacingly pulls a kerchief up over his nose.

Which leads to the third aspirational aspect of KONY 2012: Showing its viewers a future vision of themselves. I came to this idea by way of two previous research projects; the first looked at how print advertisements for designer children's clothing so often pose the young models as very serious "adults," staring straight into the camera. I argued that this offered mothers—the ostensible target of the ad—a foretaste of their children's future success.[38] The second project extended this theory through audience research and found that mothers tended to view the child models in the ads as their own, then imagine how other mothers might judge the children's clothing and, by extension, their own performance as mothers. In other words, they donned virtual goggles to audition the gaze of other mothers in a kind of vicarious dress rehearsal.[39] Therefore, I would suggest that, by pre-enacting "Cover the Night," Kony 2012 made a future promise through a vicarious gaze—this is, what you will look like through others' eyes: running around a big city at night with a bunch of your friends dressed up in cool clothes and pasting posters on walls, lighting up tunnels with flares, and dodging police.

Crucially, in addition to inspiration and aspiration, KONY 2012 got "the whole world to watch" by promoting the formation of friendships. The video explains that the "action kit" includes two bracelets ("one to keep and one to give away") with unique ID numbers that participants can input online to "join the mission to make Kony famous" and "geotag your posters and track your impact in real time."[40] In support of this idea, the video shows photographs of participants posing with their posters, selfie-style, to imply that the pictures could be posted and shared online with hashtags so as to join and connect with others putting up posters in their part of the world. The video is full of young people congregating together in groups. Whether assembled in tight formation wearing matching KONY 2012 merchandise or running around outdoors and having the time of their lives, the youth are shown in relationship with each other having fun in public—which is both inspirational and aspirational.

It is ironic that this very active vision of hundreds of young participants taking to the streets was passively watched by millions more who stayed indoors. Yet, at the same time, this is the very beauty of slacktivism—its ability to use the Internet to connect causes to marginalized supporters. In this way, friendships can be formed across divides of relative distance and perhaps interests. Even in the heyday of the anti–Vietnam War movement, there were plenty of people who showed up at the demonstration to get drugs, meet the opposite sex, or just watch; for a time, it was trendy, stylish, and even aspirational to be against the war. Some mimicked the opinions of their rock-star role models. Others were radicalized by an inspiring speaker. Still others just wanted to "Tune In, Turn On, and Drop Out"—KNOY's tagline. KONY 2012 dipped into the IC archives to compile a highlight reel of past events to persuade its young viewers to "cover the night" with an updated, decentralized protest march united online by Instagram. Even if someone only watched—and did nothing else—they still got the message loud and clear: the human rights scene is a cool place to be and to be seen.

KONY 2012's online success can be traced back to the groundwork that IC already had laid over the previous eight years. First formed in 2004 after the completion of *Invisible Children: Rough Cut*, IC debuted the film on a national tour to high schools and college campuses in 2006 and then organized the Global Night Commute, with an estimated 80,000 people in 130 American cities walking to their city centers and sleeping outdoors to show support for young Ugandan "Night Walkers" who do the same to avoid capture by the LRA.[41] In 2009, IC organized "The Rescue," a 100-city international event across 10 countries where participants "abducted" themselves and politicians and celebrities such as Representative John Lewis and Oprah Winfrey came to their rescue.[42]

The group's first legislative victory came in 2010, when President Obama deployed 100 U.S. advisers through the LRA Disarmament and Northern Uganda Recovery Act. Just a month after the launch of *KONY 2012*, Obama extended the mission. That November, IC hosted "Move DC" with thousands of volunteers joining actors from the television shows *Glee* and *Breaking Bad* to march to the White House and demand an end to LRA violence. The following January, Congress authorized $5 million for information leading to Kony's capture. That summer U.S. ambassador to the United Nations, Samantha Power, spoke at IC's Fourth Estate Summit. All in all, over the course of 10 years, IC's numbers are impressive: $32 million dollars raised; 5 million students educated about the LRA through 13,000 screenings of 12 different films; 400,000 activists attending 8 international awareness events; 3,000 in-person congressional lobby meetings; and the passage of two bipartisan bills through the U.S. Congress.[43] Although far more people might have watched versus acted, things still got done. As promised, the *KONY 2012* campaign did expire, but the ending was less than happy. "Cover the Night," the campaign to plaster KONY posters all over the world (rendered so compelling and exciting by KONY 2012's dramatic pre-enactments) was a dud.[44] Additionally, as of this writing, Joseph Kony is still at large and IC plans to shutter its doors by the end of 2015.[45]

Despite its mixed results, IC's *Kony 2012* remains a compelling case study of the promise and limitations of slacktivism. If we measure the video's worth by its ability to recruit participants for its signature culminating global event and its promise of closure through the timely capture of a notorious war criminal, then it failed. However, it also is true that this video—only the latest of many previous versions—was only able to intervene in the global conversation when it leveraged IC's existing activist network through online social media platforms. This got the ball rolling, but then the video's popularity with so many of those outside of IC's immediate network spiked the viewership and forced the issue to the top of the world's headlines.

To be sure, Jason Russell is no Martin Luther King, but his "white savior" bridge character still managed to mobilize tens of millions of young, largely white, Americans—typically thought to be too self-involved to care about world affairs[46]—to consume and spread a story about the plight of black children in a far-off land. Unlike so many activist documentaries, *KONY 2012* did not only preach to the converted. Rather, it managed to capture the attention of mostly apolitical teenagers, that most elusive of demographics, by inspiring them with a generationally specific and grandiose tone and by reframing and simplifying a complex issue as urgent, timely, and righteous by casting clear villains, victims, and heroes. It also used tactics that tapped into young aspirations; associating famous celebrities with the cause added glamour and social acceptability, co-opting the rebellious cache of street art made earnest activism more transgressive, and showing young people future visions of themselves made fighting the good fight look a lot like having fun with friends.

In the fall of 2012, a picture of president-elect Barack Obama with the caption "Four more years" became the most popular tweet in the history of Twitter. That record held for almost two years until Ellen DeGeneres snapped a selfie onstage at the Oscars with some fellow celebrities (Jennifer Lawrence, Channing Tatum, Meryl Streep, Julia Roberts, Kevin Spacey, Bradley Cooper, Brad Pitt, Lupita Nyong'o, and Angelina Jolie). DeGeneres' tweet smashed Obama's record in a mere 33 minutes. An obvious lesson to draw here is that celebrities are more popular than politicians. One might even conclude that this proves Twitter to be a trivial venue, unfit for politics or, as Gladwell put it, revolutions.

After taking a closer look at what made *KONY 2012* the kind of viral video that so much of the world chose to watch, however, I would like to suggest another moral to this story. The most popular tweet, by far, is a "selfie" of a *group*. Think about that. People love watching groups that they'd like to join, but deciding not to get up out of your chair and climb onto that particular stage does not make you a slacker or lurker; it makes you a watcher who is inspired, aspiring, and longing to build meaningful friendships. Russell understood this sensibility and his *KONY 2012* manifesto is a master class in adolescent wish fulfillment; teachers, parents, mentors, and, yes, even activists, should watch and learn.

NOTES

1. This chapter was written with support from The University of Tampa Dana Foundation grant.

2. "Invisible Children," *KONY 2102*, accessed January 2, 2014, https://www.youtube.com/watch?v=Y4MnpzG5Sqc.

3. Sam Sanders, *The "Kony 2012" Effect: Recovering from a Viral Sensation*, accessed January 2, 2014, http://www.npr.org/people/349243304/sam-sanders.

4. "Invisible Children," *KONY 2102*.

5. Teju Cole, "The White-Savior Industrial Complex," accessed January 2, 2014, http://www.theatlantic.com/international/archive/2012/03/the-white-savior-industrial-complex/254843/.

6. "Invisible Children," *KONY 2102*.

7. Sanders, *The "Kony 2012" Effect*.

8. Tom Watson, "The #StopKony Backlash: Complexity and the Challenges of Slacktivism," accessed January 2, 2014, http://www.forbes.com/sites/tomwatson/2012/03/08/the-stopkony-backlash-complexity-and-the-challenges-of-slacktivism/.

9. Barnaby Feder, "They Weren't Careful What They Hoped For," *New York Times*, accessed January 2, 2014, http://www.nytimes.com/2002/05/29/nyregion/they-weren-t-careful-what-they-hoped-for.html.

10. Amy Finnegan, "The White Girl's Burden," *Contexts* 12 (1) (2013): 30–35.

11. Malcolm Gladwell, "Small Change: Why the Revolution Will Not Be Tweeted," *New Yorker*, accessed January 2, 2014, http://www.newyorker.com/magazine/2010/10/04/small-change-3.

12. Ibid.

13. Ibid.

14. Laura Washington, The Whole World Was Watching, accessed January 2, 2014, http://inthesetimes.com/article/3876/the_whole_world_was_watching.

15. Ibid.

16. James A. Colaiaco, "Martin Luther King, Jr. and the Paradox of Nonviolent Direct Action," *Phylon* 47.1 (1986): 16–28.

17. Of course, I would soon learn that those movements, as well as most others, were not so glamorous or exciting at the time but rather could go through many fits, starts, and long slogs of stalemate, unpopularity, and failure.

18. "Invisible Children," *KONY 2102*.

19. Ibid.

20. Ibid.

21. Ibid.

22. Ibid.

23. Jessica Testa, "Two Years after KONY 2012, Has Invisible Children Grown Up?", accessed January 2, 2014, http://www.buzzfeed.com/jtes/two-years-after-kony-2012-has-invisible-children-grown-up#.hxO7rVj0A.

24. Matthew Hughey, *The White Savior Film: Content, Critics, and Consumption*. (Philadelphia, PA: Temple University Press, 2014).

25. Ibid.

26. Testa, "Two Years after KONY 2012."

27. Banksy, *Exit through the Gift Shop* (DVD) (London: Paranoid Pictures, 2010).

28. Ibid.

29. Ibid.

30. Ibid.

31. Ibid.

32. Ibid.

33. Invisible Children, *KONY 2102*.

34. Ibid.

35. Ibid.

36. Ibid.

37. Michael Renov, *Theorizing Documentary* (New York: Routledge, 1993), 203. In one of the footnotes of his book *Theorizing Documentary*, defines "pre-enactments" as "visions of what could be, presented in a documentary format." Remember that *KONY 2012* was posted to YouTube on March 5, 2012, so any talk or depiction of the April 20 "Cover the Night" event is a prediction about the future. In other words, the images portrayed in the pre-enactment scenes were shot well in advance of the actual poster-bombing event.

38. Chris Boulton, "The Mother's Gaze and the Model Child: Reading Print Ads for Designer Children's Clothing," *Advertising & Society Review* 10.3 (2009).

39. Chris Boulton, "Don't Smile for the Camera: Black Power, Para-Proxemics and Prolepsis in Print Ads for Hip-Hop Clothing,'" *International Journal of Communication* 1.1 (2007).

40. "Invisible Children," *KONY 2102*.

41. Testa, "Two Years after KONY 2012."

42. Invisible Children, International Events, accessed January 2, 2014, http://invisiblechildren.com/program/international-events/.

43. Invisible Children, Homepage, accessed January 2, 2014, http://invisible children.com.

44. Rory Carroll, "Kony 2012 Cover the Night Fails to Move from the Internet to the Streets," accessed January 2, 2014, http://www.theguardian.com/world/2012/apr/21/kony-2012-campaign-uganda-warlord.

45. Testa, "Two Years after KONY 2012."

46. Pew Research Center, "Politically Apathetic Millennials," accessed January 2, 2014, http://www.pewresearch.org/daily-number/politically-apathetic-millennials/.

BIBLIOGRAPHY

Banksy. *Exit through the Gift Shop* (DVD) (London: Paranoid Pictures, 2010).

Carroll, Rory. Kony2012 Cover the Night Fails to Move from the Internet to the Streets. Accessed January 2, 2014. http://www.theguardian.com/world/2012/apr/21/kony-2012-campaign-uganda-warlord.

Colaiaco, James A. "Martin Luther King, Jr. and the Paradox of Nonviolent Direct Action." *Phylon* 47.1 (1986): 16–28.

Cole, Teju. The White-Savior Industrial Complex. Accessed January 2, 2014. http://www.theatlantic.com/international/archive/2012/03/the-white-savior-industrial-complex/254843/.

Feder, Barnaby. They Weren't Careful What They Hoped For. Accessed January 2, 2014. http://www.nytimes.com/2002/05/29/nyregion/they-weren-t-careful-what -they-hoped-for.html.

Finnegan, Amy. "The White Girl's Burden." *Contexts* 12 (1) (2013): 30–35.

Gladwell, Malcolm. Small Change: Why the Revolution Will Not Be Tweeted. Accessed January 2, 2014. http://www.newyorker.com/magazine/2010/10/04 /small-change-3.

Hughey Matthew. *The White Savior Film: Content, Critics, and Consumption.* (Philadelphia, PA: Temple University Press, 2014).

Invisible Children. Homepage. Accessed January 2, 2014. http://invisiblechildren .com.

Invisible Children. "International Events." Accessed January 2, 2014. http:// invisiblechildren.com/program/international-events/.

Invisible Children. *KONY 2102.* Accessed January 2, 2014. https://www.youtube.com /watch?v=Y4MnpzG5Sqc.

Pew Research Center. Politically Apathetic Millennials. Accessed January 2, 2014. http://www.pewresearch.org/daily-number/politically-apathetic-millennials/.

Renov, Michael. *Theorizing Documentary* (New York: Routledge 1993), 203.

Sanders, Sam. The "Kony 2012" Effect: Recovering from a Viral Sensation. Accessed January 2, 2014. http://www.npr.org/people/349243304/sam-sanders.

Testa, Jessica. "Two Years after KONY 2012, Has Invisible Children Grown Up?" Accessed January 2, 2014. http://www.buzzfeed.com/jtes/two-years-after-kony -2012-has-invisible-children-grown-up#.hxO7rVj0A.

Washington, Laura. "The Whole World Was Watching." Accessed January 2, 2014. http://inthesetimes.com/article/3876/the_whole_world_was_watching."

Watson, Tom. "The #StopKony Backlash: Complexity and the Challenges of Slacktivism." Accessed January 2, 2014. http://www.forbes.com/sites/tomwatson/2012/03/08 /the-stopkony-backlash-complexity-and-the-challenges-of-slacktivism/.

18

Public Health's Courtship with the Internet: Slow but Steady

Samantha Lingenfelter

Health cannot be bought at the supermarket. You have to invest in health. You have to get kids into schooling. You have to train health staff. You have to educate the population.

—Hans Rosling[1]

QUESTION OF THE CENTURY: HOW DO YOU USE THE INTERNET?

(That's not your cue to skip this chapter. Keep reading, please.)

It's probably easier to list the things we do not supplement with Internet support than it is to give a timely answer to the question posed above. The simple idea of being disconnected for a weekend camping trip on the Appalachian Trail is enough to give many "Millennials" severe anxiety, or at least raise their blood pressure. A good example is the episode of *Daria*, an MTV show from the 1990s and early 2000s. In the episode, "The Teachings of Don Jake," the Morgandorfer family goes into the wild to enjoy some disconnected time away from the phones, fax machines, and computers due to Daria's father's stress-related health issues. Daria's mother cannot escape her need to be connected and smuggles a mobile phone into her backpack for the trip. This was in the 1990s, before every high school student had a cell phone. The show's creators knew we were too connected back then—and that was more than a decade before this book even was being considered and at least five years before the first iPhone was sold.

Whether arguing for its positive or the negative aspects, the Internet has really done a lot for humans as a species. Some might argue that the benefits of constantly being connected are more detrimental than beneficial but, like any tool, it is really all about how you use it. Using a cheese grater for cutting bread is just not going to be as effective as using a knife . . . unless you're making breadcrumbs, in which case that cheese grater could be useful, but still is not recommended.

Another (arguably more relevant) example could be inviting 100 people to a holiday party only using handwritten invitations sent via a postal service. Sure, it will do the job, but not as quickly or as efficiently as sending invitations through Facebook. This is a typical tradeoff that we have accepted into our developed American society in the 21st century.

As of 2013, the World Bank found that 40 out of every 100 people living on this planet use the Internet in some capacity[2]—and that number is steadily increasing. Although this accounts for less than half of the world's population, having access to the Web is slowly but surely expanding to all corners of the globe. It can be argued that this number consists mostly of those living in developed nations—a fair argument, to be sure, and even more so when compared to the 2011 statistic that only 57 percent of all roads on our planet were paved. That seems like a fairly reasonable estimation. Those living in the upper echelons of our societies typically are wealthy enough to afford Internet connectivity, both in developed nations and in parts of the Third World. Considering the amount of wealth that is distributed among those who have potential access to the Internet, and the fact that the populations of China, India, and the United States combined equal more than 2.5 billion people, it is fair to say that those countries have the lion's share of Internet users.

Why is this important? How does having Internet access really help the public in any way? Cat photos are in no short supply. Other than that, however, where are the real benefits? Internet connectivity can improve quality of life for the world's populations in countless ways, especially when paired with health initiatives and educational tools. Rapid communication opportunities and access to updated research and knowledge can expedite the process of improving quality of life for many. Sending an e-mail to someone who is 5,000 miles away and receiving a response within minutes is a level of connectivity that previous generations did not experience. This type of access enables societies to create health initiatives and programs, along with communicating with the masses as directly (and quickly) as possible.

Public health has directly benefitted from the Internet's influence during the 20th century. Although the concept of public health is broad and seemingly endless, there are numerous ways that health practitioners can use Internet communications in beneficial ways to improve quality of life locally, nationally, and globally.

This chapter reviews the important relationship shared between health, communication, the Internet, and the public. Always being connected has its benefits and its liabilities. Like many things in our society, technology can be accused of being both a blessing and a curse. Being able to protect and improve the health of the public, however, is one positive way that ubiquitous connectivity is beneficial to the world as a whole.

SO, WHAT IS PUBLIC HEALTH, EXACTLY?

It's the middle of May and you wake up to an air raid–like siren blaring from your cell phone. It is an alert from your city warning you of severe weather in your area. Perhaps a supercell thunderstorm has led to the development of a tornado. That's public health.

A local nonprofit health center is offering free HIV testing, educational seminars on safe sex, and access to free condoms for anyone in the community. Again: Public health.

A city is working to clean up pollution in a nearby lake to prevent contamination of local drinking water sources. You guessed it! Public health.

A new design for vehicle airbag systems is implemented in all new models of minivans. . . . You get the idea. Public health.

Essentially everything that can impact the quality of human life is, in one way or another, related to public health. As I said, pretty broad, right? Absolutely.

The concept of public health is rather new, although it seems like a term that should have been established as a concrete idea centuries ago. Considering that the concern for the health of the public led to the significant scientific inventions and discoveries that improved life expectancy by decades, the development of public health and public health organizations would seem to be a natural next step.

The World Health Organization (WHO) defines public health as "all organized measures (whether public or private) to prevent disease, promote health, and prolong life among the population as a whole."[3] There are numerous job positions and public interventions that could easily fall within this definition's spectrum, thus making a clear categorization simultaneously easy and difficult. Who counts as a public health practitioner when the definition incorporates such vast possibilities?

Job titles and duties that could be categorized as "public health" positions can range from the obvious—the Centers for Disease Control and Prevention (CDC) creating a more effective Ebola vaccine . . . to more subtle jobs, such as car manufacturers improving the safety of new car models through design and innovation. Public health easily creeps into many corners of the U.S. job market and potentially can be found within the job descriptions of our elected officials. Rallying for political support for certain health innovations and initiatives has an impact on constituents, and politicians prefer to keep that population healthy and happy. Our congressmen and senators might not see themselves as public health practitioners, but there are some occasions for which the case could be made.

Our planet experiences one human birth every 8 seconds and a human death every 12 seconds.[4] More than 7 billion people currently inhabit planet Earth, and there are more people are being born (at a rapid rate) than are dying. The population worldwide will continue to grow, thus resulting in increased public health demands. Considering the issues associated with sustaining a growing population with diminishing natural resources and the shrinking of habitable space, public health will need to be increasingly effective in the expectedly near future. The improvements in various health-related sectors such as sanitation have led to a longer life expectancy but result in population explosions. Birth rates have clearly increased and death rates have slowed, thus resulting in an overall "graying" of the world's population.[5]

Communications are incredibly important for disseminating key information, both for the public and for public health professionals. Considering the rapid development of communications networks and tools within the past century, there are numerous options and methods that can be used to benefit practitioners of various public health disciplines.

A BRIEF HISTORY OF PUBLIC HEALTH

How can one understand the real establishment and need for a sector like public health without some history? "Health" has no real need for a historical review, but understanding key events and time lines can illustrate where public health actually began and how we got to where we are today.

Throughout the centuries communicable diseases that have ravaged populations. Three major plague pandemics have been recorded in history, the first experienced by the Byzantine Empire. The Justinian Plague occurred in the sixth century and killed more than 25 million people.[6] The second recorded pandemic—one of the better-known pandemics—was the bubonic plague, also known as the "Black Death," which spread through Europe in the 14th century. Scholars and researchers note that 60 percent of the European population[7] died as a result of the epidemic.[8] The third pandemic on record occurred in the 1860s. The modern plague (a bacterial infection) killed more than 10 million people over a 20-year period.[9] Entire populations can be completely destroyed by a communicable disease, as evidenced by these historical examples.

Religious leaders and communities originally thought that the disease was a result of debauchery-laden behaviors and a vast lack of prayer[10] (although, ultimately, the source was found to be fleas). Tactics such as isolation, quarantine protocols for travelers, and even fumigation became tools for disease control. According to Slack, these eventual developments led to increasing government presence and essentially paved the way for interventions for public health.[11]

Public health discoveries and interventions continued into later centuries, mostly discovered through observational research.[12] Considering that many scientific standards and breakthroughs were decades—if not centuries—away, observational data were very important for the world of public health. Some of the major developments were the discovery of vaccinations, creation of sanitation services and standards, and the understanding of how infections spread. The observational research conducted by Louis Pasteur and others has led to the extended life span previously discussed.[13]

The CDC actually has a list on its Web site of what it considers to be the top-10 greatest health achievements of the 20th century. Considering the increase of the American life span that occurred due to these discoveries, it is fair to say that they were very important.

1. Vaccinations—Including school-required vaccinations for children.
2. Improvement in neonatal and OBGYN care—Reducing infant and maternal mortality rates.
3. Family planning—Better access to contraceptive methods, for example.
4. Fluoridation of drinking water—Having healthier teeth leads to fewer health issues.
5. Tobacco as a health hazard—Smoking and chewing tobacco being recognized as carcinogenic, for instance.
6. Safer and healthier foods—Read Upton Sinclair's *The Jungle* for a realistic understanding of how bad food manufacturing conditions used to be.
7. Motor-vehicle safety—Seatbelts, air bags, traffic laws.
8. Workplace safety—Decreased fatalities in the workplace, better workplace safety regulations.
9. Control of infectious diseases.
10. Declines in death from strokes and heart attacks—Healthier diets are one factor that helped this issue.[14]

Outbreaks of disease have been scattered throughout Earth's time line and humans have adapted in countless ways. The 20th century introduced a plethora of new practices and procedures that have led to an increase in the average lifespan of Americans and of people around the globe. Without vaccines, for instance, the United States might still be seeing significant outbreaks of polio and smallpox. Safer food standards and health-code regulations keep diseases under control through sanitation practices, thus resulting in a reduced risk for food poisoning. Public health's broad scope clearly made many improvements to our global way of life in the 1900s.

THINKING SMALL: COMMUNITY-LEVEL COMMUNICATIONS

Let's talk about what everybody already is talking about: Social media. The first thought that comes to mind might be annoying corporate branding Tweets or " 'Like' if you agree . . ." posts on Facebook. The (arguably) irritating parts of social media certainly exist, but the entire concept of social media changes when you put a professional, medical spin on its purpose. Community-building is a good first step.

Community-building has been an extremely beneficial side effect of social media. Those with illnesses or certain ailments can join together on Web sites such as Facebook or Reddit and have conversations about issues they face. Sometimes speaking to others is more calming than simply searching for symptoms in any search engine (Internet search diagnoses never really are a great substitute for a seeing trained medical professional). Reddit users might not have a degree in neuroscience or psychology, but the site is a free service that has the potential to help numerous people on emotional levels. Sharing experiences can help educate and spread awareness as well as create solutions that others might not be able to find. Limited access to medical care could be supplemented through social media "services" such as these. One example of an online community like this could be a Facebook group or message board for breast cancer survivors. Sharing one's story and discussing it with others could be both therapeutic and educational.

Discussions on globalization seem to always include the Internet and the communication options that go along with it. One of the biggest developments clearly is social media. Instant messenger services through AOL, Yahoo!, Google, Skype, and others keep people connected without the need for snail mail or landline telephones. It is even possible to acquire a free Internet phone number through GoogleVoice! This type of communication has been used by everyone—from elementary school–age children to leading medical professionals in the field. Always being connected and able to communicate expedites diagnoses, processes, and conditions for just about any situation.

Considering that there are links to corresponding social media accounts on many (if not most) of the U.S. government Web sites, it's easy to assume that social media networks are not going anywhere any time soon. New services can be advertised and communications with health professionals can be made potentially without fees or costs. These concepts are still in the early stages of development but there is great potential for future use of these mediums of communication.

State public health departments also use social media for communications. According to Thackeray et al., health departments use social media but the actual implementation of social media into health plans still is in its infancy.[15] Strategic communication planning must be established for the use of social media to become a more mainstream practice.[16] According to one study conducted in 2012, only 24 percent of local health departments surveyed had a Facebook page and only 8 percent had a Twitter account.[17] The same study also found information further supporting the increase use of social media by local health departments if the state health department had adopted the technology into its practice as well.[18] Because this still is in early stages for local health departments, best practices have yet to be established for effective communications to be determined. The implications of state health departments using Twitter, Facebook, and other methods of social media for information dissemination increases the ability for direct communication to the public.

Studies have found that it is possible to track illness through social media resources. Web sites have even been created to "mine" keywords related to illnesses and track them through social media accounts. HealthMap (found at HealthMap.org) conveniently places all the live-tracking information on a world map that shows various outbreaks in real time.[19] The speed that communicable diseases can travel in this world has significantly increased thanks to globalization; a case of Ebola can be caught in Africa and brought back to the United States via air travel in less than one day. Having live updates from mined information is helping public health practitioners keep ahead of illness trends, such as the Ebola outbreak in Liberia or SARS in China. Studying social media for trends within illnesses still includes caveats regarding potential accuracy of the statistics. Supporting this practice with more information could yield many positive implications for public health.

THINKING BIG: NATIONWIDE AND INTERNATIONAL COMMUNICATIONS

Community-based use of the Internet for the benefit of public health is wonderful for niche groups and local communities, but there have been beneficial uses on a much grander scale as well. Top government agencies from around the world work on public health initiatives for the benefit of their countries, and share their ideas and messages through modern-day communications. Intergovernmental organizations such as the World Health Organization (WHO) use Web sites for educational purposes and for providing real-time updates on public health crises as they occur.

The CDC was established in the 1940s to research and work toward fighting communicable diseases. The agency then expanded and focused heavily on the study of malaria and its eradication.[20] Since its inception, the CDC has grown dramatically and now works and researches all areas of public health including chronic diseases, workplace hazards, environmental health concerns, and countless other areas.[21]

The CDC supplies information to communities and states through its Web page, including a list of current outbreaks that is available on its home-page, along with access to numerous beneficial resources. One example is the page dedicated to health-related travel warnings. The list is updated as new epidemics or illnesses are experienced in different countries across all conti-nents. In early 2015, numerous African countries rose to the top of the "Avoid Nonessential Travel" list due to the recent Ebola outbreak. This travel-warning tool benefits not only those who are traveling but those who might come in contact with a person after returning from a trip to a place on the list. This is just one useful tool made available for the protection of public health.[22]

The Ebola outbreak in Liberia during the spring of 2014 is a prime example of how the Internet has improved global communications for health profes-sionals. According to the CDC, the Ebola outbreak of 2014 experienced more than 21,000 infections with more than 8,500 deaths.[23] The crisis reached the United States when a doctor unknowingly was infected with the disease, traveled back to Texas, and eventually died from the illness. Others were infected due to exposure and one exposed nurse even flew on a com-mercial jet (with CDC permission) before being tested for the illness. Once the news was released that she was infected, however, many people quaran-tined themselves as a precautionary measure if they felt they were at risk of contracting the virus. The CDC worked with communities and quarantined areas as needed. Thankfully the illness did not spread, but receiving updates during a very dramatic time was useful, particularly for understanding the needs of the health care community when such emergencies arise.

THE DANGERS OF MISINFORMATION: THE NEGATIVE RELATIONSHIP BETWEEN THE INTERNET AND PUBLIC HEALTH

There are many great things about the Internet and its effect on public health. When combined, problems can be solved, disasters potentially can be avoided, and information can be disseminated. As is the case for many good things in this world, however, there is potentially a dark side. Misinforma-tion—especially related to health—can be a dangerous thing, and the Inter-net provides access to overwhelming amounts of misinformation, all available at anyone's fingertips.

The prime example that must be discussed is the antivaccination move-ment. There has existed a group of people who are against vaccines since the days of Edward Jenner and Louis Pasteur.[24] Vaccinations are responsible for eradication and prevention of many diseases in the 19th and 20th centuries, including smallpox and polio. Many school districts require students to receive measles, mumps, and rubella vaccinations—known as MMR—before attending kindergarten.

One concept that has helped keep vaccine-preventable diseases at bay is "herd immunity." Herd immunity occurs when enough of a population is vaccinated that there is minimal risk of a susceptible individual becoming exposed to the disease in question.[25] Herd immunity has been key in keeping many diseases at bay and for protecting people who have compromised immune systems—such as chemotherapy patients, people who are immunocompromised due to medical interventions, and the elderly. A surge of misinformation on the Internet, however, has been disseminated about vaccinations in relation to risks for children, under the guise of parental autonomy,[26] civil liberties, and causes of autism. Celebrity support of the antivaccination movement also helps in spreading and publicizing the misinformation about vaccines.

The measles vaccine was introduced during the 1960s and infection rates immediately declined.[27] Measles was deemed to be "eliminated" from the United States in 2000. To give some perspective on the severity of this illness, measles was responsible for 145,000 deaths (equaling more than half of everyone infected) worldwide in 2014.[28] The Measles and Rubella Initiative was developed in 2001 with support from the CDC, the American Red Cross, the United Nations Foundation, UNICEF, and the World Health Organization. Its goal is to reducing measles-related deaths by 95 percent worldwide by 2015 and to "eliminate measles and rubella in at least five of the six World Health Organization regions by 2020."[29] This is a global effort to increase vaccinations and eliminate this disease worldwide, but many new cases still are being found in the United States in 2015.

The CDC considers the elimination of most infectious diseases to be one of the greatest public health achievements in the 20th century. The antivaccination movement is taking public health multiple steps backward, however; all because of some radical misinformation that was posted on the Internet. One "doctor" actually has written about immunizations and their effect on children but failed to cite any reliable resources for the claims made. Another doctor used various studies to show a connection between immunization and autism; however, many later studies indicated that this was not in fact the case.[30] Regardless, many people have latched on to the concept that vaccinations are the cause of conditions such as autism and Asperger's. Although on social media there is strong opposition to the antivaccination movement, the misinformation will simply not disappear. Continuous corrections of misinformation are needed to reverse the damage done by "antivaxxers."

That's one of the greatest threats of the Internet: The inability to remove inaccuracies. The freedom of the Internet is wonderful for the First Amendment and the ability to speak one's mind, but it is the virtual "wild west" of regulation—the absence of regulation enables people to run rampant. It therefore is important to always remember that anyone can post "information" on the Internet.

CONCLUDING THOUGHTS

As is the case for many things in our world, the Internet has its good, bad, and ugly sides. In the realm of public health, however, there are many benefits that have not yet been established. Reaching those people who otherwise might not receive important information has further solidified the concept of globalization and the spread of western knowledge to the most remote parts of the planet. The continuing goal should be to reach even more people—especially those who currently might not have access to the resources and tools of technology.

There definitely are ugly parts of the Internet that are related to personal health. A simple Internet search of basic symptoms can reveal a myriad of terminal illnesses that could be the cause, when the Internet user actually just has a common cold. Personal hypochondria aside, however, there are some benefits to having research readily available. Smartphone browsers and digital applications provide rapid access to basic first aid in case of an emergency, and the technology can help untrained people to save lives while waiting for responders to arrive. The benefits of this technology and interconnectivity make the quality of such experience immeasurable.

Other good things the Internet provides, however, include simple alerts from a university indicating that classes are cancelled due to weather, which benefit public health by reducing the risk of exposure to or danger from life-threatening driving. Weather alerts from community and state governments can indicate when dangerous conditions require people to evacuate or take shelter. Our technology-obsessed society is reaping benefits of public health, but more integration into public health practice is needed.

Having the foresight to know exactly where public health will go from this point is nearly impossible. There still are many lessons to be learned, challenges to be addressed, and procedures to be established. Because the categorization of public health still is fairly new, there could be significant growth through its Internet counterpart. As other nations and states increase access to resources including the Internet there will be further growth and support for public health globally. Who knows what future health discoveries and innovations will be on the CDC's illustrious list for the 21st century? Perhaps everyone will find out in 2100.

NOTES

1. Hans Rosling, "The Best Stats You've Ever Seen" (February 1, 2006) (transcript), accessed January 15, 2015, http://www.ted.com/talks/ hans_rosling_shows_the_best _stats_you_ve_ever_seen/transcript?language=en.

2. "Internet Users (per 100 People)," World Bank Data, accessed January 18, 2015, http://data.worldbank.org/indicator/IT.NET.USER.P2/countries?display=graph.

3. "Glossary—Public Health," World Health Organization, accessed January 14, 2015, http://www.who.int/trade/glossary/story076/en/.

4. United States Census Bureau, "Population Clock" (January 1, 2010), accessed January 13, 2015, http://www.census.gov/popclock/.

5. Anne Nadakavukaren, *Our Global Environment: A Health Perspective* (Waveland Press, 2011).

6. "Plague History," Centers for Disease Control and Prevention (November 18, 2014), accessed January 31, 2015, http://www.cdc.gov/plague/history/.

7. Ibid.

8. Paul Slack, "Responses to Plague in Early Modern Europe: The Implications of Public Health," *Social Research* (1988): 433–53.

9. "Plague History," Centers for Disease Control and Prevention.

10. Ibid.

11. Ibid.

12. Leon Gordis, *Epidemiology,* 5th ed. (Philadelphia, PA: Saunders, 2014).

13. Ibid.

14. "Our History—Our Story," Centers for Disease Control and Prevention (April 26, 2013), accessed January 26, 2015, http://www.cdc.gov/about/history/ourstory.htm.

15. Rosemary Thackeray, Brad L. Neiger, Amanda K. Smith, and Sarah B. Van Wagenen, "Adoption and Use of Social Media among Public Health Departments," *BMC Public Health* 12, no. 1 (2012): 242.

16. Ibid.

17. Jenine K. Harris, Nancy L. Mueller, and Doneisha Snider, "Social Media Adoption in Local Health Departments Nationwide," *American Journal of Public Health* 103, no. 9 (2013): 1700–1707.

18. Ibid.

19. Charles Schmidt, "Trending Now: Using Social Media to Predict and Track Disease Outbreaks," *Environmental Health Perspectives* (January 1, 2012), accessed January 26, 2015, http://ehp.niehs.nih.gov/120a30/?utm_source=rss&utm_medium =rss&utm_campaign=120-a30.

20. "Our History—Our Story," Centers for Disease Control and Prevention.

21. Ibid.

22. "Travel Health Notices," Centers for Disease Control and Prevention, accessed January 20, 2015, http://wwwnc.cdc.gov/travel/notices/.

23. "Travel Health Notices," Centers for Disease Control and Prevention.

24. Gregory A. Poland and Robert M. Jacobson, "The Clinician's Guide to the Anti-Vaccinationists' Galaxy," *Human Immunology* 73, no. 8 (2012): 859–66.

25. Anne Nadakavukaren, *Our Global Environment: A Health Perspective.*

26. Anna Kata. "A Postmodern Pandora's Box: Anti-Vaccination Misinformation on the Internet," *Vaccine* 28, no. 7 (2010): 1709–16.

27. Walter A. Orenstein, Mark J. Papania, and Melinda E. Wharton, "Measles Elimination in the United States," *Journal of Infectious Diseases* 189, Supplement 1 (2004): S1–S3.

28. Jason Beaubien, "Measles Is a Killer: It Took 145,000 Lives Worldwide Last Year," NPR, (January 30, 2015), accessed January 31, 2015.

29. "Learn—Measles & Rubella Initiative," Measles & Rubella Initiative, accessed January 31, 2015, http://www.measlesrubellainitiative.org/learn/.

30. Gregory A. Poland and Robert M. Jacobson, "The Clinician's Guide to the Anti-Vaccinationists' Galaxy," *Human Immunology* 73, no. 8 (2012): 859–66.

BIBLIOGRAPHY

Beaubien, Jason. "Measles Is a Killer: It Took 145,000 Lives Worldwide Last Year." NPR January 30, 2015. Accessed January 31, 2015. http://www.npr .org/2015/01/30/382716075/measles-is-a-killer-it-took-145-000-lives-worldwide -last-year.

Centers for Disease Control and Prevention. "Outbreaks Chronology: Ebola Virus Disease." January 22, 2015. Accessed January 25, 2015. http://www.cdc.gov/vhf /ebola/outbreaks/history/chronology.html.

Centers for Disease Control and Prevention. "Travel Health Notices." Accessed January 20, 2015. http://wwwnc.cdc.gov/travel/notices/.

Centers for Disease Control and Prevention. "Plague History." November 18, 2014. Accessed January 31, 2015. http://www.cdc.gov/plague/history/.

Centers for Disease Control and Prevention. "Measles History." November 3, 2014. Accessed January 31, 2015. http://www.cdc.gov/measles/about/history.html.

Centers for Disease Control and Prevention. "Our History—Our Story." April 26, 2013. Accessed January 26, 2015. http://www.cdc.gov/about/history/ourstory.htm.

Courtney, K. L. "The Use of Social Media in Healthcare: Organizational, Clinical, and Patient Perspectives." *Enabling Health and Healthcare through ICT: Available, Tailored and Closer* 183 (2013): 244.

Gordis, Leon. *Epidemiology.* 5th ed. Philadelphia, PA: Saunders, 2014.

Gore, Anita. "California Department of Public Health Confirms 59 Cases of Measles." January 21, 2015. Accessed January 26, 2015. http://www.cdph.ca.gov/Pages /NR15-008.aspx.

Harris, Jenine K., Nancy L. Mueller, and Doneisha Snider. "Social Media Adoption in Local Health Departments Nationwide." *American Journal of Public Health* 103, no. 9 (2013): 1700–1707.

Kata, Anna. "A Postmodern Pandora's Box: Anti-Vaccination Misinformation on the Internet." *Vaccine* 28, no. 7 (2010): 1709–16.

Measles & Rubella Initiative. "Learn—Measles & Rubella Initiative." Accessed January 31, 2015. http://www.measlesrubellainitiative.org/learn/.

Nadakavukaren, Anne. *Our Global Environment: A Health Perspective.* Long Grove, IL: Waveland Press, 2011.

Orenstein, Walter A., Mark J. Papania, and Melinda E. Wharton. "Measles Elimination in the United States." *Journal of Infectious Diseases* 189, Supplement 1 (2004): S1–S3.

Poland, Gregory A., and Robert M. Jacobson. "The Clinician's Guide to the Anti-Vaccinationists' Galaxy." *Human Immunology* 73, no. 8 (2012): 859–66.

Rosling, Hans. "The Best Stats You've Ever Seen." February 1, 2006. Accessed January 15, 2015. http://www.ted.com/talks/hans_rosling_shows_the_best_stats_you_ve_ever_seen/transcript?language=en.

Schmidt, Charles. "Trending Now: Using Social Media to Predict and Track Disease Outbreaks." *Environmental Health Perspectives*. January 1, 2012. Accessed January 26, 2015. http://ehp.niehs.nih.gov/120-a30/?utm_source=rss&utm_medium=rss&utm_campaign=120-a30.

Slack, Paul. "Responses to Plague in Early Modern Europe: The Implications of Public Health." *Social Research* (1988): 433–53.

Thackeray, Rosemary, Brad L. Neiger, Amanda K. Smith, and Sarah B. Van Wagenen. "Adoption and Use of Social Media among Public Health Departments." *BMC Public Health* 12, no. 1 (2012): 242.

United States Census Bureau. "Population Clock." January 1, 2010. Accessed January 13, 2015. http://www.census.gov/popclock/.

The World Bank. "Internet Users (per 100 People)." Accessed January 18, 2015. http://data.worldbank.org/indicator/IT.NET.USER.P2/countries?display=graph.

World Health Organization. "Public Health." Accessed January 14, 2015. http://www.who.int/trade/glossary/story076/en/.

Index

ABC Online Forum, 222

Acaye, Jacob, 324

accessibility, xiii–xiv, 333–335; education, 91–109; multi-platform, 3

activism, 165–176 , 267–270; hashtag, 271–277; information and communication technologies (ICTs), 165–166, 170; Internet's impact on, 183–195

Adams, Samuel, 244

Adekunle, Philip, 293

Adweek, 231

The Agency Group, 48

Agnew, Spiro, 248

AllLacqueredUp.com, 205

Amazon, 8; Prime Instant Video service, 9

American Association for the Advancement of Science, 127–128

American Council on Education (ACE), 100

Amnesty International, 323

Android, 94; Music apps, 6

Anti-vaccination movement, 340–342

Apple, 4, 6–7

Arab Spring, 165, 169, 267–268, 323

art, 3–17; accessing, 11–12; audiences and markets for, 3–17; digital, 12–14; digital; museums, 13–14; interactive, 12; internet, 12–13; internet distribution of, 11–12; internet installations, 12–13; marketing and promoting, 11–12; mobile streaming, 15; modernism, 13; net artists, 12–13; organizations, 12; post-Internet, 13; postmodernism, 11–12; shows, 12; streaming, 12–13; virtual reality and, 14–15

Arthur M. Sackler Gallery, 14

A$ASP Rocky, 48–49

Assange, Julian, 130

Astronauts Wanted, 226

athletes; advocacy and activism, engaging in, 29–32; fan advocacy, 27–29, 32; media framing, negative, 27–29; minority, 29; self-presentation, optimizing, 24–27, 31; social media and, 23–33

The Atlantic, 209–210, 277–278

Australian Recording Industry Association, 226

Babson Survey Research Group, 97

Bad Luck Brain, 230

Banksy, 325

Barnes, Nora Ganim, 206

BBC, 10, 125

Beatport, 5–6; application programming interface (API), 5; consumer-facing strategy, 4

Benedikt, Allison, 252

Benkler, Yochai, ix–x

Berens, Ricky, 30
Bernstein, Carl, 121, 245
Betaworks, 209
Big Earl's Bait House, 270–271
Bikini Kill, 40–41, 43–44, 49
bin Laden, Osama, 129
Bissinger, Buzz, 245–246
Black Eyed Peas, 187
BlackBerry riots, 169
Blackboard, 97
Blair, Jayson, 122, 246–247
The Blind Side, 324
blog(s), xii, 12, 120, 203–204, 277;
    advertising on, 214; audience,
    finding an, 212–214; demoblog, 146;
    fame and, 203–206; future of, 206–
    208, 210–212; impact of social media
    on, 208–210; micro, 307; new
    technologies and, 207–208; political,
    146, 185; video, 226–228
*A Blog About Media and Other Things
    I'm Interested In*, 250
Blogdorf Goodman, 204
Bloomberg, 4, 11
Borthwick, John, 209
Boston Journal of Occurrences, 244
Boston Red Sox, 27
Bradlee, Ben, 121
Brandenburg, Heinz, 291–292
Branding, personal, 221–235
Branzburg v. Hayes, 124
Brazil, 165
Brown, Michael, 29, 267–269
Bryant, Kobe, 29
Burr, Tanya, 228
Bush, George W., 123–124, 128, 192,
    248, 250
Butler, Shay Carl, 227–228
BuzzFeed, 61, 212, 231

camera obscura, 57
Cameron, David, 131
Camp Takota, 226
Campbell, Mel, 207
#CancelColbert, 263–265, 273

Canva, 213
Car Advice, 206
Carpini, Delli, 190
Carr, David, 129
Cashmore, Pete, 233
Castells, Manuel, ix, 287
CBS, 249
The Center for Marketing Research,
    206
Centers for Disease Control and
    Prevention (CDC), 335, 337,
    339–341
Central Intelligence Agency (CIA),
    123–124, 129
Centro de Investigaciones
    Sociológicas (CIS) Barometer,
    144–145
Champion, Edward, 250
#change, 29
Chayko, Mary, 287
Cheney, Dick, 123–124
Chicago Sun Times, 123
Child, Julia, 203
Chilean Winter, 169
China, 193
chronemics, 305; e-mail, 307–310;
    face-to-face (FTF) conversations,
    306, 308–309, 311–312; feedback,
    307; online forums, 310; role of
    technology in, 306–308, 310; text
    messaging, 307, 310
Civil Rights Movement, 322–323
Clinton, Bill, 130, 279; online strategy,
    184–185
CNN, 185, 249–250; YouTube
    debates, 187
Coates, Ta-Nahesi, 277
CODE.org, 105–106
Cohen, Dan, 121
Cohen, Roger, 63
Cohen v. Cowles Media Co., 121
Colbert, Stephen, 263–265
collective action, 165–176
Comedy Central, 263–264
Communications Decency Act, 128

community-building, xiii–xiv, 167, 338–339

Congress of Racial Equality (CORE), 322

consumers, activating, 266–267

content creation, 221, 223–224

Cooke, Janet, 122, 244–245, 247

Cooper , Anderson, 187

Cornelius, Robert, 57

Consumer Electronics Show (2015), 7

Coursera, 92, 100, 105

Craven, Kyle, 230

Cruz, Victor, 29

Cry Freedom, 324

C-SPAN, 185

Cupcakes and Cashmere, 214

cybercrimes, 127, 166

cyberspace, corporate colonization of, xii

cyber-utopianism. 167

da Vinci, Leonardo, 57

Daguerre, Louis, 57

The Daily Dot, 231

Daily Grace, 225

The Daily What, 231

Dances with Wolves, 324

Daria, 333

data mining, 195

Davies, Ian, 230

Daydream Nation, 47

De Tocqueville, Alexis, 243

Deadspin, 277

Dean, Howard, 185–186

Dean, Jodi, 290

DeFranco, Philip, 226–228

DeGeneres, Ellen, 55, 329

Democracy in America, 243

Democratic National Convention (1968) 322

democratization, xiii–xiv

Denton, Nick, 278

Derringer, Nancy, 250–251

Dewey, John, 288

DeYoung, Karen, 249

digital democracy, 141–154; downside of, 144–145

digital divide, xii, 93, 143

digital media platforms, 243

District Lines, 229

Dole, Bob, 185

Donovan, Lisa, 228

Dorfman, Dan, 122

Drake, Thomas A., 129

DuBravac, Shawn, 58

E! Entertainment Television, 226

e-mail conversations, 307–308

East Carolina University, 227

Easy Gourmet, 206

Eco, Umberto, 144

education, xiv; definition of, 92–93; distance, 92; face-to-face, 98; formal, 83; higher, 91, 97–98, 101–102; informal, 83; K–12, 98–99; massive open online courses (MOOC), 91, 99–102; nonprofit organizations, 105–106; nonprofit resources, 106–107; online, 91–109; online, benefits of, 108–109; online, how we're learning, 97–99; peer learning, 85–86

edX, 104–105

Electric Zoo, 5

Eler, Alicia, 60, 65

Ellsberg, Daniel, 121

England, 165

Ephron, Nora, 203

ePolicy, 146

Erdely, Sabrina Rubin, 252

Erdogan, Recep, 165

Espionage Act, 120, 129

Etsy, 12

European Parliament, 145

Exit Through the Gift Shop, 325–326

Extra TV, 205

Facebook, 27–28, 55, 61–63, 107, 208, 211, 213, 221, 248, 267, 277; artists pages, 12; communities, 338–339;

Fortune 500 companies and,
    207–208; iTunes, 4; NYPD, 272–273;
    political movements and, 146,
    193–194; terms and Conditions, ix
Fairey, Shepard, 325–326
Falkner, Nickolas, 103
Falkner, Vivian, 103
Fallah, Alborz, 206
fameballs, 232–234
Federal Rules of Evidence, 124
Felt, Mark, 121
*Feminists Read Habermas: Gendering
    the Subject of Discourse*, 290
#Ferguson, 29
Fernald, Anne, 290
15M Spanish Movement, 165,
    169–170, 174
Fisher, Ian, 206–207
Fitzgerald, Patrick, 124
Flickr, 222
Florida Marlins, 26–27
Food & Wine Magazine, 205
The Food Illusion, 204
food standards, 337–338
Forbes, 227
Forbes Digital Tool, 245–246
Ford Fiesta Movement, 228–229
Fort Wayne (IN) News–Sentinel, 250
Franco, James, 64
Frankfurt School, 288
Fraser, Nancy, 288–290, 297
Free Flow of Information Act, 130–131
Freedom of Information Act, 128
Freemasons, 288
Freer Gallery of Art, 14
French, Antonio, 268
Friedlander, Lee, 57–58
Fulfillment Fund, 105–106
Fullscreen, 229

Galley Café, 270–271
Gallup, 241
Garner, Eric, 267–269
Gates, William Henry "Bill" III, 210
Gawker Media, 277–278

Geocities, 206
Gharaibeh, Kassem, 228
Gladwell, Malcolm, 250, 322
Glamour, 205
Glass, Stephen, 245–247
Global Night Commute, 328
Globalization, xiii–xiv
Goeglein, Tim, 250–251
Goffman, Erving, 60
Google, 6, 108, 193, 248, 250–251;
    Adsense, 206, 228–229; Art Project,
    14; Chrome Cast, 4, 7; Fiber, 10;
    Hummingbird, 206
Gordon, Kim, 42
Gore, Al, 184
government accountability, 119–132
GPS technology, 13
GQ magazine, 42
*Grace's Guide: The Art of Pretending To
    Be a Grown-Up*, 226
The Grace Helbig Project, 226
Gran Torino, 324
Grantland, 251
Green, Bill, 244
Greenpeace, 323
The Guardian, 263–264
Gumuchian, Marie-Louise, 250

Habermas, Jürgen, x, 288–291
hacking, 128–131
Hamilton, Josh, 28
Hanna, Kathleen, 40–41
Hannan, Caleb, 251
Hart, Mamrie, 226
Harvard University, 211, 221; Harvardx,
    101, 104–105; Neiman Lab, 208
hashtag activism; audience,
    understanding, 272–273; education
    through widespread awareness, 276;
    national events become personal,
    275–276; subverting mainstream
    awareness through minority opinion,
    273–275
HealthMap, 339
Hearst, William Randolph, 244

Hecox, Ian, 224
Heiderman, Marvin, 59
Helbig, Grace, 225–226
The Help, 324
Hernandez, Macarena, 246–247
HeyUSA, 226
Hochman, Jessa, 212
Holder, Eric, 131
the Horrors, 42–43, 51
Hotmail, 208
The Huffington Post, 233
Hughey, Matthew, 324
humiliation culture, 279
Hussein, Saddam, 123

Imagine: How Creativity Works, 250
In Defense of Food, 204
In Rainbows, 50
information and communication
    technologies (ICTs), 165–166, 172,
    193; debates and challenges,
    166–169
Instagram, 55, 61, 208, 222; filters,
    61; posting teasers on, 214
Internet; civic mobilization and, 192–
    194; crafting legislation for, 191–192;
    persuasion and, 190–191; political
    discussion and, 183–195; political
    knowledge and, 188–190; service
    providers (ISPs), 128
Invisible Children (IC), 321–329
Invisible Children: Rough Cut, 328
iPhone, 94; music apps, 6
Iran, 322
Iraq War, 130, 192
It'sGrace, 226
iTunes, 4, 43–45, 209, 226; Facebook,
    4; iTunesU, 107; Radio, 4, 7; Store,
    7, 43–44
ITV, 125

J-Phone, 58
Jefferson, Thomas, 243
Jezebel, 231, 277
Jordan, Michael, 29

journalists/journalism; accountability,
    241–254; accountability, historical
    lack of, 243–248; accountability in
    the digital age, 248–253; anonymous
    sources, verifying, 122–123, 126–
    127; anonymity, promise of, 119–
    122, 127–128; anonymity versus
    transparency, 120–125; citizen, 120;
    Fifth Estate, 120, 125–128; legal
    protections for, 120; mistrust of, 242;
    online sources and, 126–127;
    plagiarism, 122–123, 246–247, 250;
    plagiarism-flagging software for, 250;
    scandals, 242, 246–253; "sock
    puppet," 191; technology and,
    125–128
Julia & Julia, 203
Julia and Julia: 365 Days, 524 Recipes, 1
    Tiny Apartment Kitchen, 203
The Jungle, 337

Kahlo, Frida, 56–57, 60, 64
Kahrl, Christina, 251
Kang, Jay Caspian, 264
Karim, Jawed, 226
Kaskade, 4
Katz, Scott, 228
Kelly, Jack, 122, 246–247
Kelner, Douglas, 292
Kendall, Mikki, 273–274, 276
Khan, Philippe, 58
Khan Academy, 105–107
Kindle, 15–16; Unlimited Program,
    16–17
Kinja, 277
Kirakou, John, 129
Kjellberg, Felix Arvid Ulf "PewDiePie,"
    225
Kluge, Alexander, 291
Knox, Shelby, 276
Kony, Joseph, 321–329
KONY 2012, 321–329
Koselleck, Reinhart, 288
Kovach, Bill, 242
Kurutz, Steven, 214

Kurtz, Howard, 248
Kwantlen Polytechnic University, 223

Landis, Floyd, 27
Lane, Charles, 246
The Last Samurai, 324
Le, Stephanie, 205–206
Le Cordon Bleu, 203
leak(s), government, 119–132; morality
    and legality of, 128–131; planned, 12
Leake, Lisa, 204
Lehrer, Jonah, 250
Lewis, John, 328
Lewinsky, Monica, 130, 279
Libby, Lewis "Scooter, " 123
Liberia, 339–340
Libya, 165, 249
Lieberman, Matthew D., 63
Life in Color, 5
Little Green Footballs, 248
Llauradó, Josep M., 145
Lochte, Ryan, 30
Lodwick, Jakob, 233
Long. Live. A$AP, 48
Loomio, 148–149
Lord's Resistance Army (LRA),
    321–329
LRA Disarmament and Northern
    Uganda Recovery Act, 328
Los Angeles Times, 191
Lustig, Andrea, 204–205

Maker Studios, Inc., 228–229
Mandela, Nelson, 60
Manning, Bradley, 130
Manz, Brianne, 205
Marche, Stephen, 57, 63
Marcus, Greil, 42
Marie Claire, 205
MarksDailyApple.com, 205
Martin, Trayvon, 29, 267
Mashable, 230–231, 233
Massachusetts Institute of
    Technology (MIT), 104-105;
    MITx, 100–101

massive open online courses (MOOC),
    91, 99–102; badge-style format,
    103; categories of, 103; cMOOC,
    100–101; definition of, 101; degree
    programs, 102–103; developing,
    102–103; future of, 104; in 2012,
    101–102; in 2013, 102; providers,
    104–105; purpose of, 101;
    xMOOC, 100–101
Match.com, 307
Mayes, Stephen, 59
McCain, John, 185–186
McKessen, DeRay, 268–270
McLaughlin, Neil, 290
Measles and Rubella Initiative, 341
media bias, 267–270
media overexposure, 232–234
Media Research Center, 241
Media Richness Theory (MRT),
    306–307
Medium, 208
memes, 230
Mexico, 165, 171–173
Microsoft, 210
Mikkelson, Barbara, 321
Mismas, Michelle, 204–205
Mississippi Burning, 324
mobile devices; 4k on, 11; high-
    definition link (MHL) 3, 11;
    subscribers, 204; UHD video, 11
mobile networks, 10
mobile phones, 58, 94; camera, 125
The Modest Man, 205
Moldova, 322
Monash University, 207
Money Magazine, 122
Moodle, 97
Morris, Laina, 230–231
Morrison, Logan, 27
Moss, Benjie, 208
movies, 3–17; audiences and markets
    for, 3–17; 4K content, 8–10;
    streaming, 8–10
Moynihan, Michael, 250
Mukasey, Michael B., 130–131

Murphy, Eileen, 207
music, 3–17, 39–52, 209; audiences and markets for, 3–17, 48–49; digital, benefits for listeners, 44–46; distribution, 40–41; distribution, new approaches to, 4–8; DJs, 5; downloads, 43–44; economics, 7–8; festivals, 5; geography and, 46–48; imports, 45–46; Internet and, 39–40; local scenes, 41–42; on-demand, 5; pre-Internet, 43; royalties, 8; sampling, 44–45; streamed, 3–4
Myanmar, 193
MyDamnChannel, 225
#MyNYPD, 272–273
MySpace, 42–43, 45, 51, 221

Napster, 43
National Association for the Advancement of Colored People (NAACP), 322
National Mobilization Committee to End the War in Vietnam (MOBE), 322
National Moratorium Antiwar Demonstrations, 323
National Security Agency, 129
NBC News, 124
Negt, Oskar, 291
Netflix, 8
The New Inquiry, 271
The New Republic, 191, 245–246
New York Journal, 244
New York Police Department, 272
New York Times, 63, 121, 206–207, 214, 226, 246, 278, 321
The New Yorker, 63, 250, 264
News sites, 242, 277
Nguyen, Dao, 231
Nielsen SoundScan, 3
Niépce, Joseph Nicéphore, 57
Nigeria, 287–298
Nigerian Village Square, 287; as a collaborative online community, 293–296; citizen journalism on,

296–298; guidelines, 293–294; i-Witness, 294; Publishing Guide, 294
Nirvana, 47
Nixon, Richard, 121, 248
#NotYourAsianSidekick, 263–264
Novak, Robert, 123

Obama, Barack, 55, 60, 129, 131, 328–329; online strategy, 186–188; Twitter, 186; YouTube, 186–187; Web ads, 188
Obasanjo, Olusegun, 294–295
Occupy Wall Street, 165, 169, 174, 194
Oculus VR, 14–15
O'Donnell, Susan, 293
OK Cupid, 308
online demographics, 94–96, 333–335; age, 95–96; ethnicity, 94–95
Orta, Ramsey, 269
O'Sullivan, Patrick, 314–315
Oxford Dictionaries, 55, 222

Pablo Honey, 50
Padillo, Anthony, 224
Page Street Publishing, 206
Pandora, 4, 7–8, 47
Pantaleo, Daniel, 269
Park, Suey, 263–265, 273
Parmigianino, 222–223
Pasteur, Louis, 337
Patriot Act, 120, 128
PC Magazine, 11
Pentagon Papers, 121
Petrusich, Amanda, 61
Pew Research Center, 55, 93–96, 189, 248
PhilipDeFranco, 227
photography; digital, 58–59; history of, 57
PicMonkey, 213
Ping, 4
Pinterest, 211, 213, 267
Pitchfork.com, 51
Plame, Valerie, 123–124

Plenty of Fish, 308
#POC4CulturalEnrichment, 263
Podemos, 142–153; digital revolution and, 145–146; Facebook and, 150–151; fan pages, 151–152; online circles, 148–150, 153; online tools used by, 146–148; worldwide debates on, 152–153
political campaigns, history of online, 184–186
political mobilization, 142–144; pseudo-participation, 143
politics, future of, 195
Pollan, Michael, 204
Powell, Julie, 203
Power Line, 248
The Poynter Institute, 206, 250
public health, xiv; definition of, 335–336; history of, 336–338; Internet and, 333–342; misinformation and, 340–342; social media and, 339
public spheres, 287–298; definition of, 289; the Internet and, 291–293
Pulitzer, Joseph, 244

Racists Getting Fired, 270
radical transparency, 119–132
Radiohead, 49–50
Rajoy, Mariano, 145
Rather, Dan, 249
Recording Industry Association of American (RIAA), 43–44
Reddit, 147, 153, 208, 230, 338
Reluctant Habits, 250
REM, 323
Republican National Convention (2004), 248
Resisting the Virtual Life, 293
response delays; asynchronous channels, 311; changing how we interpret, 315–316; e-mail, 309, 310; expectations for responses and, 315–316; face-to-face (FTF) conversations, 309, 311; impression

management, 311–312; negative perceptions of, 308–310; nonverbal, 309, 312–313; online forums, 310; persuasion and, 314–315; positive interpretations of, 310–311; reconceptualizing, 305–316; self-disclosure and, 312–313; social support and, 313–314; text messaging, 310
Rettberg, Jill Walker, 223–224
Rheingold, Howard, 292
Rice, Janay, 275–276
Rice, Ray, 275
Roig-Franzia, Manuel, 247
Rolling Stone, 252
Romenesko, Jim, 250
Romney, Mitt, 188; Facebook, 188; Twitter, 188
Ronstein, Richard, 5
Rose, Don, 322
Rosenstiel, Tom, 242
Rosin, Hannah, 252
Rosling, Hans, 333
Rough Trade, 51
Rountree, Elspeth, 231
Rucker, J. D., 214–215
Russell, Jason, 323–325, 329

Salon, 265
San Antonio Express-News, 246
San Diego Zoo, 226
sanitation practices, 337–338
Saveur, 205
Schilling, Curt, 27
Schmidt, Eric, 108
Schonfeld, Erick, 209
Schuman, Emily, 214
Schumer, Charles, 131
Schwartz, Peter, 48–49
Schwyzer, Hugo, 273–274
Seeing Ourselves Through Technology: How We Use Selfies, Blogs and Wearable Devices to See and Shape Ourselves, 223–224
Self Portrait in a Convex Mirror, 222–223

self-presentation, 31, 61; gender and, 25–26, 31; optimizing, 24–27

self–promotion, 62, 221–235

selfies, 222–223; as a communication tool, 55–65; definition of, 55; history of, 56–59; rise of modern, 57–59; self-portraiture before the camera phone, 56–57; why people post, 62–64

SFX Entertainment, 5

Sheppard, SI, 243

Sherman, Cindy, 58, 60

Siddiqi, Ayesha, 271

Siemens, George, 104

Silverman, Craig, 250–251

Simmons, Bill, 251

Sinclair, Upton, 244, 337

Sission, Mark, 205

Sivan, Troy, 226

60 Minutes Wednesday, 248–250

Skoler, Michael, 231

Sky News, 125

Skying, 51

slacktivism, 271–277; defense of, 321–329

Slate, 252

smartphones, 94–95, 222, 308

Smithsonian Institution, 14

Smosh.com, 224–225

Snapchat, 11, 208, 307

Snowden, Edward, 130

Snyder, Daniel, 263

social media, 12, 62–64, 222; as an agent of change, 263–279; athletes and, 23–33; blogging and, 208–210; collective action, 166, 267–270; community building through, 338–339; directions for future research, 31–32; filters, 61; mechanisms, 263–279; missteps, 32; news and, 231–232; positive side of, 23–33; practitioner implications, 32–33

Social Media Examiner, 211

Social Media Today, 204, 206

social mobility, xiv

social networks/networking, xii, 60, 221, 242; analysis, 32; collective action and, 169–172; self-worth and, 63–64

societal transformation, Internet's impact on, 183–195

#solidarityisforwhitewomen, 273–274, 276

Sonic Youth, 42, 47

Sony Entertainment, 4, 10, 131, 203

Soshable, 214–215

SoundCloud, 4, 224

Spain, 141–154; government transparency, 141; *Podemos*, 142–153; political parties in, 141

Spotify, 4, 6–7, 47, 51; music economics and, 7–8

Spreadshirt, 229

Sprout Social, 212

Stack Overflow, 108

Stelzner, Michael, 211

Stewart, Potter, 124

Stroller in the City, 205

The Structural Transformation of the Public Sphere, 288

Sundin, Per, 7

surveillance, 189, 192

Swift, Taylor, 6

sxephil, 227

Tablet, 250

Taksim Dayanismasi (Taksim Solidarity), 174

Tarbell, Ida, 244

Tea Party, 193

Technology, Entertainment, and Design (TED) Talks, 106

Telecommunications Act, 128

television, streaming

Texas Rangers, 28

third-party media, 229–231

Time, 124, 225

#tippingpoint, 29

titanpad, 148

Toledo Blade, 245

Tomorrowland, 5

Tour de France, 27
Trip, Linda, 130
Tumblr, 194, 208, 267, 270
Turkey, 165
Twitter, 55, 107, 172, 211–212,
    221–222, 266–267, 278, 339; activism
    and, 165–176, 193–194; athletes on,
    26–27, 29–30; brand–building on, 27;
    #CancelColbert, 263–265, 273;
    #change, 29; iTunes, 4; #Ferguson,
    29; Fortune 500 companies and,
    207–208; #MyNYPD, 272–273;
    #NotYourAsianSidekick, 263–264;
    #POC4CulturalEnrichment, 263;
    political movements and, 146;
    revolutions, 165, 169, 322; scheduling,
    212; #solidarityisforwhitewomen,
    273–274, 276; #tippingpoint, 29;
    #WhyILeft, 276; #WhyIStayed,
    275–276

Udacity, 100, 105
United Nations; Education for All
    program, 92–93; Foundation, 341
United States; Anti-Doping Agency,
    27; Congress, 129; Department of
    Commerce, 93; Department of
    Education, 97; Supreme Court, 121
U.S. Constitution; First Amendment,
    124, 270, 342; Fifth Amendment,
    119; Sixth Amendment, 124
U.S.S. *Maine*, 244
Universal Music Group, 4; Sweden, 7
University of Bergen, Norway,
    223–224
University of New Orleans, 46
University of North
    Carolina–Wilmington, 30
Upworthy, 212
Urban Music, 48–49
USA Today, 122, 247
U2, 323

Valley Wag, 277
van Rijn, Rembrandt, 56–57

Vanderbilt, Essay Anne, 251
Vanity Fair, 279
video blogging/vlogging, 226–228
Vietnam War, 121, 248, 322
Villarrasa, Colombo, 142
Vlogging Here, 227
Volokh, Eugene, 252
The Volokh Conspiracy, 252

Wall Street Journal, 250
Walther, Joseph, 311
Warfield, Katie, 223
Warner, Michael, 291
Warner Music Group, 4
Washington Post, 121–122, 124,
    244–245, 248–249, 252
Washington Redskins, 263
Washington Redskins Original
    Americans Foundation, 263
Watergate scandal, 121, 245
Web 2.0, 168, 170, 175
webcams, 222
Webdesigner Depot, 208
What a Brand, 205
WhatsApp, 11
whistle-blowers, 119–132; protection
    of, 120
White, Byron, 124
White, Roddy, 29
White House Internet page, 184
The White Savior Film, 324
#WhyILeft, 276
#WhyIStayed, 275–276
WikiLeaks, 120, 130
Wikipedia, 105–107, 129–130
Wilson, Darren, 267–269
Wilson, Joseph C., 123–124
Winfrey, Oprah, 328
Wired, 59, 250
Woods, Tiger, 27–28
Woodward, Bob, 121, 245
WordPRess, 13
World Bank, 334
World Health Organization (WHO),
    335, 339, 341

World Social Forum (WSF), 165
World Trade Organization, 192
written word, the Internet and, 15–16

#YoSoy132, 165, 171–174
Yorke, Thom, 49–50
YouTube, 8, 48, 92, 193, 234, 266, 323;
    advertisements on, 84; careers in,
    228–229; Creator Academy,
    227–228; critical ambivalence and,
    84–85; debates, 187; Edu, 107–108;
    frequently asked questions (FAQs),
    78–80; FAQ for life, 81–84;

instructional videos, 77–87;
instructional videos, analyzing,
80–81; life hacking, 80; Music Key,
4, 6; non-neutrality of, 84–85;
political economy of expertise,
85–86; revenue opportunities,
228–229; videos, 224–226;
walkthroughs, 78–80

Zapatista Army, 165, 192
Zappin, Danny, 228
Zimmerman, George, 29, 267
Zoosk, 308

About the Editors and Contributors

EDITORS

DANIELLE SARVER COOMBS (PhD) is an associate professor in the School of Journalism and Mass Communication at Kent State University. Her research primarily focuses on sports, politics, and the intersection of the two. Danielle co-edited two anthologies for Praeger: *We Are What We Sell: How Advertising Shapes American Life. . . and Always Has* (2014), and *American History through American Sports* (2013). Dr. Coombs is coauthor of *Female Fans of the NFL: Taking their Place in the Stands* (2015), and the author of *Last Man Standing: Media, Framing, and the 2012 Republican Primaries* (2014). Danielle also has published in major international journals, including *International Journal of Sport Communication*, the *Journal of Public Relations Research*, and *Sport in Society: Cultures, Commerce, Media, Politics*. Dr. Coombs has been invited to provide expert commentary on sports fandom and political affairs around the world.

SIMON COLLISTER is a senior lecturer at the London College of Communication, University of the Arts London in the United Kingdom, where he teaches strategic communication, social media, and critical approaches to public relations. Collister currently is completing doctoral research into strategic political communication and digital media at Royal Holloway, University of London's New Political Communication Unit. His current research interests include strategic communication, big data, computational aspects of communication, algorithms, the mediation of power, 21st century organizational models, and the future of the public relations industry.

Simon has recently authored and coauthored articles in leading journals, including *Ephemera: Theory & Politics in Organization* and the *International Journal of Communication* and authored and coauthored book chapters on PR and big data in *Share This Too* (2013) and social media and text-mining in *Innovations in Digital Research Methods* (2015). He is cofounder of the research hub, The Network for Public Relations and Society, and also a founder member of the UK Chartered Institute of Public Relation's (CIPR) Social Media Advisory Panel. Before academia, Simon worked with some of the world's leading public relations consultancies, including We Are Social, Edelman, and Weber Shandwick.

CONTRIBUTORS

LÁZARO M. BACALLAO-PINO is a postdoctoral fellow at the University of Chile (Fondecyt Postdoctoral Fellowship Programme). He earned his PhD (Summa Cum Laude) in sociology at the University of Zaragoza (Spain) in 2012. Dr. Bacallao-Pino has been a professor and a researcher at the University of Zaragoza and the University of Havana (Cuba). His main research interests include communication and power relationships and social movements, specifically their communication dimension and the use of information and computer technologies (ICTs). He has published more than 30 articles, books, and chapters on these topics.

EVAN BAILEY is an assistant professor of advertising and public relations at Kent State University. Prior to Kent State, Evan spent nearly a decade working in newsrooms, design studios, and advertising agencies in northeastern Ohio. Bailey holds a bachelor's degree in advertising and a master's degree in media management from Kent State. Mr. Bailey's research interests include graphic design, audience analysis, new media technology, and live entertainment marketing. In his spare time, Bailey works with SFX Entertainment—one of the world's leading live entertainment producers—developing marketing strategies for concerts in more than 100 markets in the United States, Canada, Mexico, and Latin America.

RICHARD (RICK) J. BATYKO, APR, Fellow PRSA, President Regional Marketing Alliance of Northeast Ohio. Mr. Batyko has more than 25 years of Fortune 100 and nonprofit communications, marketing, and brand strategy experience at organizations including Babcock & Wilcox, AlliedSignal, Honeywell International, and The Cleveland Foundation. Rick is a graduate of Ohio University's E. W. Scripps School of Journalism, where he majored in public relations, and received his Master of Arts in public relations from Kent State University. He holds his accreditation with the Public Relations Society of America (PRSA), and in 2009, he was inducted into PRSA's College of Fellows. Batyko is an adjunct faculty member at Kent State University's College of Journalism and Mass Communication. He has published work in the *PR Journal*, in the ABC-CLIO book *We Are What We Sell*, and in other outlets. His thesis, "The Impact of Corporate and Country Culture on Public Relations Crisis Response in Japan: A Case Study Examining Tokyo Electric Power Company's Response to the Fukushima Nuclear Power Plant Disaster," was published to OhioLink in December 2012.

CHRISTINA BEST is a graduate of the University of Mount Union (2011) and holds a master's degree from the School of Journalism and Mass Communication at Kent State University. Ms. Best currently lives in Nashville, Tennessee, with her husband. She does freelance social media consulting and graphic design.

CHRISTOPHER BOULTON is both a veteran and critic of the creative industries. After working in public and cable television, Boulton earned his doctorate and is now an assistant professor at the University of Tampa, where he teaches critical media studies and production. Boulton's research focuses on the intersection of communication, inequality, and activism, and his writing has appeared in the *International Journal of Communication, The Communication Review, Advertising & Society Review, The Routledge Companion to Advertising and Promotional Culture,* and *New Views on Pornography.*

KRISTEN CHORBA is an instructional designer at Kent State University, where she also earned a PhD in educational psychology and a master's degree in higher education administration. The focus of Chorba's dissertation was a peer mentoring project, originally created to support undergraduate teacher education majors. Her dissertation research incorporated reflecting processes, photo elicitation, and phenomenological interviewing to describe the experiences of the mentors who participated in this mentoring project and to continue the conversation regarding what it is to be a mentor. This research project is ongoing and continues to evolve. Dr. Chorba's research interests include teacher education, mentoring, relational learning, and online education.

KATHRYN CODUTO is a junior strategist at an advertising agency based in Cleveland, Ohio. She graduated from Kent State University's School of Journalism and Mass Communication in 2014 with a graduate degree in media management. While at Kent State, Coduto focused her research on the music industry, specifically examining how listeners choose to access music, and the differences between physical and digital interactions in the music industry. Kathryn herself is an avid collector of vinyl records, as well as a daily user of online streaming services, which she uses to listen to whatever music she chooses all of the time.

ANDREW FRISBIE is a doctoral student at Ohio University. Frisbie earned a master's degree in communication with a certificate in conflict and dispute resolution at Missouri State University. His general research interests include information processing, self-construction, and inquiry methodology. More specifically, he is interested in the use of new media for self-expression and identity construction. One of his current research projects explores the use of long-duration latencies with asynchronous media. Frisbie's nonacademic interests include kayaking, cycling and folk music.

ZAC GERSHBERG is an assistant professor of journalism and media studies in the Department of Communication, Media, and Persuasion at Idaho State

University. Dr. Gershberg earned a PhD in communication studies from Louisiana State University, and his research interests include the history of journalism, media ecology, and political communication. Before entering academia, Gershberg was a reporter for the *Honolulu Weekly* among other publications, and he served as an adviser to *The Signal*, the student newspaper at California State University, Stanislaus, while teaching there. In addition to his current position, he contributes a media-industry column to the *Idaho State Journal*, the daily newspaper of Pocatello, Idaho.

R. BENJAMIN (BEN) HOLLIS is a senior instructional designer at Kent State University (KSU). Dr. Hollis has worked with KSU faculty to design more than 125 online courses, a fully online master's degree program in public relations, and fully online degree concentrations in health informatics, museum studies, and scholastic education. He also teaches online workshops in the Instructional Technology program and graduate courses for in-service educators in the scholastic education concentration at KSU. Hollis won national awards for innovation and interactive multimedia for his instructional design contributions to the online course Media, Power and Culture offered by the School of Journalism and Mass Communication at KSU. He holds a master's degree in instructional technology and a doctoral degree in educational psychology, both from KSU. In his dissertation, Dr. Hollis examined mind wandering in online learning, an intersection of technology learning and cognitive psychology. In this study, he investigated how individual differences in interest, working memory capacity, and motivation impacted off-task thinking and academic performance. Hollis aims to continue researching the practical approaches to improving student achievement in the classroom based on findings in technology-enhanced learning and cognitive psychology.

KEVIN HULL holds an MAT from North Carolina A&T University and is a PhD candidate in the College of Journalism and Communications at the University of Florida. Hull's research focuses on how sports broadcasters are using social media to engage viewers, sports organizations, and athletes. Before returning to academia, he spent nearly 10 years as a television sports broadcaster and news reporter in Wilmington, North Carolina. Hull will start his next career as a journalism professor at the University of South Carolina in fall 2015. His work has appeared in outlets such as *International Journal of Sport Communication* and *Journal of Sport and Social Issues*.

FAROOQ KPEROGI is an assistant professor of journalism and citizen media in the Department of Communication at Kennesaw State University, Georgia. Dr. Kperogi received his PhD in communication from Georgia State University. A former Nigerian newspaper journalist and presidential researcher and speech writer, Dr. Kperogi has written several journal articles and book chapters on a wide range of communication topics, including alternative and citizen journalism, online journalism, journalistic objectivity, public sphere theory, indigenous language newspaper publishing in Africa, cybercrime, international English usage, and media theory. His work has appeared in *The Review of Communication, New Media & Society, Journal of Global Mass Communication,* and *Asia Pacific Media Educator.* Kperogi was managing editor of the *Atlanta Review of Journalism History,* and blogs at www.farooqkperogi.com.

JUSTIN LAGORE, a central Ohio native, holds a bachelor of science degree from Kent State University, where he completed both the advertising and public relations sequences in the School of Journalism and Mass Communication. During his academic career, Lagore studied abroad short-term in London, where he examined variances in social media's applications to marketing and customer service in the United States and the United Kingdom, specifically within the hospitality industry. His other research interests include online community management, digital media applications for reputation management, online identity and self-esteem, and content creation and consumption in the digital age.

SAMANTHA LINGENFELTER is a graduate student in the College of Public Health at Kent State University working toward a master's degree in public health with a focus on health policy and management. She earned a bachelor's degree in political science with a focus on international relations from Kent State University and a Certificate of Nonprofit Business Management in 2010. Lingenfelter's research interests include substance abuse, health literacy, and communicable diseases. When she's not writing book chapters or reading for school, Lingenfelter can be found reading, playing video games, and hanging out with her corgi.

JAMES D. PONDER holds a PhD from Kent State University and is an assistant professor in the School of Communication Studies in the College of Communication and Information at Kent State University. His primary area of research is mass communication, with a secondary specialization in political communication. In particular, Ponder's research focuses on the uses and effects of political media by citizens and how this use can influence conversations with similar and dissimilar others, political knowledge, and other politically relevant outcomes. He has also examined how different social movements have used social media to connect with others and seek social and political change. Ponder has begun to examine how people use financial information available to them and how this use affects issues related to credit and knowledge and, ultimately, how this affects debt.

LINA RAHM is a doctoral student in the Department of Behavioural Sciences and Learning at Linköping University, Sweden. She holds a BSSc in gender studies and an MSSc in welfare studies. Rahm's current research includes citizenship in adult education and socio-material aspects of survivalism and prepping.

KIRAN SAMUEL is a masters student in the Media, Culture and Communication department at NYU and award-winning digital strategist who's worked on some of the most culturally relevant brands today in entertainment, music, technology, fashion, and others, including Beats by Dre, TNT/TBS, Google, YouTube, Reebok, Playtex, and Armani Exchange. She's a research junkie, relying on insights to create rich, dynamic stories for her clients. Kiran has also guest lectured classes on digital strategy and strategic storytelling at Columbia University and New York University. Continually amazed by the ways digital media—social media especially—has affected society and behavior, Samuel hopes to use that curiosity within her work as an academic at the Masters level and beyond.

JIMMY SANDERSON holds a PhD from Arizona State University and is an assistant professor and director of the Sports Communication Program in the Department of Communication Studies at Clemson University. Dr. Sanderson's research interests center on the influence of social media on sport media, sport organizational governance, fan-athlete interaction, advocacy, and identity expression. His research has appeared in outlets such as *Communication & Sport*, *Journal of Sport & Social Issues*, *Journal of Sports Media*, and *Mass Communication and Society*. He is the author of *It's a Whole New Ballgame: How Social Media Is Changing Sports* (2011).

REKHA SHARMA holds an MA and an MS from Kent State University and is a doctoral candidate and an assistant professor in the School of Communication Studies in the College of Communication and Information at Kent State University. Her primary area of research is mass communication, with a secondary specialization in political communication. Sharma's educational background is in journalism and information use. She has examined a diverse array of topics in mass media and computer-mediated communication, including portrayals of issues and people in news and film, messages about war, and about consumerism in animated cartoons, potential knowledge gains from infotainment, motives and outcomes of political social media use, history and applications of viral marketing, and case studies of television fandom. Her research has been published in academic journals such as the *Ohio Communication Journal*, *Mass Communication & Society*, *Electronic News*, *Global Media Journal—Canadian Edition*, *Media, War, & Conflict*, and the *Journal of Fandom Studies*. Sharma has also contributed to the anthologies *War and the Media: Essays on News Reporting, Propaganda and Popular Culture*; *We Are What We Sell: How Advertising Shapes American life . . . and Always Has*; and *Heroines of Film and Television: Portrayals in Popular Culture*.

JÖRGEN SKÅGEBY is an associate professor at the Department for Media Studies at Stockholm University, Sweden. Skågeby's research interests include material interactions theory, post-social media, and feminist design.

SALOMÉ SOLA-MORALES holds a PhD in media, communication and culture from the Autonomous University of Barcelona, Spain (2012). Dr. Sola-Morales has worked as a researcher and assistant professor at the Autonomous University of Barcelona (2007–13) and at the International University of Catalonia (2013).

She is currently an associated professor at the School of Journalism, Universidad de Santiago de Chile (Chile), and chair of Communication Theory and Thesis Seminar I and II. Dr. Sola-Morales is principal investigator of a project on Identity Politics, Participation and Youth and co-investigator on the international "World of Journalism Project." Her research interests include communication theory, anthropology of communication, identity processes, political participation, and the Internet.

STEPHANIE A. TIKKANEN is an assistant professor in the School of Communication Studies at Ohio University. She earned her MA and PhD in Communication at the University of California, Santa Barbara. Dr. Tikkanen's research program focuses on the growing role of new media (e.g., social networking sites, mobile phones) in interpersonal relationships. Specifically, she takes a theoretical approach to understanding the way in which channel features interact with individual and relational motivations to affect interpersonal processes across relational types. Dr. Tikkanen's work has been published in journals including the *Journal of Communication* and *Communication Research Reports*. She enjoys both research and teaching, but spends the rest of her time baking, reading, and secretly analyzing all of her friends' posts on Facebook.